AF352773

Water Quality of Freshwater Ecosystems in a Temperate Climate

Water Quality of Freshwater Ecosystems in a Temperate Climate

Editors

Ryszard Gołdyn
Piotr Klimaszyk

MDPI • Basel • Beijing • Wuhan • Barcelona • Belgrade • Manchester • Tokyo • Cluj • Tianjin

Editors
Ryszard Gołdyn
Adam Mickiewicz University
Poland

Piotr Klimaszyk
Adam Mickiewicz University
Poland

Editorial Office
MDPI
St. Alban-Anlage 66
4052 Basel, Switzerland

This is a reprint of articles from the Special Issue published online in the open access journal *Water* (ISSN 2073-4441) (available at: https://www.mdpi.com/journal/water/special_issues/Water_Quality_Freshwater).

For citation purposes, cite each article independently as indicated on the article page online and as indicated below:

LastName, A.A.; LastName, B.B.; LastName, C.C. Article Title. *Journal Name* **Year**, *Article Number*, Page Range.

ISBN 978-3-03943-414-5 (Hbk)
ISBN 978-3-03943-415-2 (PDF)

Cover image courtesy of Ryszard Gołdyn.

Contents

About the Editors

Ryszard Gołdyn (Prof. Dr.) is a hydrobiologist who deals with the functioning of aquatic ecosystems. He obtained his PhD based on phytoplankton research in dam reservoirs; he obtained his postdoctoral degree on the influence of preliminary reservoirs on the main reservoir, used for recreation and water sports. He is interested in the relationships between various groups of organisms, as well as the mutual relations between abiotic conditions and organisms, especially in the role of internal nutrient loading from bottom sediments and the influence of external loading from the catchment area. Recently, he has been working on the possibility of using nature-based solutions for lake restoration. His research covers several lakes that are subject to sustainable restoration. According to the Web of Science database, he is an author of 65 published papers and many more published in non-indexed journals and book chapters. He belongs to many scientific societies; he is the vice president of the Polish Hydrobiological Society and the national representative of SIL.

Piotr Klimaszyk (Dr.) obtained a PhD degree in freshwater ecology from the Adam Mickiewicz University of Poznań, Poland. His dissertation concerned the functioning of small water bodies in diverse landscapes. He also conducted research on the influence of colonies of water birds on the transfer of nitrogen and phosphorus between aquatic and terrestrial ecosystems. He analyzed the evolution of soft-water lakes with isoetides and problems related to the reaction of water ecosystems to human impact. He is interested in environmental toxicology. According to ResearchGate he is an author of over 70 research articles and 12 chapters. Currently, he is an associate professor and the head of the Department of Water Protection at the Faculty of Biology of Adam Mickiewicz University in Poznań, Poland.

Editorial

Water Quality of Freshwater Ecosystems in a Temperate Climate

Piotr Klimaszyk and Ryszard Gołdyn *

Department of Water Protection, Adam Mickiewicz University, Uniwersytetu Poznańskiego 6,
61-614 Poznań, Poland; pklim@amu.edu.pl
* Correspondence: rgold@amu.edu.pl

Received: 7 September 2020; Accepted: 18 September 2020; Published: 22 September 2020

Abstract: Water is the substance that made life on Earth possible. It plays a key role in both the individual and population development of all species. Water is also a critical resource for humans as populations continue to grow and climate change affects global and local water cycles. Water is a factor limiting economic development in many regions of the world. Under these conditions, good water quality becomes an extremely important factor that determines its economic utility, including water supply, recreation, and agriculture. Proper water quality maintenance of freshwater ecosystems is also very important for preserving biodiversity. The quality of water depends on many factors, the most important of which are related to human impact on water ecosystems, especially the impact of various pollutants from municipal economy, industry and agriculture. Hydrotechnical changes, such as river damming, drainage processes and water transport between catchments also have a significant impact. Water quality is also dependent on the impact of natural conditions connected, e.g., with climate, catchment, water organisms and their interactions within the food-webs, etc. This Special Issue consists of fourteen original scientific papers concerning different problems associated with the water quality of freshwater ecosystems in a temperate climate. Most of the articles deal with the relations between water quality and the structure of ecosystem biocenoses. The conclusion of these articles confirms the fact that the deterioration of water quality has a direct impact on the quantitative and qualitative structure of biocenoses. This is accompanied by a decline in biodiversity and the disappearance of rare plant and animal species. They also draw attention to the particular importance of internal physical and chemical differentiation within the aquatic ecosystem, both in horizontal and vertical dimensions. The problem of ensuring proper ecological conditions and good quality of water in freshwater aquatic ecosystems is also raised, and methods for the restoration of water bodies are presented. The majority of the research presented in this Special Issue was carried out in Central Europe, and one of the papers concerns the area of West Africa—the edge of temperate climate zone.

Keywords: water quality; eutrophication; biodiversity; aquatic plants; plankton assemblages; water cycle; European Water Framework Directive; freshwater ecosystems restoration

---

1. Introduction

Earth water resources contain a huge amount of water (about 1.4 billion cubic kilometers). However, only 3% of this amount is freshwater. It is surprising, but only 1% of the total amount of freshwater is stored in surface freshwater ecosystems (lakes, rivers, ponds and reservoirs) [1]. Proper water quality enables the use of many ecosystem services by human beings. It also maintains the stable functioning of water ecosystems and ensures high biodiversity, both within the waterbody and in the entire landscape. Freshwaters are strictly connected with their surrounding terrestrial ecosystems and cumulate (lakes, ponds, reservoirs) or transport (streams and rivers) matter and energy from catchments. Natural catchment area traits, e.g., geology, morphology, climate, determine the strength and scale of the impacts. Undisturbed catchment areas usually deliver insignificant amounts of pollutants, so deterioration of water quality and rise in trophy are slow and imperceptible. Occasionally, natural phenomena may cause rapid deterioration of the water quality of ecosystems, but most often they are local or limited in time [2–4]. A rapid increase of human population leads to a shift in water and chemical elements within the landscape and ecosystems that finally triggers freshwater eutrophication and deterioration of water quality. Deforestation and transformation of natural catchments into agriculture and industrial areas lead to changes in the water cycle. A decrease of water retention in the catchment accelerates water outflow and increases the loads of nitrogen, phosphorus, carbon, calcium and other chemical elements reaching the freshwaters [5,6]. The increase of heavy rainfall events associated with global climate change is causing intensified erosion and washing of nutrients and other chemical substances from catchments that affect the water quality of waterbodies. Another serious threat to water quality is industrial origin atmospheric deposition of nitrogen and sulfur that acidify freshwaters. Loads of N and S reaching lakes and rivers not only decrease water pH but also increase the bioavailability of C and N [7,8].

External nitrogen and phosphorus loads frequently reach surface waters with treated municipal sewage, discharged directly or flowing with ground waters [4,9]. Very large loads also originate from agriculture; they can be particularly intensive in the case of large farms using high doses of mineral fertilizers. They are non-point sources, reaching the waterbody along the entire shoreline [10,11]. Excessive nutrient concentration causes a hypereutrophic state of waterbodies, which is manifested by an intensive growth of phytoplankton, loss of submerged vegetation and a decrease in the biodiversity of benthic macroinvertebrates [12]. Recent decades have given rise to emerging pollutants, i.e., pharmaceuticals, which affect water quality [13] and modify the functioning of water ecosystems [14].

Degraded aquatic ecosystems require protective measures that eliminate or limit the inflow of external pollutants. Often, however, these treatments are not sufficient and must be supplemented with restoration treatments. Their main goal is to improve water quality in the waterbody, but for this to be possible, it is necessary to change the functioning of the entire ecosystem. Restoration treatments are usually very expensive, so nature-based solutions that use natural processes taking place in the ecosystem to improve water quality are increasingly preferred [9,15,16]. It is very advantageous to use several restoration methods simultaneously, thus reducing the possibility of feedback effects that can eliminate the positive effects of individual methods [17].

The problem of the bad ecological status of water resources—groundwater, lakes, rivers and other fresh and coastal waters—is of particular interest to the European Commission. In 2000, the Commission implemented the Water Framework Directive (WFD) to ensure sustainable water management based not on political or administrative boundaries but on basins units. The WFD provides a legal framework that obliges EU countries to prevent the deterioration of the aquatic environment and to achieve a good status of all water bodies.

The WFD identifies the most important causes of water degradation, prescribes the reduction and control of pollution and recommends sustainable water use. It also establishes requirements for the assessment and monitoring of water quality. In the case of surface waters, the WFD pays more attention to the ecological state of the ecosystem's biocenoses than to water quality, defined by physical, chemical and hydromorphological indicators, which are of auxiliary nature. The main aim of the WFD is to achieve good ecological status for natural waters and the ecological potential of artificially changed water bodies. The first deadline set for EU countries to meet this target was 2015. However, due to unsatisfactory progress, the Commission decided to derogate, and eventually set 2027 as the date. However, there are serious concerns that many countries will not be able to achieve this goal by that date either [18]. Therefore, protective and restoration measures should be widely implemented to improve the current ecological status of freshwater.

We hope that this Special Issue discussing the problems related to the water quality of freshwater ecosystems in a temperate climate will contribute to a better understanding of the causes of water quality deterioration, the mechanisms responsible for the functioning of ecosystems and the possibilities of improving their condition.

2. Summary of the Issue

The articles published in this Special Issue cover a wide range of aspects connected with the water quality of freshwater ecosystems in a temperate climate.

Zieliński et al. [4] used the chemical tool: strontium isotopes ($^{87}Sr/^{86}Sr$) to identify water supply and track the source of pollution of meromictic Lubińskie Lake (N Poland). They found that the main factor affecting the lake's water quality is the nutrient inflow from the catchment. Despite a wide ecotone zone that acts as a sink for non-point sources of pollutants, nitrogen and phosphorus migrate from croplands, infiltrate the first aquifer and enter the lake through springs. The deep aquifer and good quality water supply is of secondary importance in the Lubińskie Lake water budget. In meromictic lakes, the lack of total horizontal mixing prevents the water quality of the upper layer (mixolimnion) from deteriorating because the deep layer (monimolimnion) acts as a trap for both autochthonous and allochthonous pollutants [19]. The authors found that several subsequent heavy storms that occurred in 2008 had broken the multiannual stratification of the lake. The consequence of these phenomena was an upwelling of the nutrient rich water of the monimolimnion and a deterioration of the surface water of the lake.

Two further articles [10,20] deal with the problem of water quality changes in soft-water lakes with isoetids. Soft water lakes with isoetids are considered as valuable natural habitats and are included in the Natura 2000 protected area network as habitat 3110–"Oligotrophic waters containing very few minerals of sandy plains (*Littorelletalia uniflorae*)". They are prone to degradation due to the low buffer capacity of their waters [20].

Zawiska et al. [10] performed a paleolimnological assessment of the functioning of the soft water Lake Sekšu in Latvia, which is under long term stress due to the artificial supply of eutrophic pumped water. Before the commencement of water pumping (1950s) Lake Sekšu was a low productive water body with some symptoms of eutrophication. It was inhabited by characteristic plant species: *Lobelia dortmanna*, *Isoëtes lacustris* and *Juncus bulbosus*. Diatom-based total phosphorus (TP) reconstruction indicates between 25 and 50 µg L^{-1}. After 10 years of water pumping, lake level had risen and total P concentration had increased to over 100 µg L^{-1}. Closing the inflow of eutrophic waters led to a slight decrease of the total P concentration. The rapid increase trophic state of the lake increased the abundance of diatoms causing a decrease in water transparency, which induced hypolimnetic oxygen deficits. This caused the vanishing of characteristic plant species and the emergence of eutrophic species.

Klimaszyk et al. [20] describe long-term changes in the water chemistry and transformation of biocenosis of one of the largest Polish soft water lake with isoetids, Lake Jeleń. The lake is located within the borders of the city of Bytów and is under recreational pressure. Since the beginning of 1990s, a gradual increase in the trophic state of the lake has been observed as indicated by increased nutrient availability, deterioration of deep-water oxygen conditions and a decrease in water transparency. The alterations in water chemistry induced biological transformations, in particular, an increase in phytoplankton abundance (4-fold increase of biomass in the epilimnion) as well as a gradual reduction in the range of the phytolittoral (from 10 to 6 m), a decrease in the frequency of isoetids, and expansion of plant species characteristic of a eutrophic state. At the same time, a very rare charophyte species *Nitella translucens* has been observed in the lake since the beginning of the 20th century [21].

Babko et al. [22] present the results of studies on the structure of the ciliate population in a floodplain eutrophic lake. The authors show that the main factor responsible for the spatial distribution of ciliates in a floodplain eutrophic waterbody is the pattern of habitat oxygenation. According to results, the authors distinguished four characteristic habitats: (i) habitat with permanent high oxygen concentration, (ii) habitat with permanent low oxygen concentration, (iii) permanent anoxic habitat, and (iv) habitat with high diurnal fluctuations of oxygen concentration. They concluded that a high diversity of habitats ensures high biodiversity in water bodies.

The authors of the next article [23] come to similar conclusions to those of Babko et al. [22]. Antonowicz and Kozak [23] study the concentration of chemical parameters and biocenoses occurring in the surface microlayer, i.e., the ecotone, between the hydrosphere and the atmosphere. They prove that the surface microlayer is characterized by a specific chemical and biological composition. Compared to the deeper zone, the surface microlayer accumulates more trace metals, nutrients and magnesium. The authors also found a higher abundance of heterotrophic bacteria and phytoplankton in the surface microlayer. Taxonomic analysis of the phytoneuston showed that the surface microlayer is a separate habitat.

Kirichenko-Babko et al. [24] deal with the problem of the impact of river pollution and transformation of riverbanks of floodplain lakes on the assemblages of riparian ground beetles. The authors state that rivers with their floodplains are some of the most diverse and biologically productive ecosystems, but at the same time, they are among the most prone to human impact. The obtained results show that riverbanks and floodplain lakes are habitats with different characteristics. They identified 95 species of ground beetles, of which 81 species were found on riverbanks, and 51 species were found on the shores of floodplain lakes. The authors found that the expansion of vegetation on the open banks of the river (resulting from the increase of river eutrophication) significantly affects the species composition and spatial distribution of the riparian ground beetles. The number and density of riparian species decreases with increasing development of the vegetation cover.

Czerniawski et al. [12] describe the impact of human activity on the structure of invertebrates in a small urban stream. The composition of invertebrates was strongly influenced by the human impacts occurring in both the streambed and catchment. The authors found that spatial differentiation of the richness and abundance of zooplankton in the studied stream increased with an increase in the distance from the river source. The highest richness of zooplankton was stated in river pools where the water current was slower. Only a small number of macroinvertebrate taxa were found in all sections of the studied river. The authors hypothesized that this is the result of the regulation of stream banks, deforestation of the catchment and water pollution.

Kuczyńska-Kippen [11] presents the results of studies on the response of zooplankton community indices to abiotic, biotic, and habitat traits in two types of ponds differing in the level of human stress. Ponds of low human alterations characterize richer communities and a higher share of littoral zooplankton, whose occurrence was associated with higher water transparency and complex macrophyte habitats. Plankton communities predominated in the highly human-influenced ponds. Their distribution was mainly related to the open water area and fish presence. Anthropogenic disturbance was also reflected in the frequency of rare species, which were more related to low impacted ponds. Despite the fact that the majority of rare species are littoral-associated, they had no prevalence towards a certain ecological type of plant, which suggests that any kind of plant cover, even macrophytes typical for eutrophic waters, will create a valuable habitat for pond conservation. The author postulates that a complex and dense cover of submerged macrophytes ought to be maintained in order to improve the ecological value of small water bodies.

Napiórkowski et al. [25] studied the relationship between zooplankton communities of floodplain lakes with the degree of their connection with rivers and flood-pulse dynamics. They found higher diversity and abundance of zooplankton in floodplain lakes connected with river compared to permanently isolated ones. The authors also noticed that the river flood-pulse has an impact on the zooplankton diversity in the studied lakes. The high level of the river and therefore the studied lakes increased zooplankton diversity and the predomination of rotifer density. When the level of a river is low, total diversity decreased, but crustacean density rose. Research shows that the stable functioning and high diversity of floodplains requires a mosaic of river habitats. The regulation of rivers and cutting off floodplain lakes from their connection with rivers disturbs specific hydrological conditions and leads to an impoverishment of water biocenoses.

Nouaceur and Murarescu [26] studied rain variability in West Africa. To analyze rainfall trends and variability, the authors used long term data from 27 Sahelian climatic stations in Burkina Faso, Mauritania and Senegal and a chronological graphic method of information processing (MGCTI) of the "Bertin Matrix" and continuous wavelet transformation (CWT). The results of the analysis of rainfall trend evolution showed that, following a long Sahelian drought, rains returned to this part of West Africa. The change from the dry to the wet period began in the second half of the 1990s; however, the level of rainfall did not reach the maximum values noted in the 1950s. The authors argue that this return of rainfall is related to the surface temperature of the Atlantic Ocean.

The European Water Framework Directive requires that all European waterbodies should achieve at least good ecological quality. Rücker et al. [18] analyzed the effects of long-term measures to achieve a good ecological status of North German lowland lakes. The authors studied the influence of morphometry, land use, catchment size, and hydrology on water quality. They noticed that despite great efforts, 70% of German lakes have not achieved proper ecological status. They identified excessive nutrient emissions from agriculture as the main cause of this failure. Based on long-term data, they analyzed the response of two morphometrically differentiated lakes to the reduction of nutrient loads. In the case of shallow polymictic lakes, even significant long-term reduction of external loads of nutrients did not result in changes in biocenoses or satisfactory improvement of ecological status. Therefore, catchment size in relation to lake area or volume becomes a major predictor for the probability of lakes reaching at least a good ecological quality.

Undoubtedly, achieving an improvement in ecological status requires, in the case of many aquatic ecosystems, not only a decrease in the inflow of nutrients from the catchment area but also a number of technical treatments within the waterbody itself [17]. Dondajewska et al. [15] present data on the sustainable restoration of a dimictic lake. The strongly eutrophicated lake was treated for almost 10 years with hypolimnetic aeration, phosphorus inactivation using small doses of chemicals and top-down biomanipulation. During that period, the authors noticed a gradual improvement of oxygen conditions in the lake hypolimnion. Induced nitrogen transformation in the deep-water zone resulted in a 30% decrease in the concentration of ammonium nitrogen and a decrease in phosphorus content. The reduction of nutrient content lead to a lower phytoplankton biomass, expressed in a 55% decrease of chlorophyll-a concentrations and a two-fold increase in water transparency.

One of the expected effects of lake restoration is the reduction of phytoplankton abundance and the elimination of potentially toxic taxa of cyanobacteria. Kozak et al. [16] present data on changes in the functional groups of phytoplankton in a long-term restored lake. The small postglacial eutrophic lake was initially treated with iron sulfate to inactivate phosphorus. This treatment did not bring about the expected outcome. Some rebuilding of phytoplankton functional groups was observed; however, continuous domination of cyanobacteria species such as *Aphanizomenon gracile*, *Planktolyngbya limnetica*, and *Limnothrix redekei* indicated a highly eutrophic and turbid state of the lake. The second stage of restoration consisting in the delivery to the over-bottom zone of the lake water of nearby streams rich in oxygen and nitrate led to a significant change in the phytoplankton structure. Species of groups characteristic for eu-mesotrophic waters started to dominate.

The control of phytoplankton blooms is one of the greatest environmental engineering challenges. The addition of hydrogen peroxide to the aquatic ecosystem and the generation of reactive oxygen species has been shown to selectively target cyanobacteria that create harmful water blooms. Thoo et al. [27] studied the effect of increasing the concentration of reactive oxygen species on zooplankton as a result of adding sodium percarbonate. They found that the genus *Daphnia* shows moderate sensitivity to sodium percarbonate, although the size of specimens affects the probability of survival. The lower likelihood of survival of larger individuals may be related to a higher relative filtration rate. The results of the studies show that a safe concentration of sodium percarbonate for *Daphnia* individuals would be below 10.0 mg L^{-1} of sodium percarbonate (2.8 mg L^{-1} of hydrogen peroxide). The presence of zooplankton, especially large daphnids, is a key factor in controlling the biomass and phytoplankton structure. Therefore, the use of too high a dose of oxidants may cause side effects and lead to the failure of restoration treatment.

Author Contributions: Conceptualization and writing, P.K.; text completion and editing, R.G. All authors have read and agreed to the published version of the manuscript.

Funding: This research received no external funding.

Conflicts of Interest: The authors declare no conflict of interest.

References

1. Holden, J. Water fundamentals. In *Water Resources: An Integrated Approach*; Holden, J., Ed.; Routledge Taylor and Francis Group: London, UK; New York, NY, USA, 2014; p. 371.

2. Spencer, C.N.; Gabel, K.O.; Hauer, F.R. Wildfire effects on stream food webs and nutrient dynamics in Glacier National Park, USA. *For. Ecol. Manag.* **2003**, *178*, 141–153. [CrossRef]

3. Klimaszyk, P. May a cormorant colony be a source of coliform and chemical pollution in a lake? *Oceanol. Hydrobiol. Stud.* **2012**, *4*, 67–73. [CrossRef]

4. Zieliński, M.; Szczucińska, A.; Drożdżyński, M.; Frankowski, M.; Pukacz, A. Water quality assessment of a meromictic lake based on physicochemical parameters and strontium isotopes (87Sr/86Sr) analysis: A Case Study of Lubińskie Lake (Western Poland). *Water* **2019**, *11*, 2231. [CrossRef]

5. Barałkiewicz, D.; Chudzińska, D.; Szpakowska, B.; Świerk, D.; Gołdyn, R.; Dondajewska, R. Storm water contamination and its effect on the quality of urban surface waters. *Environ. Monit. Assess.* **2014**, *186*, 6789–6803. [CrossRef]

6. Klimaszyk, P.; Rzymski, P. Catchment vegetation can trigger lake dystrophy through changes in runoff water quality. *Ann. Limnol. Int. J. Limnol.* **2013**, *49*, 191–197. [CrossRef]

7. Klimaszyk, P.; Rzymski, P. Surface runoff as a factor determining trophic state of midforest lake. *Pol. J. Environ. Stud.* **2011**, *20*, 1203–1210.

8. Shigaki, F.; Sharpley, A.; Prochnow, L.I. Rainfall intensity and phosphorus source effects on phosphorus transport in surface runoff from soil trays. *Sci. Total Environ.* **2017**, *373*, 334–343. [CrossRef]

9. Dondajewska, R.; Kozak, A.; Budzyńska, A.; Kowalczewska-Madura, K.; Gołdyn, R. Nature-based solutions for protection and restoration of degraded Bielsko Lake. *Ecohydrol. Hydrobiol.* **2018**, *18*, 401–411. [CrossRef]

10. Zawiska, I.; Dimante-Deimantovica, I.; Luoto, T.P.; Rzodkiewicz, M.; Saarni, S.; Stivrins, N.; Tylmann, W.; Lanka, A.; Robeznieks, M.; Jilbert, T. Long-term consequences of water pumping on the ecosystem functioning of Lake Sekšu, Latvia. *Water* **2020**, *12*, 1459. [CrossRef]

11. Kuczyńska-Kippen, N. Response of zooplankton indices to anthropogenic pressure in the catchment of field ponds. *Water* **2020**, *12*, 758. [CrossRef]

12. Czerniawski, R.; Sługocki, Ł.; Krepski, T.; Wilczak, A.; Pietrzak, K. Spatial changes in invertebrate structures as a factor of strong human activity in the bed and catchment area of a small urban stream. *Water* **2020**, *12*, 913. [CrossRef]

13. Rzymski, P.; Drewek, A.; Klimaszyk, P. Pharmaceutical pollution of aquatic environment: An emerging and enormous challenge. *Limnol. Rev.* **2017**, *17*, 92–107. [CrossRef]

14. Klimaszyk, P.; Rzymski, P. Water and aquatic fauna on drugs: What are the impacts of pharmaceutical pollution? In *Water Management and the Environment: Case Studies. WINEC 2017. Water Science and Technology Library*; Zelenakova, M., Ed.; Springer: Cham, Switzerland, 2018; Volume 86, pp. 255–278. [CrossRef]

15. Dondajewska, R.; Kowalczewska-Madura, K.; Gołdyn, R.; Kozak, A.; Messyasz, B.; Cerbin, S. Long-term water quality changes as a result of a sustainable restoration—A case study of dimictic Lake Durowskie. *Water* **2019**, *11*, 616. [CrossRef]

16. Kozak, A.; Budzyńska, A.; Dondajewska-Pielka, R.; Kowalczewska-Madura, K.; Gołdyn, R. Functional groups of phytoplankton and their relationship with environmental factors in the restored Uzarzewskie Lake. *Water* **2020**, *12*, 313. [CrossRef]

17. Gołdyn, R.; Podsiadłowski, S.; Dondajewska, R.; Kozak, A. The sustainable restoration of lakes—Towards challenges of the water framework directive. *Ecohydrol. Hydrobiol.* **2014**, *14*, 68–74. [CrossRef]

18. Rücker, J.; Nixdorf, B.; Quiel, K.; Grüneberg, B. North German lowland lakes miss ecological water quality standards—A lake type specific analysis. *Water* **2019**, *11*, 2547. [CrossRef]

19. Klimaszyk, P.; Heymann, D. Vertical distribution of benthic macroinvertebrates in a meromictic lake (Lake Czarne, Drawienski National Park). *Oceanol. Hydrobiol. Stud.* **2010**, *39*, 99–106. [CrossRef]

20. Klimaszyk, P.; Borowiak, D.; Piotrowicz, R.; Rosińska, J.; Szeląg-Wasielewska, E.; Kraska, M. The effect of human impact on the water quality and biocoenoses of the soft water lake with isoetids: Lake Jeleń, NW Poland. *Water* **2020**, *12*, 945. [CrossRef]

21. Rosińska, J.; Piotrowicz, R.; Celiński, K.; Dabert, M.; Rzymski, P.; Klimaszyk, P. The reappearance of an extremely rare and critically endangered *Nitella translucens* (Charophyceae) in Poland. *J. Phycol.* **2019**, *55*, 1412–1415. [CrossRef]

22. Babko, R.; Kuzmina, T.; Danko, Y.; Szulżyk-Cieplak, J.; Łagód, G. Oxygen gradients and structure of the ciliate assemblages in floodplain Lake. *Water* **2020**, *12*, 2084. [CrossRef]

23. Antonowicz, J.P.; Kozak, A. Phytoneuston and chemical composition of surface microlayer of urban water bodies. *Water* **2020**, *12*, 1904. [CrossRef]

24. Kirichenko-Babko, M.; Danko, Y.; Franus, M.; Stępniewski, W.; Babko, R. Riparian ground beetles (Coleoptera) on the banks of running and standing waters. *Water* **2020**, *12*, 1785. [CrossRef]

25. Napiórkowski, P.; Bąkowska, M.; Mrozińska, N.; Szymańska, M.; Kolarova, N.; Obolewski, K. The effect of hydrological connectivity on the zooplankton structure in floodplain lakes of a regulated large river (the Lower Vistula, Poland). *Water* **2019**, *11*, 1924. [CrossRef]

26. Nouaceur, Z.; Murarescu, O. Rainfall variability and trend analysis of rainfall in West Africa (Senegal, Mauritania, Burkina Faso). *Water* **2020**, *12*, 1754. [CrossRef]

27. Thoo, R.; Siuda, W.; Jasser, I. The effects of sodium percarbonate generated free oxygen on *Daphnia*—Implications for the management of harmful algal blooms. *Water* **2020**, *12*, 1304. [CrossRef]

Article

Water Quality Assessment of a Meromictic Lake Based on Physicochemical Parameters and Strontium Isotopes (^{87}Sr/^{86}Sr) Analysis: A Case Study of Lubińskie Lake (Western Poland)

Mateusz Zieliński [1], Anna Szczucińska [2], Mateusz Drożdżyński [3], Marcin Frankowski [4] and Andrzej Pukacz [5,*]

[1] Institute of Geoecology and Geoinformation, Adam Mickiewicz University, Bogumiła Krygowskiego 10, Pl-61-680 Poznań, Poland; mateusz.zielinski@amu.edu.pl
[2] Institute of Physical Geography and Environmental Planning, Adam Mickiewicz University, Bogumiła Krygowskiego 10, 61-680 Poznań, Poland; szana@amu.edu.pl
[3] ul. Zbietka 8, PL-62-290 Mieścisko, Poland; mateuszdrozdzynski1@gmail.com
[4] Faculty of Chemistry, Adam Mickiewicz University Poznań, Poland, Uniwersytetu Poznańskiego 8, PL-61-614 Poznań, Poland; marcin.frankowski@amu.edu.pl
[5] Collegium Polonicum, Adam Mickiewicz University, Kościuszki 1, PL-69-100 Słubice, Poland
* Correspondence: pukacz@europa-uni.de

Received: 19 September 2019; Accepted: 21 October 2019; Published: 25 October 2019

Abstract: In 2017, hydrochemical surveys of meromictic Lubińskie Lake (W Poland) and its water inflows were carried out. The lake experienced complete mixing in 2008 due to a series of orkan winds, and since 2015, intensifying worsening of water quality in the lake has been observed. Our aim was to determine the degree of transformation of Lubińskie Lake based on water chemistry and to identify the source of pollution of the lake using strontium isotopes (^{87}Sr/^{86}Sr) as a new chemical tracking tool. The physicochemical analysis confirmed the meromictic character of the lake. The comparison with previous studies (2003 and 2008) showed significant year-to-year differentiation, indicating intensifying eutrophication of the lake's water, both in the epilimnion and the hypolimnion. Nine spring niches, directly supplying the lake, provide water with very high phosphorus and nitrogen concentrations (up to 10 kg of nitrogen and 0.9 kg of phosphorus daily). The strontium isotopes (^{87}Sr/^{86}Sr) analysis indicated that the lake's water was supplied mostly by the springs, and recharge from deep aquifers is of secondary importance. Moreover, strontium isotope data and the relationship between Sr and Cl content support the finding that the high load of nutrients is of anthropogenic origin and reaches the lake through springs.

Keywords: water quality; strontium isotopes; springs; meromictic lake; eutrophication

---

1. Introduction

Climate changes triggered by human activity have caused profound alterations in lake ecosystems, including imbalances in the water budget, eutrophication, and increased sediment resuspension, which lead to the worsening of water quality and the loss of habitats [1–3]. These negative changes have affected both large water bodies, such as the Great Lakes or Lake Chad [4,5], and small lakes in remote areas [6,7]. Anthropogenically driven eutrophication is considered as one of the most critical threats for water quality in lakes, resulting in plankton blooms, decreased water clarity, oxygen deficiencies, and finally, loss of habitats [8]. It is mostly induced by an increased supply of nutrients coming from manure and mineral fertilizers applied on crop fields. Nutrients reach the lake through ground water recharge and surface runoff [9]. They can also be delivered to the lake by sublacustrine springs supplying some lakes of early post-glacial areas [10,11]. This applies in particular to the springs whose alimentation area is located within the zone of strong anthropogenic influence [12].

In addition to external supply, an important source of nutrients is also internal loading from sediments [13]. This is particularly important in deep, stratified lakes, in which the decomposition of organic matter under anaerobic conditions causes the release of various substances, such as ammonia, sulfur, or heavy metals, which endanger the quality of water [14]. Thus, stratification and the associated cyclic mixing regime are important for the functioning of lake ecosystem and water quality as well.

The meromictic type is very special and rare with regards to natural lakes. Unlike dimictic lakes (a type of holomixis), in these lakes the deepest stratum of water (monimolimnion) remains isolated from the upper mixing stratum (mixolimnion) by a stratum called chemocline [15]. Hutchinson [16] distinguished three types of meromictic lakes: (i) Ectogenic meromictic lakes with meromixis caused by external events (e.g., the inflow of saline water), (ii) crenogenic meromictic lakes (inflow of saline spring water at a constant, low temperature), and (iii) biogenic meromixis, where differences in water density leading to limited mixing result from the decomposition of organic matter at the bottom. In all these types, stratification may be aided by the morphometric characteristics of the lake as well as shelter from the prevailing wind by surrounding hills or tall vegetation at the shore [9].

Due to permanent density stratification within the monimolimnion, there is no direct supply of oxygen from the atmosphere and the vertical transport of solutes is limited. As a result, sedimentary substances are trapped and accumulated for a long time in anoxic conditions. In holomictic lakes there is a systematic exchange between surface and bottom waters. In meromictic lakes hypolimnion water accumulates huge amounts of ammonia, carbon dioxide, and hydrogen sulfide, which may pose an internal threat to the stability of the ecosystem [17]. Consequently, if the meromixis is disturbed, substances accumulated in the monimolimnion for a long time, including substantial amounts of nutrients, can be rapidly released and cause negative changes within the whole ecosystem. This negative process may be accelerated by external supply from the catchment, e.g., via springs or surface runoffs [13].

Nowadays, water quality assessment of the lakes, mostly based on trophy-related parameters, has been replaced by ecological status assessment, whose main tools are bioindicators [18]. However, the parameters that directly determine water quality (i.e., visibility, nutrient concentration, and chlorine concentration) are still widely used and largely reflect the quality and functioning of the ecosystem as a whole. This is particularly important for assessing the degree and sources of pollution, given the simple and relatively quick analytical methods [19].

Numerous studies show that radiogenic strontium isotopes ($^{87}Sr/^{86}Sr$) have the potential to fingerprint anthropogenic additives (typically, mineral fertilizers, municipal sewage, and industrial emissions) in aquatic systems [20–26]. Moreover, strontium isotopes proved to be a reliable tool for the identification of water reservoirs and interactions between water bodies in various aquatic systems [27–31]. They are also a valuable tracer of water-rock interactions and are widely applied in weathering studies [32–37].

The strontium isotope method utilizes differences in relative concentration of two strontium isotopes, ^{87}Sr, and ^{86}Sr. As the amount of ^{87}Sr increases with time due to the decay of rubidium isotope ^{87}Rb, rocks of different age develop different $^{87}Sr/^{86}Sr$ ratios. Strontium is released from rocks by weathering into water and soil and becomes available for further uptake by plants and animals [38]. Since the strontium isotope composition of surface water results from the mixing of strontium derived from various natural (bedrock, soil, precipitation, sea spray) and anthropogenic (industrial emissions, sewage, mineral fertilizers) sources with different isotopic ratios, the $^{87}Sr/^{86}Sr$ ratio might be applied to trace strontium sources [39–41]. Although strontium isotopes are commonly applied in studies of lakes, works investigating meromictic lakes are scarce [42].

In 2004, Pełechaty et al. [43] described a new meromictic lake in mid-Western Poland: Lubińskie Lake. According to Pełechata et al. [44], this lake experienced complete mixing in 2008 due to a series of storm winds. Since then, intensifying worsening of water quality (i.e., increase in nutrient concentration, phytoplankton blooms, decrease of water visibility) in the lake has been observed.

Additionally, in 2015, increased fertilization of fields, mainly with liquid manure, occurred within the catchment area.

Our aim was to determine the degree of transformation of Lubińskie Lake based on water chemistry and to identify the source of pollution of the lake using strontium isotopes ($^{87}Sr/^{86}Sr$) as a new chemical tracking tool. We hypothesized that complete mixing of the water column after storm winds triggered the eutrophication process, which was then reinforced by the inflow of nutrients from the catchment.

2. Study Object

Lubińskie Lake is a small lake with the surface area of 22.7 ha located in western Poland (52°18′32″ N, 14°54′25″ E), in Lubuskie Lakeland (Figure 1). Its catchment area only covers 5.5 km². It is one of 21 lakes located in the post-glacial tunnel valley running by Gronów, Rzepin, and Torzym towns. The valley spreads NW-SE but in the southern part, its direction changes to W-E, so the basin of the water body has an arched shape. In the southern part of the lake, a bay has formed, separated from the main pool with a 60 m wide isthmus [45]. The maximum depth of the lake is 21.6 m [46,47], whereas the mean depth is 8.8 m [47]. This results in steep slopes of the lake bottom. Detailed parameters and morphometric indices of the lake are presented in Table 1. The west and north-east slopes are particularly steep. The slopes of the bay, however, are gentle, and its depth does not exceed 5 m. The north and west shores of the lake are directly adjacent to the edge of the valley and have the form of steep scarps, over a dozen meters high. The south and north sides are milder.

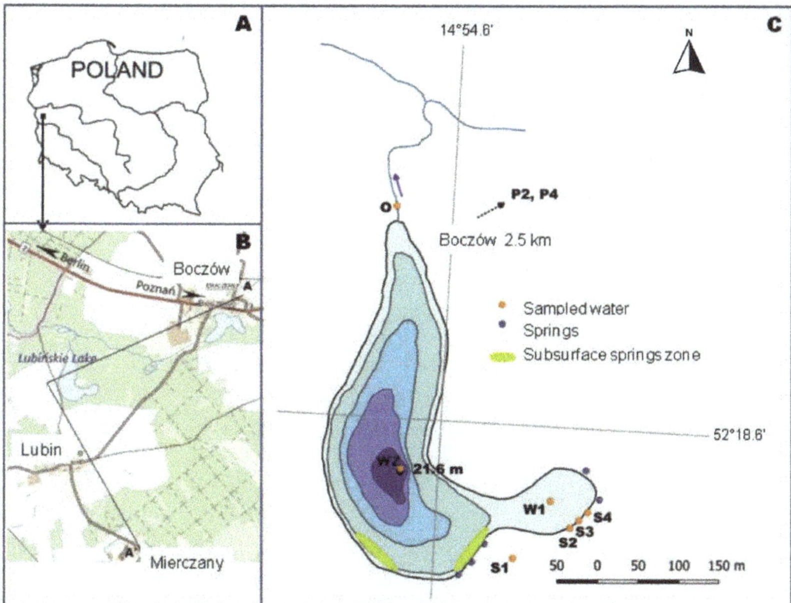

Figure 1. Location of Lubińskie Lake and the points of water samples intake from: The lake (W1 and W2), outflow (O), springs (S1–S4), and wells P2 and P4 (**A,C**). The line A–A′ shows the line of hydrogeological cross-section (**B**).

Table 1. Parameters and morphometric indices of Lubińskie Lake (according to Jańczak [47]).

Parameters and Indices	Unit	Value
Surface area	ha	22.7
Maximum length	m	1130
Maximum width	m	270
Mean width	m	200
Length ratio	-	4.2
Shoreline length	m	2950
Shoreline development	-	1.75
Volume	10^3 m^3	1997.6
Maximum depth	m	21.6
Mean depth	m	8.8
Relative depth *	-	0.0453
Depth ratio	-	0.41
Exposure ratio	-	2.6

* Maximum depth (m)/surface area (km^2).

The bottom of the lake is mostly made of fine- or medium-grained sands, and at some points also of gravels. In the bay and in the north part of the lake, the bottom is mostly covered with gyttjas [45]. The lake basin cuts through alternating deposits of fluvial sands and gravels [48] whose thickness fluctuates between several and over a dozen meters, which facilitate the infiltration of water from the catchment. In the lower part, the basin probably also cuts through the deposits of post-glacial clay (Figure 2), whose thickness in the open pit located approximately 1 km north of the lake is up to 35 m. Above and below the layer of clay there are approximately 10 m thick deposits of fine- and medium-grained sands, whose waters directly feed the lake in the form of groundwater upwellings [48].

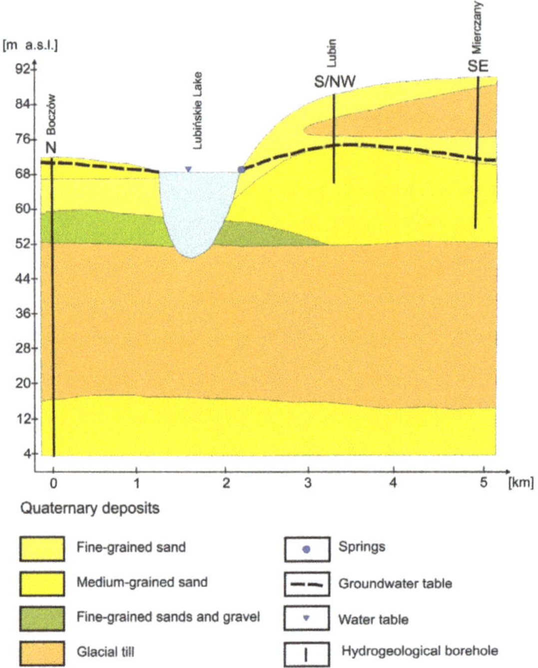

Figure 2. Hydrogeological cross-section of Lake Lubińskie drainage basin along the line Boczów—Lubińskie Lake—Mierczany.

As evidenced on an orthophotomap made in 2010 [49], woodlands and farmlands prevail in the catchment, covering the area of 2.7 and 2.6 km^2, respectively (Figure 3). The remaining area is covered by waters and the urban territory of Lubin town. The lake is surrounded by forests from all sides, but on the west and south side the forest buffer is very narrow, only between about a dozen and several dozens of meters from the lake shore. Woodlands predominantly occur in the eastern part of the catchment, while arable lands are located in the western part.

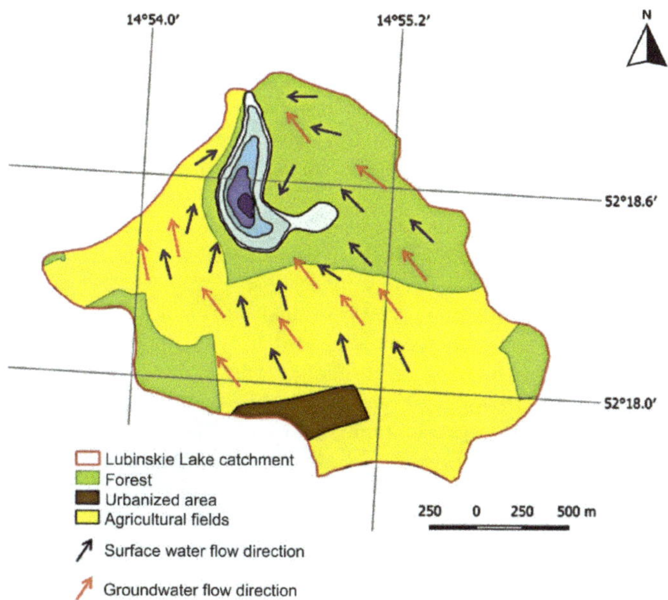

Figure 3. Structure of land use in the catchment of Lubińskie Lake based on the orthophotomap.

3. Materials and Methods

The lake was investigated in April and August 2017, both for comparison with the previous results in 2003 [43] and 2008 (unpublished data) and with regard to the potential meromictic nature of the lake. In both months, the electrolytic conductivity (EC) and pH were measured in the deepest part of the lake, using an Elmetron CX-701 meter (Elmetron Sp.j.). The Secchi disk was used for visibility. Oxygen concentration and temperature were measured in the vertical profile using an Elmetron CX-401 meter. In order to confirm the mixing type, additional measurements for oxygen and temperature were taken at the end of November 2017.

For further chemical analysis, the samples were collected in 1-l plastic bottles (one from the surface and one from the above-bottom layer—using an electric pump) and kept in a portable refrigerator. Then, the samples were kept in a refrigerator (at 4 °C) until the remaining chemical analyses were performed at the Department of Water and Soil Analysis, Faculty of Chemistry, Adam Mickiewicz University in Poznań.

Concentrations of total nitrogen (TN) were determined in the filtered samples by high-temperature combustion (HTC) using a Shimadzu TOC-L Total Organic Carbon analyzer (Shimadzu, Japan). For TN determination, the analyzer was equipped with a TNM-L TN unit (Shimadzu, Japan). The concentrations of NO_2^-, NO_3^-, and NH_4^+ ions were determined by the ion chromatography method with a conductivity detector (Shimadzu, Japan), using Dionex AS22 with AG22 columns. Total water hardness was determined by the versenate method. In order to determine Ca^{2+} and Mg^{2+} ion concentrations, a Shimadzu ion chromatograph with conductivity detection was applied, using Dionex IonPac SCS 1 Silica Cation Separator Column.

The inductively coupled plasma optical emission spectrometer (ICP-OES) ICPE-9820 (Shimadzu, Japan) was used for the qualitative and quantitative detection of total phosphorus (TP). Prior to the analysis, the Sigma-Aldrich (USA) containing 1000 mg L^{-1} P in 10% nitric acid (comprising hydrofluoric acid traces) was used for the calibration of ICP-OES. In order to preserve the standard/sample conditions, the matrix match method was used.

Soluble reactive phosphorus (SRP) in the water was determined by the molybdate method with ascorbic acid as a reducer. Prior to the analyses, the water samples were filtered through a rinsed 0.45 μm pore size filter (Macherey-Nagel, Germany).

Based on Secchi disc visibility and the concentration of TP in water, Carlson's trophic state index (TSI) was calculated [50]. The TN and TP concentrations were used for the N:P ratio.

The same methodology as above was applied in the case of the analysis of physicochemical parameters of water in springs feeding the lake, also carried out in April and August 2017. Methods described by Pełechaty et al. [43] were applied with regard to the compared years 2003 and 2008.

The flow and discharge of springs flowing out of the lake were measured three times: On 15 March, 15 April, and 15 August, using an electromagnetic flow meter Valeport model 801 with measurement accuracy of ± 0.5% of reading plus zero stability < 0.005 m s^{-1}.

Separate samples were taken for $^{87}Sr/^{86}Sr$ analysis of the dissolved strontium load. Additionally, rain water was collected at Słubice, about 30 km to the west, and groundwater was collected at Boczów (about 2.5 km to the north-east) in order to evaluate its contribution to the Sr budget of the lake. A total of 16 samples were taken. Grab samples were taken from the springs. Lake water was sampled with an electric pump from the deepest part of the lake and from the south-eastern bay (Figure 1). In the former, it was gathered at points from the surface up to the depth of 20 m in 4 m intervals, whereas in the latter, samples were collected from the surface and the depth of 4 m. The rain was caught on the roof of Collegium Polonicum of Adam Mickiewicz University to a precleaned rain collector. The obtained sample encompass rainwater from several rain events during two weeks. The groundwater was collected from water intakes. All the samples were taken unfiltered to precleaned polyethylene bottles (250 mL) after flushing the bottles with the sampled water. The samples were filtered through Rotilabo cellulose syringe filters (0.45 μm), acidified with Merck sub-boiled ultrapure HNO_3 (1 mL of acid per 100 mL of water), and stored at 4 °C immediately after returning to the laboratory.

The analytical part of this work, including chemical separation of Sr and the measurements of Sr isotope ratios, were carried out in the Isotope Laboratory of the Adam Mickiewicz University in Poznań. About 15 mL of water was evaporated on a hot plate in PFA vials. Subsequently, a 340 μL portion of 2N HNO_3 was added to the sample and left overnight to equilibrate. The Sr was separated from matrix elements on PFA columns filled with Sr.Spec resin using the miniaturized chromatographic technique described by Pin et al. [51]. Some modifications in the column size and concentration of reagents were introduced by Dopieralska [52]. Strontium was loaded with a $TaCl_5$ activator on a single W filament and analysed in the dynamic collection mode on a Finnigan MAT 261 mass spectrometer. During the course of this study, the NBS 987 Sr standard yielded $^{87}Sr/^{86}Sr$ = 0.710210 ± 10 (2σ mean on thirteen analyses). The measured ratios were normalized to the nominal value of 0.710240 for the standard NBS 987. Total procedure blanks were below 83 pg.

Statistical analyses were performed using STATISTICA 12 (StatSoft, Kraków, Poland) software. The normality of distributions of the analyzed variables and the homoscedasticity of the samples were tested with the Shapiro-Wilk and Levene tests, respectively. The significance of differences was tested by ANOVA with the Kruskal-Wallis test. $p < 0.05$ was accepted as being statistically sound. The principal components analysis (PCA) was performed for the statistical analysis of the physicochemical parameters data set. Prior to this analysis, the data (except pH values) was subject to logarithmic transformation (log (1 + x)) to minimize the discrepancies between the empirical distribution of the variables and their theoretical normal variation. Out of the studied environmental variables, speciation forms of nutrients that were highly interrelated were reduced to TP and TN and included in the PCA. Additionally, to reduce the number of variables, a simple factor analysis was applied. Covariation was tested using non-parametric Spearman rank correlation. The significant levels were set to $\alpha = 0.05$.

4. Results

4.1. Lake Physio-Chemical Characteristics

The thermal and oxygen profiles (Figure 4) performed during the spring overturn and summer stagnation and repeated during the autumnal overturn indicate the meromictic character of Lubińskie Lake. In all seasons, the monimolimnion zone began at 13 m, which was indicated both by a constant temperature (approx. 4.5 °C) and by anoxia.

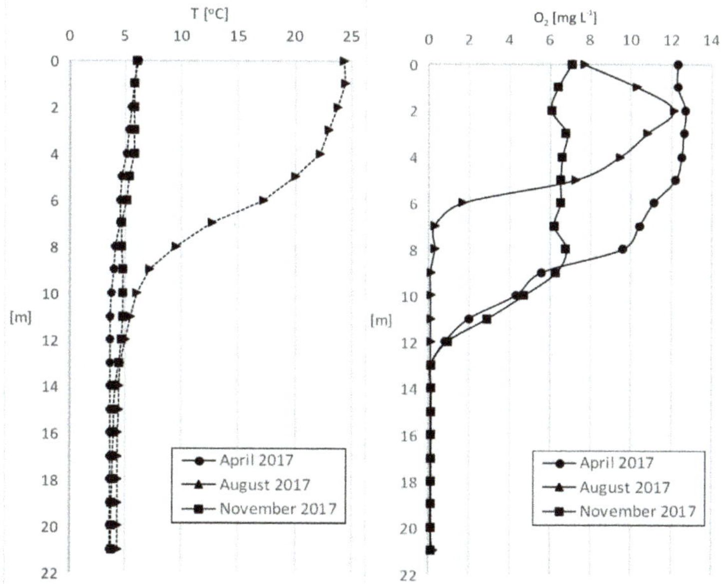

Figure 4. Vertical variability of temperature and oxygen concentration in Lubińskie Lake studied in 2017.

The analyzed water properties were typical for eutrophy (TSI = 61), with high nutrients concentration (TN = 3.81 mg N L^{-1} and TP = 0.13 mg P L^{-1}, on average), high water mineralization (mean EC = 380 µS cm^{-1}), and poor light conditions (mean SD visibility = 2 m; Table 2).

Comparison with the previous studies showed significant ($p < 0.05$) year-to-year differentiation for three parameters only: Hardness, TN, and NO$_3^-$. For all of them, a significant increase in values was observed between 2003, 2008, and 2017. A similar pattern of increase was also observed for TP concentration and trophic indicators: TSI and N:P. The results indicate intensifying eutrophy, both in the epilimnion and hypolimnion waters. Interestingly, in the epilimnion, the N:P ratio increases with increasing TSI (phosphorus limitation), while in the hypolimnion, N:P decreases (nitrogen limitation). Although similar differentiation between the surface (epilimnion) and the above-bottom (hypolimnion) layer was observed in all the study years, the results from 2017 are the most divergent, indicating the highest mineralization and trophy.

The factor analysis showed that pH, EC, hardness, Ca^{2+}, Mg^{2+}, TN, and TP were the variables significantly differentiating the physicochemical composition of water within the studied periods. Based on these variables, the PCA analysis was performed (Figure 5). The first two main components explained almost 80% of the observed variance. The PCA output proved pH, hardness, Mg^{2+}, TN, and TP (correlated with the first axis, r > 0.80) as well as EC (correlated with the second axis, r > 0.80) to be primarily responsible for the observed variance.

Table 2. Physicochemical characteristic of Lubińskie Lake epilimnion (E) and hypolimnion (H) water over three years of study: 2003 [43], 2008 (unpublished data) and 2017. The mean value ± standard deviation is given for each year.

Parameter	(unit)	2003		2008		2017	
		E	H	E	H	E	H
SD	(m)	3.2 ± 0.5		2.5 ± 0.9		2 ± 0.1	
Temp.	(°C)	14.4 ± 10.7	4.4 ± 0.1	13.4 ± 9.8	4.3 ± 0.8	15.3 ± 12.7	4 ± 0.5
EC	(μS cm^{-1})	416 ± 11.3	452 ± 19.8	442 ± 15.6	485 ± 28.3	380 ± 7.1	460.5 ± 0.7
pH	(mg L^{-1})	8.4 ± 0.6	7.5 ± 0	8.6 ± 0.3	7.7 ± 0.6	8.4 ± 0.2	7.2 ± 0
O_2	(mg L^{-1})	14 ± 2.2	0 ± 0	12.2 ± 1.5	2.8 ± 3.9	8.2 ± 0.6	0.2 ± 0.1
Ca^{2+}	(mg L^{-1})	70 ± 8.06	76.4 ± 7.07	77.3 ± 13.72	67.45 ± 0.49	74.16 ± 2.76	90.26 ± 1.75
Mg^{2+}	(mg L^{-1})	4.3 ± 0	5.05 ± 2.19	1.91 ± 1.33	3.29 ± 0.79	3.54 ± 0.15	6.81 ± 0.11
Hard.	(dH)	10.8 ± 1.13	12.2 ± 0	9.05 ± 1.49	10.2 ± 0.14	12 ± 0.14	12.8 ± 0.14
TP	(mg L^{-1})	0.08 ± 0.025	0.199 ± 0.106	0.088 ± 0.071	0.834 ± 0.069	0.13 ± 0.02	0.998 ± 0.04
TN	(mg L^{-1})	1.524 ± 1.561	4.357 ± 1.008	1.964 ± 0.31	4.008 ± 3.029	3.812 ± 0.173	5.844 ± 0.047
$N_{org.}$-N	(mg L^{-1})	1.252 ± 1.177	3.194 ± 0.504	1.335 ± 0.064	2.605 ± 1.563	3.029 ± 0.033	3.804 ± 0.157
NO_3-N	(mg L^{-1})	0.1 ± 0.141	0 ± 0	0.199 ± 0.177	0.148 ± 0.209	0.467 ± 0.195	0.656 ± 0.109
NO_2-N	(mg L^{-1})	0.002 ± 0.003	0.004 ± 0.005	0.005 ± 0.004	0.003 ± 0.004	0.001 ± 0.001	0 ± 0
NH_4-N	(mg L^{-1})	0.17 ± 0.24	1.16 ± 0.509	0.426 ± 0.555	1.253 ± 1.679	0.316 ± 0.013	1.384 ± 0.218
SRP	(mg L^{-1})	0.011 ± 0.016	0.127 ± 0.051	0.036 ± 0.009	0.126 ± 0.096	0.089 ± 0.012	0.858 ± 0.084
N:P *		6.8	18.5	15.8	7.8	33.9	6.0
TSI		52.8		59.5		61	

* For epilimnion water only.

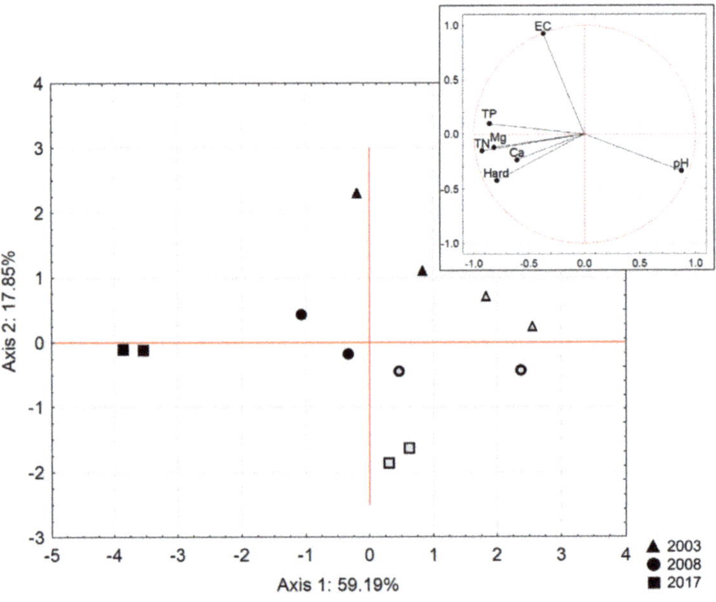

Figure 5. The principal components analysis (PCA) for water analyses in Lubińskie Lake performed in 2003 [43], 2008 (unpublished data), and 2017—differentiation across the selected hydrochemical characteristics (preceded by factor analysis). Gray color indicates samples taken in August, and black—those taken in April.

4.2. Site-to-Site Differentiation of Water Chemistry

The water analyses showed that springs provide water with a significantly lower temperature (from about 10 °C in winter to 12 °C in summer) and much higher mineralization (about 50 μS cm^{-1} higher in spring and about 100 μS cm^{-1} in summer) than in the lake's water (Figure 6). This was also reflected in the values of total hardness and concentrations of Ca^{2+} and Mg^{2+}. In the case of pH and oxygen, however, the lowest values were found in spring waters and the highest in the outflow waters.

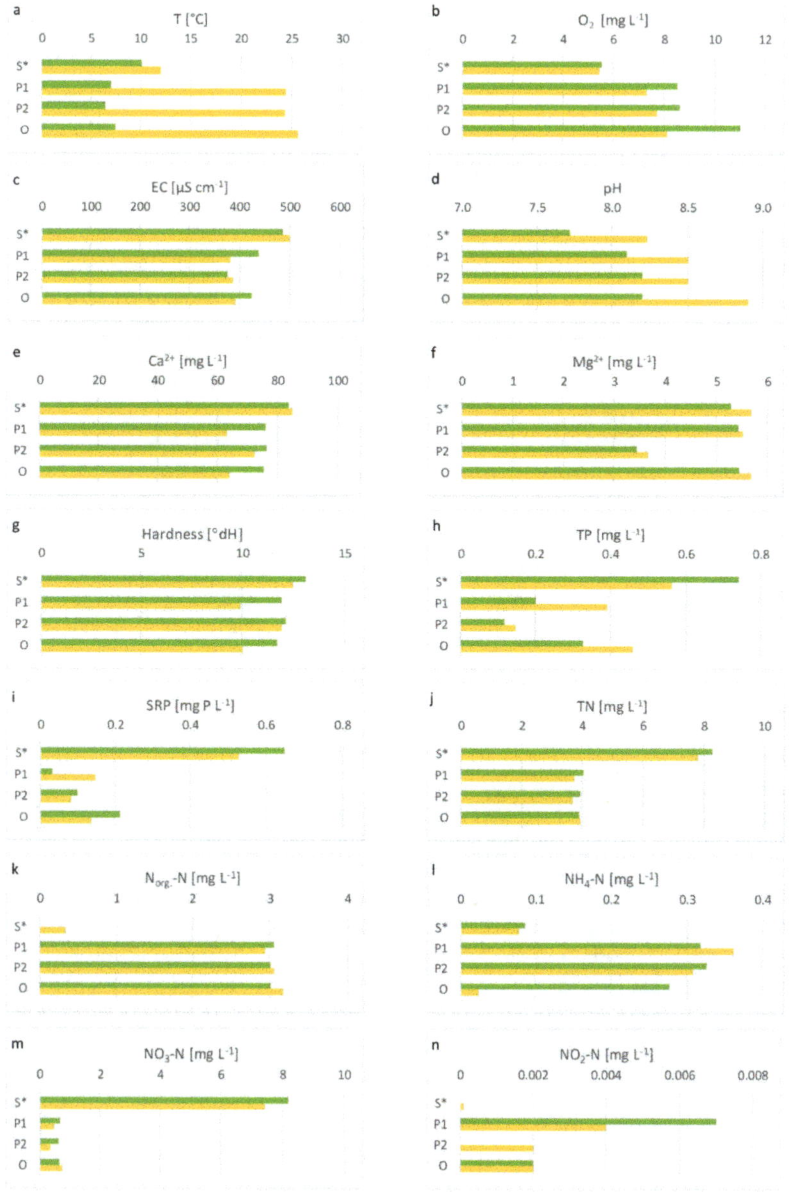

Figure 6. Differentiation of water properties between the studied sites. P1—pelagial 1, P2—pelagial 2, O—outflow, S—spring, * — mean values for all springs. Green bars—April, yellow bars—August.

The spring waters were characterized by much higher nutrient concentrations. This is particularly noticeable in the case of TP and SRP as well as TN and NO_3-N concentrations (Figure 6). It is worth noting that the waters of the bay directly fed by springs (site P1) had higher values than the waters in site P2. Even higher values were found in outflow waters (site O). Total nitrogen concentrations in spring were almost twice as high (up to 8.2 mg L^{-1} on average) as in lake waters. The majority of this load was NO_3-N, the highest values of which were found in spring S1, both in the spring and the summer (24.13 and 22.86 mg L^{-1}, respectively). The exceptions were: NO_2-N and $N_{org.}$-N, which was found to be almost non-existent in spring waters, and lower concentrations of NH_4-N.

4.3. Water Inflow to the Lake

Groundwater upwellings occur both in the littoral zone and in the pelagic zone (Figure 1). They are fed from the shallow aquifer at the depth of 2 to 5 m below ground level and from the aquifer occurring at the depth below 10 m and isolated from the surface with post-glacial clay deposits. The gravity springs feeding the lake are permanent. In the littoral zone of the lake, nine spring niches were recorded, from which waters are directly supplied to the lake. The flow discharge rate of the four measured springs was in the range of 1 to 8 dm^3 s^{-1} (Table 3). The other flow discharge rates were estimated at below 0.5 dm^3 s^{-1}. The total flow discharge rate of all the springs is on average 20 dm^3 s^{-1}. In the southern part of the lake, two zones of sublacustrine springs were also recorded. They are approximately 10–20 m from the shoreline. However, it is difficult to measure their flow discharge. It is estimated to be about 50% of the inflow from overground springs, but this value may be underestimated. The discharge rate of the stream flowing out of the lake in the spring, directly after the disappearance of the ice cover was 40 dm^3 s^{-1}, and in the summer, 21 dm^3 s^{-1}. The difference between the amount of water feeding from the springs and the outflow from the lake indicates additional feeding with groundwater from aquifer outcrops on the bottom of the lake.

Table 3. Efficiency of the springs and discharge of the outflow from the lake in 2017.

Discharge	Q (dm^3 s^{-1})		
Date	15.03	12.04	02.08
Spring (S1)	6.0	4.0	4.0
Spring (S2)	8.0	8.0	7.0
Spring (S3)	1.0	5.0	4.0
Spring (S4)	1.0	2.0	1.0
Outflow (O)	29.0	40.0	21.0

4.4. Strontium Isotope Composition in Water

The profiles of $^{87}Sr/^{86}Sr$ and Sr content for Lubińskie Lake are shown in Figure 7, and Sr isotope and elemental characteristics for components of the hydrologic system of the lake are given in Table 4. Since the bay yields a uniform Sr isotope composition and Sr concentration throughout the water column, it will not be discussed in detail. The lake is characterized by a narrow range of $^{87}Sr/^{86}Sr$ values, between 0.710741 and 0.710798. Thus, it is isotopically homogeneous throughout the water column (Figure 7a). Similarly, Sr content of lake water is highly uniform in the profile and changes from 0.124 to 0.148 mg L^{-1} (Figure 7b). It is worth mentioning that the highest concentration of dissolved Sr occurs in the lower part of the water column. As can be expected, the stream draining the lake has the $^{87}Sr/^{86}Sr$ ratio close to that of surface lake water, but is enriched in Sr (Figure 7).

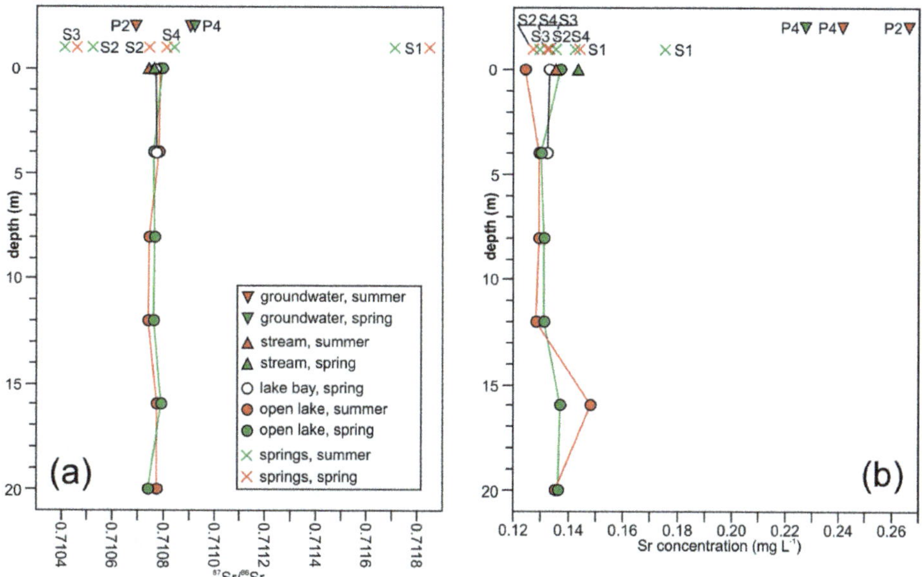

Figure 7. Depth profiles for (**a**) ^{87}Sr/^{86}Sr and (**b**) Sr content of the lake water. Data for the springs, the stream, and the groundwater is also shown for comparison. Explanations for panel (**a**) also refer to panel (**b**).

Table 4. Strontium isotope composition (^{87}Sr/^{86}Sr) and Sr content of components of the Lubińskie Lake hydrologic system.

Component	^{87}Sr/^{86}Sr		Sr Content (mg L^{-1})	
	Spring	Summer	Spring	Summer
Lake water–open lake				
0 m	0.710798 ± 10	0.710790 ± 10	0.137	0.124
4 m	0.710764 ± 10	0.710784 ± 9	0.130	0.129
8 m	0.710770 ± 10	0.710748 ± 11	0.131	0.129
12 m	0.710762 ± 10	0.710744 ± 8	0.131	0.128
16 m	0.710791 ± 12	0.710775 ± 9	0.137	0.148
20 m	0.710741 ± 9	0.710776 ± 10	0.136	0.135
Lake water–bay				
0 m	0.710770 ± 10	ND	0.133	ND
4 m	0.710771 ± 10	ND	0.132	ND
Stream				
	0.710765 ± 10	0.710741 ± 9	0.143	0.135
Springs				
S1	0.711851 ± 10	0.711710 ± 10	0.175	0.144
S2	0.710746 ± 10	0.710525 ± 10	0.135	0.127
S3	0.710463 ± 11	0.710412 ± 9	0.129	0.133
S4	0.710813 ± 11	0.710844 ± 16	0.142	0.132
Groundwater				
P2	ND	0.710693 ± 10	ND	0.266
P4	0.710922 ± 10	0.710906 ± 10	0.227	0.241
Rain				
	0.708879 ± 11	ND	ND	ND

ND—no data.

The springs show a wide range of $^{87}Sr/^{86}Sr$ values from 0.710412 to 0.711851 and relatively diverse Sr dissolved load (0.127–0.175 mg L^{-1}). The strontium isotope composition of the groundwater falls within the range for the springs ($^{87}Sr/^{86}Sr$ values of 0.710693–0.710906), yet it has a clearly higher Sr content (0.241–0.266 mg L^{-1}). Finally, the rain is the least radiogenic component (with the lowest $^{87}Sr/^{86}Sr$ ratio) and yields $^{87}Sr/^{86}Sr$ value of 0.708879.

5. Discussion

5.1. Water Chemistry Analysis

The conducted research showed that Lubińskie Lake is currently in the state of higher trophy, which is indicated by the concentrations of nutrients and low Secchi disk visibility. The comparison with previous years of research shows clearly worsening of the water quality. In 2003, despite high nutrient concentrations, the lake maintained the clean water state (visibility over 4 m) and was classified as a charophyte lake [43,53].

The results of the physicochemical and algological tests showed that the meromictic character of the lake maintained since 2003 was disturbed between January and March 2008 [44]. At the end of February 2008, the waters in the lake mixed completely. At that time, a series of storm winds went across the western part of Poland ("Paula", "Zizi", and "Emma"). The complete restoration of the meromictic conditions lasted until June 2008 [54] (Pukacz et al. 2010). Although the complete mixing of lake waters lasted only a few months (which is confirmed by measurements carried out in the subsequent years), it probably contributed to the release of nutrients and other substances (e.g., methane and hydrogen sulfide) deposited for years in the monimolimnion [55]. This resulted in gradual worsening of the condition of the whole ecosystem. One of the visible effects was the worsening of light conditions and gradual disappearance of submerged vegetation (Pukacz 2008–2018, unpublished data).

Although meromictic stratification was restored in the lake, further increase in nutrients concentration was observed. The study showed that an important source of phosphorus and nitrogen (especially in the form NO_3-N and NH_4-N) supplied to the lake is the spring waters. Based on the average concentrations and water supply from the springs, it can be estimated that each day, approximately 10 kg of nitrogen and 0.9 kg of phosphorus can be introduced to the lake. However, these values may be underestimated by approximately 25%, because they do not take into account the load supplied with discharges for which the volume was estimated, and the load supplied through the springs in the bottom of the lake.

The increased supply of those elements must be associated with the change in the use of land in the catchment area of Lubińskie Lake. Since 2015, in the fields located in the SE part of the catchment, corn, and rapeseed have been cultivated, which are not only fertilized with mineral fertilizers but also, in the spring, with organic fertilizers (own observations). Taking into account the directions of inflow of groundwater and the very high concentrations of nutrients in spring waters, it is a very probable cause of the increased inflow of nutrients to the lake.

Interestingly, the values of the studied parameters in the SE bay, where most springs are located, were not significantly higher than in the central basin. This may result from the small surface area of the lake and from feeding it with inflows located in the slope of the basin. Feeding the lake through springs located in the bottom of the lake basin may have an impact, not only on the lower quality of the lake waters, but also on disturbed stratification and crenogenic meromixis [56].

5.2. Strontium Isotopes Perspective

Strontium isotope composition coupled with Sr content enables the identification of reservoirs controlling the $^{87}Sr/^{86}Sr$ ratio of water and mixing processes involved [27,36,57]. There is a strong correlation between $^{87}Sr/^{86}Sr$ and 1/Sr for the springs dataset, which points to mixing between the two Sr reservoirs (Figure 8). The first reservoir is characterized by a relatively radiogenic $^{87}Sr/^{86}Sr$ ratio and moderately low 1/Sr (high Sr concentration), whereas the second reservoir has a less radiogenic $^{87}Sr/^{86}Sr$ ratio and higher 1/Sr (lower Sr content). Since the springs emerge from shallow Quaternary aquifers, the first reservoir is represented by silicates, mostly glacial deposits. It controls the isotopic composition of the S1 spring. Since the investigated springs are charged by atmospheric waters [58], we identify the second reservoir with rainwater, which strongly influences the Sr isotope budget of the S2-S4 springs. As shown by Raiber et al. [59], rain can be an important source of Sr to shallow groundwater. Strontium isotope composition of rainwater at Słubice ($^{87}Sr/^{86}Sr$ value of 0.708879) is less radiogenic than that of modern seawater ($^{87}Sr/^{86}Sr$ about 0.70917; [60]) being the source of rainwater, likely due to the addition of atmospheric dust. Inland deviation of precipitation $^{87}Sr/^{86}Sr$ ratio is a well-documented phenomenon [59,61,62]. Typically, rainwater is not rich in Sr [59,61,63]. Zieliński et al. [25] provide Sr concentration of about 0.001 mg L^{-1} for rain and snow at Poznań, about 140 km east of Lubińskie Lake. However, the admixture of dust elevates the Sr content [59]. Therefore, Sr dissolved load of tens of μg L^{-1} for the rain at Słubice can reasonably be assumed.

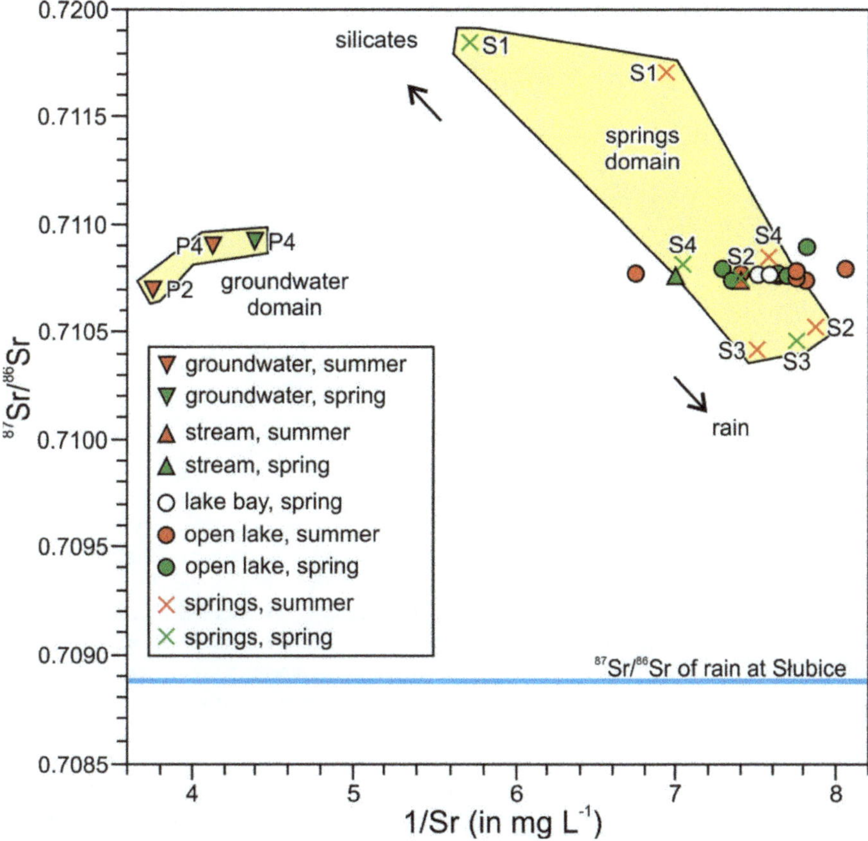

Figure 8. $^{87}Sr/^{86}Sr$ versus 1/Sr mixing diagram of the Lubińskie Lake hydrologic system. Note a binary mixing trend for the springs dataset and the deviation of data points for lake water from mixing paths between the springs.

The role of rain as one of the major sources of Sr for the springs is supported by a clear relationship between Cl and Sr in the springs dataset (Figure 9), indicating their common origin. Chloride ions are a useful atmospheric input reference to the hydrosystem, because in the dissolved load of the hydrosystem they originate from seasalt and saltrock dissolution [64,65].

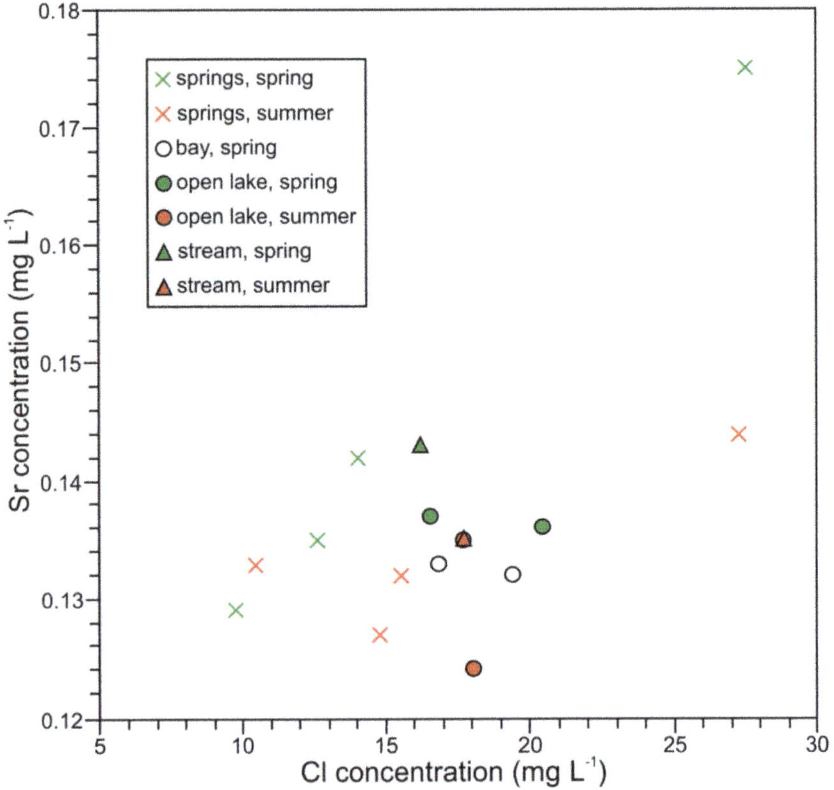

Figure 9. Relationship between Cl and Sr in samples collected from the springs and Lubińskie Lake. Chlorine content in the lake was measured only for surface and near bottom waters (see Table 4).

However, they can also result from various human activities (domestic sewage, fertilizers, and deicing salts) [66]. Since the Lubińskie Lake drainage basin is free of evaporates, they do not participate in the weathering reactions. Consequently, the general trend of relationships between Cl and Sr can be explained by the contribution of Cl from the atmospheric and anthropogenic sources. Since crop fields occur in hinterland of the spring area, the latter is likely represented by mineral fertilizers. This assumption is supported by the high load of nitrates (NO_3) carried by the springs (Figure 6). Additionally, sewage leaking from surrounding villages may also provide some load of Cl.

Recently, Zieliński et al. [25] provided evidence for the common lowering of the present-day $^{87}Sr/^{86}Sr$ values of Noteć River water in western Poland due to the impact of fertilizers used in agriculture. The estimated scale of the lowering was not higher than 0.0006 to 0.0007 of the observed $^{87}Sr/^{86}Sr$ ratios in the downstream portion of the Noteć River. Given that the input of fertilizers-related Sr to the springs waters was recognized, the lowering of the lake water isotopic ratio should be assumed.

Figure 8 shows that data points for Lubińskie Lake in general plot in the springs domain on the $^{87}Sr/^{86}Sr$ versus 1/Sr diagram; thus, crucial contribution of spring-derived Sr is identified. Surprisingly, spring S1, the second largest yet richest in Sr, has a negligible impact on the Sr isotopic composition of the lake. Instead, springs S2–S4 with lower flow rates and Sr contents seem to contribute to it the most. This suggests additional input of shallow groundwater with $^{87}Sr/^{86}Sr$ ratios similar to these springs below the water table due to the lack of other springs along the lake's shore. Indeed, we found several seepages at the bay's bottom.

If the isotopic composition of Lubińskie Lake was solely related to the springs-derived Sr, data points for the lake water would lie on paths between the springs. Most of the data points, however, are moved aside the mixing trends, mainly towards lower Sr concentrations but without noticeable changes in $^{87}Sr/^{86}Sr$, and evidence the contribution of less concentrated waters. Apart from rainwater supplied to Lubińskie Lake through the springs, as inferred above, it also reaches the lake directly. Therefore, it is likely responsible for the dilution of the lake's Sr dissolved load.

The input of dust-derived Sr can account for differences in Sr content between Lubińskie Lake and the stream originating from it. Graustein and Armstrong (1983) [39] show that coniferous foliage is more effective in dust trapping than deciduous foliage. As the stream flows in a coniferous forest surrounding the northern fringe of the lake, dust deposited on the foliage and then incorporated into the soil by throughfall seems to charge soil waters and hence the stream by additional Sr.

Although the lake basin cuts deeply into the aquifer (Figure 2), and the recharge of the lake by groundwater arguably occurs, its impact on Sr isotope composition of the lake water seems negligible. Nonetheless, this impression is only apparent, because $^{87}Sr/^{86}Sr$ ratios of the groundwater fall within the $^{87}Sr/^{86}Sr$ range for the springs, so the contribution of the former will be masked. The input of groundwater-derived Sr to Lubińskie Lake is suggested by the rise of Sr concentration several meters above the bottom, which is consistent with the lateral input of groundwater enriched in Sr and its downward descend. Due to the meromictic character of the lake, it is, however, limited to its deepest part.

6. Conclusions

The results of the study indicate that Lubińskie Lake has regained its meromictic character after complete mixing in 2008. That sudden event probably led to the release of nutrients from the monimolimnion layer, which resulted in an increase in trophy and deterioration of water quality in subsequent years. An additional, significant factor affecting the lake's water quality is the nutrient inflow from the catchment. Although the lake is insulated with a belt of trees, water polluted with fertilizers infiltrates into the first aquifer and enters the lake through springs. The difference between the lake supply from springs and the outflow from the lake indicates lake charging with groundwater from underwater outcrops of aquifers. That was also confirmed by strontium isotopes ($^{87}Sr/^{86}Sr$) analysis. It indicates that water is supplied to Lubińskie Lake mostly by the springs, and recharge from deep aquifers is of secondary importance. Moreover, strontium isotope data and the relationship between Sr and Cl content support finding that high load of nutrients is of anthropogenic origin and reach the lake through springs. We show that the $^{87}Sr/^{86}Sr$ composition of lake water supplemented by major ion chemistry is a valuable tool in the investigation of meromictic lakes.

Author Contributions: Conceptualization, A.P. and A.S.; Methodology and investigation, M.Z., A.S., M.D., M.F., A.P.; Data analysis, M.Z., A.S., M.D., M.F., A.P.; Writing—original draft preparation, M.Z., A.S., A.P.; Writing—review and editing, M.Z., A.S., A.P.

Funding: The APC was founded by Adam Mickiewicz University and the research received no external funding.

Acknowledgments: We wish to thank Z. Belka (Poznań) for the constructive criticism of the MS as well as comments of two anonymous reviewers and journal editor help to improve the manuscript.

Conflicts of Interest: The authors declare no conflict of interest.

References

1. Evans, D.O.; Nicholls, K.H.; Allen, Y.C.; McMutry, M.J. Historical land use, phosphorus loading, and loss of fish habitat in Lake Simcoe, Canada. *Can. J. Fish. Aquat Sci.* **1996**, *53* (Suppl. 1), 194–218. [CrossRef]
2. Callisto, M.; Molozzi, J.; Barbosa, J.L.E. Eutrophication of lakes. In *Eutrophication: Causes, Consequences and Control*, 1st ed.; Ansari, A.A., Gill, S.S., Eds.; Springer: Dordecht, The Netherlands, 2014; pp. 55–71.
3. Hijosa-Valsero, M.; Bécares, E.; Fernández-Aláez, C.; Fernández-Aláez, M.; Mayo, R.; Jiménez, J.J. Chemical pollution in inland shallow lakes in the Mediterranean region (NW Spain): PAHs, insecticides and herbicides in water and sediments. *Sci. Total Environ.* **2016**, *544*, 797–810. [CrossRef] [PubMed]

4. Gao, H.; Bohn, T.J.; Podest, E.; McDonald, K.C.; Lettenmaier, D.P. On the causes of the shrinking of Lake Chad. *Environ. Res. Lett.* **2011**, *6*, 034021. [CrossRef]

5. Gronewold, A.D.; Stow, C.A. Water loss from the Great Lakes. *Science* **2014**, *343*, 1084–1085. [CrossRef]

6. Kang, W.; Chen, G.; Wang, J.; Huang, L.; Wang, L.; Li, R.; Hu, K.; Liu, Y.; Tao, J.; Blais, J.M.; et al. Assessing the impact of long-term changes in climate and atmospheric deposition on a shallow alpine lake from southeast Tibet. *Sci. Total Environ.* **2019**, *650*, 713–724. [CrossRef]

7. Wan, D.; Mao, X.; Jin, Z.; Song, L.; Yang, J.; Yang, H. Sedimentary biogeochemical record in Lake Gonghai: Implications for recent lake changes in relatively remote areas of China. *Sci. Total Environ.* **2019**, *649*, 929–937. [CrossRef]

8. Welch, E.B.; Jacoby, J.M. *Pollutant Effects in Freshwater: Applied Limnology*, 3rd ed.; CRC Press: London, UK, 2004; pp. 1–520.

9. Wetzel, R.G. *Limnology, Lake and River Ecosystems*, 3rd ed.; Elsevier: Amsterdam, The Netherlands, 2001; pp. 1–1006.

10. Choiński, A.; Ilyin, L.; Marszelewski, W.; Ptak, M. Lakes supplied by springs: Selected examples. *Limnol. Rev.* **2008**, *8*, 145–150.

11. Cieśliński, R.; Piekarz, J.; Zieliński, M. Groundwater supply of lakes: The case of Lake Raduńskie Górne (northern Poland, Kashubian Lake District). *Hydrol. Sci. J.* **2016**, *61*, 2427–2434. [CrossRef]

12. Scott, J.T.; Marcarelli, A.M. Cyanobacteria in freshwater benthic environments. In *The Ecology of Cyanobacteria*; Whitton, B.A., Ed.; Springer: Durham, UK, 2012; pp. 271–289.

13. Dodds, W.K. Freshwater ecology. In *Concepts and Environmental Applications*, 1st ed.; Academic Press: San Diego, CA, USA, 2002; pp. 1–591.

14. Last, W.M.; Smol, J.P. An Introduction to Physical and Geochemical methods Used in Paleolimnology. In *Tracking Environmental Change Using Lake Sediments*; Last, W.M., Smol, J.P., Eds.; Kluwer Academic Publishers: Dordrecht, The Netherlands, 2001; pp. 1–504.

15. Lampert, W.; Sommer, U. *Ecology of Inland Waters*; PWN Scientific Press: Warsaw, Poland, 2001; pp. 1–415.

16. Hutchinson, G.E. A contribution to the limnology of arid regions: Primarily founded on observations made in the Lahontan basin. *Trans. Conn. Acad. Arts Sci.* **1937**, *33*, 47–132.

17. Göltenboth, F.; Timotius, K.H.; Milan, P.P.; Margraf, J. *Ecology of Insular Southeast Asia*; Elsevier: Amsterdam, The Netherlands, 2006; pp. 1–568.

18. Bis, B. Assessing the Ecological Status Assessment of Freshwaters. In *Freshwater Ecosystems in Europe–An Educational Approach*; Voreadou, C., Ed.; Natural History Museum of Crete, Selena Press: Heraklion, Greece, 2008; pp. 56–69.

19. Imteaz, M.A.; Asaeda, T.; Lockington, D.A. Modelling the effects of inflow parameters on lake water quality. *Environ. Model. Assess.* **2003**, *8*, 63–70. [CrossRef]

20. Böhlke, J.K.; Horan, M. Strontium isotope geochemistry of groundwaters and streams affected by agriculture, Locust Grove, MD. *Appl. Geochem.* **2000**, *15*, 599–609. [CrossRef]

21. Hosono, T.; Nakano, T.; Igeta, A.; Tayasu, I.; Tanaka, T.; Yachi, S. Impact of fertilizer on a small watershed of Lake Biwa: Use of sulfur and strontium in environmental diagnosis. *Sci. Total Environ.* **2007**, *384*, 342–354. [CrossRef] [PubMed]

22. Chapman, E.C.; Capo, R.C.; Stewart, B.W.; Hedin, R.S.; Weaver, T.J.; Edenborn, H.M. Strontium isotope quantification of siderite, brine and acid mine drainage contributions to abandoned gas well discharges in the Appalachian Plateau. *Appl. Geochem.* **2013**, *31*, 109–118. [CrossRef]

23. Vengosh, A.; Jackson, R.B.; Warner, N.; Darrah, T.H.; Kondash, A. A critical review of the risk to water resources from unconventional shale gas development and hydraulic fracturing in the United States. *Environ. Sci. Technol.* **2014**, *48*, 8334–8348. [CrossRef] [PubMed]

24. Sherman, L.S.; Blum, J.D.; Dvonch, J.T.; Gratz, L.E.; Landis, M.S. The use of Pb, Sr, and Hg isotopes in Great Lakes precipitation as a tool for pollution source attribution. *Sci. Total Environ.* **2015**, *502*, 362–374. [CrossRef]

25. Zieliński, M.; Dopieralska, J.; Belka, Z.; Walczak, A.; Siepak, M.; Jakubowicz, M. Sr isotope tracing of multiple water sources in a complex river system, Noteć River, central Poland. *Sci. Total Environ.* **2016**, *548*, 307–316. [CrossRef]

26. Zieliński, M.; Dopieralska, J.; Belka, Z.; Walczak, A.; Siepak, M.; Jakubowicz, M. Strontium isotope identification of water mixing and recharge sources in a river system (Oder River, central Europe): A quantitative approach. *Hydrol. Process.* **2018**, *32*, 2597–2611. [CrossRef]

27. Négrel, P.; Fouillac, C.; Brach, M. A strontium isotopic study of mineral and surface waters from the Cézallier (Massif Central, France): Implications for mixing processes in areas of disseminated emergences of mineral waters. *Chem. Geol.* **1997**, *135*, 89–101. [CrossRef]

28. Négrel, P.; Casanova, J.; Aranyossy, J.P. Strontium isotope systematics used to decipher the origin of groundwaters sampled from granitoids: The Vienne Case (France). *Chem. Geol.* **2001**, *177*, 287–308. [CrossRef]

29. Négrel, P.; Petelet-Giraud, E.; Barbier, J.; Gautier, E. Surface water-groundwater interactions in an alluvial plain: Chemical and isotopic systematics. *J. Hydrol.* **2003**, *277*, 248–267. [CrossRef]

30. Ojiambo, S.B.; Lyons, W.B.; Welch, K.A.; Poreda, R.J.; Johannesson, K.H. Strontium isotopes and rare earth elements as tracers of groundwater-lake water interactions, Lake Naivasha, Kenya. *Appl. Geochem.* **2003**, *18*, 1789–1805. [CrossRef]

31. Flusche, M.A.; Seltzer, G.; Rodbell, D.; Siegel, D.; Samson, S. Constraining water sources and hydrologic processes from the isotopic analysis of water and dissolved strontium, Lake Junin, Peru. *J. Hydrol.* **2005**, *312*, 1–13. [CrossRef]

32. Blum, J.D.; Gazis, C.A.; Jacobson, A.D.; Chamberlain, C.P. Carbonate versus silicate weathering in the Raikhot watershed within the High Himalayan Crystalline Series. *Geology* **1998**, *26*, 411–414. [CrossRef]

33. Jacobson, A.D.; Blum, J.D.; Walter, L.M. Reconciling the elemental and Sr isotope composition of Himalayan weathering fluxes: Insight from the carbonate geochemistry of stream water. *Geochim. Cosmochim. Acta* **2002**, *66*, 3417–3429. [CrossRef]

34. Jin, Z.; Wang, S.; Zhang, F.; Shi, Y. Weathering, Sr fluxes, and controls on water chemistry in the Lake Qinghai catchment, NE Tibetan Plateau. *Earth Surf. Process.* **2010**, *35*, 1057–1070. [CrossRef]

35. Lyons, W.B.; Welch, K.A.; Priscu, J.C.; Tranter, M.; Royston-Bishop, G. Source of Lake Vostok cations constrained with strontium isotopes. *Front. Earth Sci.* **2016**, *4*, 78. [CrossRef]

36. Zieliński, M.; Dopieralska, J.; Belka, Z.; Walczak, A.; Siepak, M.; Jakubowicz, M. The strontium isotope budget of the Warta River (Poland): Between silicate and carbonate weathering, and anthropogenic pressure. *Appl. Geochem.* **2017**, *81*, 1–11. [CrossRef]

37. Négrel, P.; Pauwels, H.; Chabaux, F. Characterizing multiple water-rock interactions in the critical zone through Sr-isotope tracing of surface and groundwater. *Appl. Geochem.* **2018**, *93*, 102–112. [CrossRef]

38. Capo, R.C.; Stewart, B.W.; Chadwick, O.A. Strontium isotopes as tracers of ecosystem processes: Theory and methods. *Geoderma* **1998**, *82*, 197–225. [CrossRef]

39. Graustein, W.C.; Armstrong, R.L. The use of strontium-87/strontium-86 ratios to measure atmospheric transport into forested watersheds. *Science* **1983**, *219*, 289–292. [CrossRef]

40. Åberg, G. The use of natural strontium isotopes as tracers in environmental studies. *Water Air Soil Poll.* **1995**, *79*, 309–322. [CrossRef]

41. Bain, D.C.; Midwood, A.J.; Miller, J.D. Strontium isotope ratios in streams and the effect of flow rate in relation to weathering in catchments. *Catena* **1998**, *32*, 143–151. [CrossRef]

42. Faure, G.; Mensing, T.M. *The Transantarctic Mountains. Rocks, Ice, Meteorites and Water*; Springer: Dordrecht, The Netherlands, 2010; pp. 1–832.

43. Pełechaty, M.; Pukacz, A.; Pełechata, A. Diversity of micro- and macrophyte communities in the context of the habitat conditions of a meromictic lake on Lubuskie Lakeland. *Limnol. Rev.* **2004**, *4*, 209–214.

44. Pełechata, A.; Pełechaty, M.; Pukacz, A. Complete mixing of a meromictic lake as a result of extreme weather events and its impact on phytoplankton community. In Proceedings of the 23rd Congres of Polish Hydrobiologists, Koszalin, Poland, 8–12 September 2015.

45. Zieleniewski, W. Fisherman report for waters of circumference I.5.8. In *The Fishing Region of the Lubińskie Lake on an Unnamed Watercourse Flowing into the Rzepia River-No. 1 in the Ilanka River Basin in the Lower Oder and Western Catchment Water Region*; Polish Fishermans Association: Zielona Góra, Poland, 2007.

46. Choiński, A. *Outline of Polish Physical Limnology*; Adam Mickiewicz University Press: Poznań, Poland, 1995; pp. 1–298. (In Polish)

47. Jańczak, J. *Atlas of Polish Lakes*; Bogucki Press: Poznań, Poland, 1996; pp. 1–256.

48. Stupnicka, E. *Regional Geology of Poland*; University of Warsaw Press: Warsaw, Poland, 2007; pp. 1–346. (In Polish)

49. Main Office of Geodesy and Cartography. Available online: http://www.gugik.gov.pl/ (accessed on 15 September 2019).

50. Carlson, R.E. A trophic state index for lakes. *Limnol. Oceanogr.* **1977**, *22*, 361–369. [CrossRef]
51. Pin, C.; Briot, D.; Bassin, C.; Poitrasson, F. Concomitant separation of strontium and samarium-neodymium for isotopic analysis in silicate samples, based on specific extraction chromatography. *Anal. Chim. Acta* **1994**, *298*, 209–222. [CrossRef]
52. Dopieralska, J. Neodymium Isotopic Composition of Conodonts as a Palaeoceanographic Proxy in the Variscan Oceanic System. Ph.D. Thesis, Justus-Liebig-University, Giessen, Germany, 2003.
53. Pełechata, A.; Pełechaty, M.; Pukacz, A. An attempt to the trophic status assessment of the lakes of Lubuskie Lakeland. *Limnol. Rev.* **2006**, *6*, 239–246.
54. Pukacz, A.; Pełechaty, M.; Pełechata, A.; Sękowska, N. The influence of winter windstorms on the disorder of meromixia of Lubińskie Lake. In Proceedings of the 14th Polish Limnologic Conference Natural and Anthropogenic Transformations of the Lakes, Szczecin-Stare Drawsko, Poland, 20–23 September 2010. (In Polish).
55. Boehrer, B.; von Rohden, C.; Schultze, M. Physical features of Meromictic Lakes: Stratification and circulation. In *Ecology of Meromictic Lakes*; Gulati, R.D., Zadereev, E., Degermendzhy, A.G., Eds.; Springer: Dordrecht, The Netherlands, 2017; pp. 1–405.
56. Hakala, A. Meromixis as a part of lake evolution—Observations and a revised classification of true meromictic lakes in Finland. *Boreal Environ. Res.* **2004**, *9*, 37–53.
57. Shand, P.; Darbyshire, D.P.F.; Love, A.J.; Edmunds, W.M. Sr isotopes in natural waters: Applications to source characterisation and water-rock interaction in contrasting landscapes. *Appl. Geochem.* **2009**, *24*, 574–586. [CrossRef]
58. Szczucińska, A. Spatial distribution and hydrochemistry of springs and seepage springs in the Lubuska Upland of western Poland. *Hydrol. Res.* **2014**, *45*, 379–390. [CrossRef]
59. Raiber, M.; Webb, J.A.; Bennetts, D.A. Strontium isotopes as tracers to delineate aquifer interactions and the influence of rainfall in the basalt plains of southeastern Australia. *J. Hydrol.* **2009**, *367*, 188–199. [CrossRef]
60. McArthur, J.M.; Howarth, R.J.; Bailey, T.R. Strontium isotope stratigraphy: LOWESS version 3: Best fit to the marine Sr-isotope curve for 0-509 Ma and accompanying look-up table for deriving numerical age. *J. Geol.* **2001**, *109*, 155–170. [CrossRef]
61. Négrel, P.; Guerrot, C.; Millot, R. Chemical and strontium isotope characterization of rainwater in France: Influence of sources and hydrogeochemical implications. *Isot. Environ. Health S* **2007**, *43*, 179–196. [CrossRef] [PubMed]
62. Xu, Z.; Han, G. Chemical and strontium isotope characterization of rainwater in Beijing, China. *Atmos. Environ.* **2009**, *43*, 1954–1961. [CrossRef]
63. Nakano, T.; Morohashi, S.; Yasuda, H.; Sakai, M.; Aizawa, S.; Shichi, K.; Morisawa, T.; Takahashi, M.; Sanada, M.; et al. Determination of seasonal and regional variation in the provenance of dissolved cations in rain in Japan based on Sr and Pb isotopes. *Atmos. Environ.* **2006**, *40*, 7409–7420. [CrossRef]
64. Stallard, R.F.; Edmond, J.M. Geochemistry of the Amazon, 1. Precipitation chemistry and the marine contribution to the dissolved load at the time of peak discharge. *J. Geophys. Res.* **1981**, *86*, 9844–9858. [CrossRef]
65. Meybeck, M. Atmospheric inputs and river transport of dissolved substances. In Proceedings of the International Association of Hydrological Sciences Conference Dissolved load of rivers and surface water quantity/quality relationships, Hamburg, Germany, 15–27 August 1983; pp. 173–192.
66. Sherwood, W.C. Chloride loading in the South Fork of the Shenandoah River, Virginia, U.S.A. *Environ. Geol. Water S* **1989**, *14*, 99–106. [CrossRef]

Article

Long-Term Consequences of Water Pumping on the Ecosystem Functioning of Lake Sekšu, Latvia

Izabela Zawiska [1,*], Inta Dimante-Deimantovica [2,3], Tomi P. Luoto [4], Monika Rzodkiewicz [5], Saija Saarni [6], Normunds Stivrins [7,8,9], Wojciech Tylmann [10], Anna Lanka [2,11], Martins Robeznieks [2,12] and Tom Jilbert [6]

[1]   Institute of Geography and Spatial Organisation, Polish Academy of Sciences, Twarda 51/55, PL-00818 Warsaw, Poland
[2]   Latvian Institute of Aquatic Ecology, Agency of Daugavpils University, Voleru 4, LV-1007 Riga, Latvia; inta.dimante-deimantovica@lhei.lv (I.D.-D.); anna.lanka0@gmail.com (A.L.); martinsrobez@gmail.com (M.R.)
[3]   Norwegian Institute for Nature Research, Gaustadalléen 21, NO-0349 Oslo, Norway
[4]   Faculty of Biological and Environmental Sciences, Ecosystems and Environment Research Programme, University of Helsinki, Niemenkatu 73, FI-15140 Lahti, Finland; tomi.luoto@helsinki.fi
[5]   Institute of Geoecology and Geoinformation, Adam Mickiewicz University, Bogumiła Krygowskiego 10, PL-61680 Poznań, Poland; monika.rzodkiewicz@amu.edu.pl
[6]   Faculty of Biological and Environmental Sciences, Aquatic Biogeochemistry Research Unit, Ecosystems and Environment Research Program, University of Helsinki, P.O. Box 65, FI-00014 Helsinki, Finland; saija.saarni@helsinki.fi (S.S.); tom.jilbert@helsinki.fi (T.J.)
[7]   Department of Geography, University of Latvia, Jelgavas 1, LV-1004 Riga, Latvia; normunds.stivrins@lu.lv
[8]   Department of Geology, Tallinn University of Technology, Ehitajate tee 5, EST-19086 Tallinn, Estonia
[9]   Lake and Peatland Research Centre, LV-4063 Puikule, Aloja, Latvia
[10]   Faculty of Oceanography and Geography, University of Gdańsk, Bażyńskiego 4, PL-80309 Gdańsk, Poland; wojciech.tylmann@ug.edu.pl
[11]   Faculty of Biology, University of Latvia, Jelgavas 1, LV-1004 Riga, Latvia
[12]   Department of Geology, University of Latvia, Jelgavas 1, LV-1004 Riga, Latvia
*   Correspondence: izawiska@twarda.pan.pl

Received: 4 April 2020; Accepted: 16 May 2020; Published: 20 May 2020

Abstract: Cultural eutrophication, the process by which pollution due to human activity speeds up natural eutrophication, is a widespread and consequential issue. Here, we present the 85-year history of a small, initially *Lobelia–Isoëtes* dominated lake. The lake's ecological deterioration was intensified by water pumping station activities when it received replenishment water for more than 10 years from a eutrophic lake through a pipe. In this study, we performed a paleolimnological assessment to determine how the lake's ecosystem functioning changed over time. A multi-proxy (pollen, Cladocera, diatoms, and Chironomidae) approach was applied alongside a quantitative reconstruction of total phosphorus using diatom and hypolimnetic dissolved oxygen with chironomid-based transfer functions. The results of the biotic proxy were supplemented with a geochemical analysis. The results demonstrated significant changes in the lake community's structure, its sediment composition, and its redox conditions due to increased eutrophication, water level fluctuations, and erosion. The additional nutrient load, particularly phosphorus, increased the abundance of planktonic eutrophic–hypereutrophic diatoms, the lake water's transparency decreased, and hypolimnetic anoxia occurred. Cladocera, Chironomidae, and diatoms species indicated a community shift towards eutrophy, while the low trophy species were suppressed or disappeared.

Keywords: eutrophication; water level fluctuation; multi-proxy approach; Cladocera; Chironomidae; diatoms; Northern Europe

---

1. Introduction

Eutrophication strongly influences the functioning of freshwater ecosystems by changing their water qualities, such as oxygen availability, light conditions, and increasing the production of algae, which results in a reduction of the water's self-purification capacity. The process of nutrient enrichment of water bodies is a part of a lake's natural lifecycle. However, the introduction of sewage water, fertilizer, and detergent to lake systems greatly accelerates eutrophication and results in significantly increased biological productivity. Therefore, cultural eutrophication continues to be ranked as the most common water-quality problem in the world [1–5].

Not only intensified cultivation and clear-cut logging, but also activities such as artificial water replenishment into lakes, can strongly affect lake ecosystems. There are several reasons for taking such actions, such as increasing the flow of water into lakes for restoration (particularly relevant in arid regions due to intense evaporation), in connection to the hydropower industry [6,7], or increasing the groundwater level to secure the operations of water pumping stations. Such activities can cause significant environmental issues since the physical, chemical, and biological characteristics of the lake can be changed.

Water was artificially replenished to increase the water level in Lake Sekšu, located in the vicinity of Riga city (Latvia, Baltic Region, Northern Europe). This lake is a part of the drinking water supply system and enriches the groundwater horizons near the "Baltezers" drinking water pumping station. The pumping station "Baltezers" began to operate in 1904 as an extension to the pre-existing water supply system meant to solve the problem of water shortages in the Riga city. In addition to the existing drinking water source (the river Daugava), the new water pumping station was designed to use groundwater. However, alongside the construction of residential and public buildings, the area's population and industrial activity increased rapidly after World War I. The size of the water supply network developed proportional to urban growth, but during World War II, the water supply system was seriously damaged. However, by 1948, the pre-war level of industrial activity was again reached [8]. In the late 1950s, the suburbs of Riga were developed for housing purposes. Natural biotopes in the vicinity of the city were replaced by dense residential areas and small kitchen gardens [9]. The consumption of water continued to increase, and in 1953, an artificial groundwater recharge system went into operation. Between 1953 and 1965, the water supply from an adjacent eutrophic lake to Lake Sekšu was established through a pipe and ditch [10].

In this study, we hypothesized that even the relatively short-term pumping of water from the nearby eutrophic lake could have changed the trophic conditions in Lake Sekšu and could have led to persistent shifts in the lake's ecosystem functioning. We analyzed sediment core representing the time period ~1935–2018 to investigate how water replenishment affected the lake's ecosystem functioning and to determine if the lake system showed recovery after the water replenishment activities were terminated. We applied a multi-proxy approach and developed a quantitative reconstruction of the total amount of phosphorus using diatoms and hypolimnetic dissolved oxygen with chironomid-based transfer functions along with the indicative properties of Cladocera. The plant succession in the lake and catchment was reconstructed using pollen analysis. The results were supplemented with a geochemical analysis, which helped to detect changes in the relative supply of organic and inorganic sediment material, as well as variability in the organic matter sources. The geochemical analysis also provided further evidence of changes in the lake's trophic state.

2. Materials and Methods

2.1. Study Site

Lake Sekšu (57°03′ N, 24°35′ E, Figure 1) is a small (surface area 7.9 ha), shallow (the average and maximum depth is 2.5 m and 6 m, respectively) lake at an elevation of 2.5 m a.s.l., located in the vicinity of the capital city of Riga, central Latvia, Northern Europe [11,12]. The average annual air temperature (1981–2010) in areas close to the Baltic Sea coast is +6.8 to +7.4 °C. July is the warmest month of the year, with an average air temperature +17.4 °C (average maximum +22.3 °C). The coldest month of the year is February, with an average air temperature of −3.7 °C. The annual rainfall is 692 mm. According to the data on climate change, the air temperature and precipitation in the area are increasing [13].

The lake is mostly surrounded by inland dune forest and dominated by pine growing on sandy soil. The southern part of the lake is dominated by birches growing on organic soils [14]. The lake has no runoff apart from a small, shallow ditch that transports humic substances to the lake. In this area, there is also a peat soil-based, meliorated forest [9,13].

According to the EU Water Framework Directive criteria (total phosphorus, Secchi depth, and chlorophyll *a*), the lake's current ecological status is good. The latest macrozoobenthos studies confirmed the lake's high biodiversity. Following Carlson's trophic state index, in terms of water transparency (SD), Lake Sekšu is a mesotrophic lake, while its chlorophyll *a* concentration (CA) indicates a eutrophic lake status. The average value of the indices corresponds to a eutrophic lake. Nevertheless, the lake still features low eutrophication. The main problem for such lakes is a loss of transparency [15].

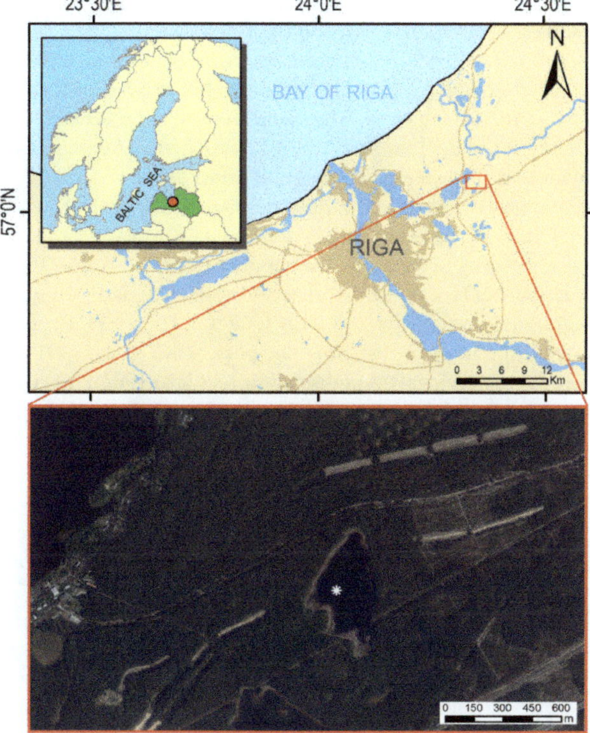

Figure 1. Lake Sekšu location on map. Asterisk indicates deepest part of the lake where sampling occurred.

2.2. Materials and Methods

2.2.1. Lake Sediment Coring

A 46 cm sediment core from the deepest part of Lake Sekšu was taken on 13th February 2019 using a Kayak/HTH gravity-type corer. The sediment core was divided in the field into 1 cm sections and stored in a cold room. In the laboratory, 1 cm^3 subsamples of fresh sediment from each section were taken for analysis of the biological (Cladocera, Chironomidae, diatoms, pollen) and spheroidal carbonaceous particles (SCPs). Dried material was used for $^{137}Cs/^{210}Pb$ dating, as well as chemical and physical analysis.

2.2.2. Core Chronology

A sediment core from Lake Sekšu was dated with ^{137}Cs and ^{210}Pb at the Geochronology Laboratory at the Gdańsk University according to the standard procedure [16]. The activity of ^{137}Cs and ^{226}Ra was determined directly by gamma-ray spectrometry. Gamma measurements were carried out using a HPGe well-type detector (GCW 2021, Canberra) coupled to a multi-channel analyzer and shielded by a 15 cm thick layer of lead. Counting efficiency was determined using reference materials (CBSS-2 for ^{137}Cs at 661.6 keV, and RGU-1 for ^{226}Ra via ^{214}Pb at 352 keV) with the same measurement geometry as the samples. The counting time for each sediment sample was 24 h.

The activity of total ^{210}Pb was determined indirectly by measuring ^{210}Po using alpha spectrometry. Dry and homogenized sediment samples of 0.2 g were transferred into Teflon containers, spiked with a ^{209}Po yield tracer, and digested with concentrated HNO_3, $HClO_4$, and HF at a temperature of 100 °C using a CEM Mars 6 microwave digestion system. After 24 h, the solution was transferred into a Teflon beaker, evaporated with 6 M HCl to dryness, and then dissolved in 0.5 M HCl. Polonium isotopes were spontaneously deposited within 4 h on silver discs. After deposition, the discs were analyzed for ^{210}Po and ^{209}Po using a 7200-04 APEX Alpha Analyst integrated alpha-spectroscopy system (Canberra) equipped with PIPS A450-18AM detectors. The samples were counted for 24 h. A certified mixed alpha source (^{234}U, ^{238}U, ^{239}Pu, and ^{241}Am; SRS 73833-121, Analytics, Atlanta, Georgia, USA) was used to check the detector counting efficiencies, which varied from 30.9% to 33.9% for the applied geometry.

In addition, an abundance of spheroidal carbonaceous particles (SCPs) was estimated throughout the sediment sequence following the methodology of Rose [17] and Alliksaar [18]. The analysis was performed in the Department of Geography, University of Latvia. According to the black carbon combustion continuum model of Hedges et al. [19] and Masiello [20], SCPs only form during industrial fuel combustion at high temperatures (greater than 1000 °C). Therefore, the peak followed the fuel combustion pattern: 1950—the rise of SCPs, 1982—the peak of SCPs, and 1991—the decrease of SCPs. The peak SCP emissions were previously established for Latvia at 1982 ± 10 years [21].

In the final step, we combined the results from the radionuclides and the SCP analyses to build an age-depth model using the Clam 2.2 deposition model [22] with a 95.4% confidence level in the R environment [23].

2.2.3. Physical and Chemical Sediment Analyses

Sediment geochemical characteristics were determined using loss on ignition (LOI) combustion analysis and inductively coupled plasma-optical emission spectrometry (ICP-OES). Altogether, 45 subsamples at 1 cm intervals were analyzed, but only the topmost two sediment samples were merged into one sample representing depths from 0 to 2 cm. The sediment organic matter and carbonate content were investigated using the LOI method [24]. A measure of 0.1–0.2 g of fresh sub-sample sediment was weighed in a crucible and dried at 105 °C for 12 h; then, it was combusted at 550 °C for 4 h and, finally, at 950 °C for 2 h. Between each step, the samples were cooled with an exicator and weighed. The organic matter (OM) content was measured as the LOI from the combusted samples at 550 °C, and the carbonate matter (CM) was calculated as the difference between the LOI at 950 °C and the LOI at 550 °C multiplied by 1.36 [24]. Non-carbonate siliciclastic matter, here referred to as minerogenic matter (MM) content, was obtained by subtracting OM and CM from the total sample weight after final combustion. MM also includes biogenic silica (opal). However, siliciclastic matter here mainly represents terrigenous clastic matter.

A set of homogenized sub-samples was freeze dried and powdered using an agate mortar. The total carbon (TC) and total nitrogen (TN) contents were analyzed using thermal combustion elemental analyses (Element Analyzer Vario EL III) with an uncertainty of ±5%. Approximately 7–8 mg of prepared material was weighed into tin cups for analyses that were performed at an accredited laboratory (ISO/IEC/17025).

For the ICP-OES analyses, 0.1–0.2 g of dry sediment powder was weighed in Teflon vials, and 5 mL of 65% HNO_3 was added. The vials were closed loosely, and digestion was carried out at 160 °C for 30 min. Gases and volatiles from the digestion were allowed to escape through the cap. Following digestion, 10 mL H_2O was added to dilute the remaining acid, and the vials were weighed to determine the dilution factor. A measure of 1 mL of the resulting solution was then pipetted into 15 mL centrifuge tubes, and 9 mL 0.5 mol/l HNO_3 was added. These steps were performed to give the samples a final acid concentration of approximately 1 mol/l HNO_3 and to have the element concentrations in a suitable range for the ICP-OES analysis. The sulfur (S) concentration was determined by ICP-OES (Thermo Scientific iCAP 6000) using a concentric nebulizer. Certified control samples and blank samples were used to ensure the quality of the analytical process.

2.2.4. Pollen and Non-Pollen Palynomorphs

Samples of known volume for the pollen analysis were processed using standard procedures [25]. Known quantities of *Lycopodium* spores were added to each sample to allow the calculation of pollen concentrations [26]. At least 500 terrestrial pollen grains per sample were counted under a light microscope (400× magnification). Taxa were identified to the lowest possible taxonomic level using the reference collection at the Department of Geography at the University of Latvia along with published pollen keys [27]. The percentage of dry-land taxa was estimated using arboreal and non-arboreal pollen sums (excluding the sporomorphs of aquatic and wetland plants). Counts of spores were calculated as the percentages of the total sum of terrestrial pollen. Non-pollen palynomorphs were recorded throughout the pollen analysis and identified using the published literature listed in Miola [28], as well as from the descriptions of Sweeney [29] and Finsinger and Tinner [30]. Non-pollen palynomorphs were expressed as presence or concentrations. The pollen diagram was compiled using TILIA software [31].

2.2.5. Diatom Analysis

Samples were prepared according to the standard methods [32]. The material was treated with 10% HCl to remove calcareous matter, washed with distilled water, and then treated with 30% H_2O_2 in a water bath to remove organic matter. The material was repeatedly washed with distilled water, and a known number of microspheres in a solution (concentration 8.02×10^6 microspheres/cm^3) was added to the diatom suspensions to estimate the diatom concentrations [33]. A few drops of diatom suspension were dried on a cover glass. At least 300 diatom valves per sample were analyzed using oil immersion at 1000× magnification under a light microscope. For identification, a selection of published keys was used [34–41].

The diatom ecological groups were determined using the OMNIDIA software (Version 4.2) [42]. Next, the resulting groups were distinguished according to Denys [43] and van Dam [44]. We considered the following indicator parameters: habitat category [43], dominant taxa (abundance over 2%) and preference for pH, saprobic level (OM contamination), and trophy [44]. The percentage diatom diagram was prepared with the Tilia software [31].

The total phosphorus (TP) concentration was reconstructed based on changes in the diatom species composition (DI-TP). The reconstruction was performed using the European Diatom Database (EDDI) in the ERNIE software [45]. A model based on inverse regression had a root mean square error of prediction (RMSEP) of 0.33 μg L^{-1} and a coefficient of determination (r^2) of 0.64. The reconstruction of the TP was based on the diatom taxa present at more than 2% abundance. The DI-TP was calculated using the combined TP dataset (derived from nine datasets with 347 samples in total), covering a TP range of 2–1189 μg L^{-1}, with a mean of 98.6 μg L^{-1}. The weighted averaging (WA) method with good empirical predictive ability was used [45].

2.2.6. Cladocera Analysis

The 46 fresh sediment samples were prepared in a laboratory according to the standard procedures [46], heated in 10% KOH, and sieved using a 38 µm mesh size. Microscope slides were prepared from 0.1 mL of each sample and examined with a light microscope under magnifications of ×100, ×200, and ×400. For each sample, 1–3 slides were scanned, and all skeletal elements (head shields, shells, and postabdomens) were counted until 70–100 individuals were found, which is regarded as an adequate number to characterize the assemblages [47]. Identification of the cladoceran remains was based on the key by Szeroczyńska and Sarmaja-Korjonen [46]. The stratigraphic diagrams presenting the results were prepared using C2 freeware [48].

The Cladocera composition is presented in the stratigraphic diagrams with percentage values. For trophic state reconstruction, we used changes in the percentage of species regarded as indicators for the eutrophic state, which, according to Flössner [49], are *Alona rectangula* and *Chydorus* cf. *sphaericus*. Lake water level changes were reconstructed by using the added percentage values of planktonic and littoral cladocerans.

2.2.7. Chironomidae Analysis

Standard methods were applied in the fossil Chironomidae analysis [50]. The wet sediment was sieved through mesh (100 µm), and the residue was examined under a stereomicroscope for larval head capsule extraction using a target counting sum of 50 per sample. The head capsules were mounted with Euparal on microscope slides for taxonomic identification following Brooks et al. [50] under a light microscope (400× magnification).

Hypolimnetic dissolved oxygen (DO) was reconstructed using a Finnish 30-lake chironomid-based calibration model [51,52]. The calibration sites range from anoxic ($O_2 = 0.5$ mg L^{-1}) to hypersaturated sites ($O_2 = 18.1$ mg L^{-1}). The weighted averaging partial least squares (WA-PLS) model had an r^2 (leave-one-out cross-validation) of 0.74 and an RMSE P of 2.3 mg L^{-1}.

3. Results

3.1. Core Chronology

The depth profile of ^{137}Cs activity is long and smooth, with only one wide maximum between a 28 and 38 cm sediment depth (Figure 2). The sharp decrease below 41 cm indicates the presence of sediments older than 1950 below this depth. The lack of two independent peaks indicating global fallout during 1961–1965 and the Chernobyl peak in 1986 might suggest the deep mixing of surface sediments.

However, the excess ^{210}Pb activities decrease regularly with mass depth, which demonstrates no significant disturbance of the sediment column and relatively stable mass accumulation rates. Thus, we used the CFCS (Constant Flux Constant Sedimentation) model to calculate the mean value of the mass accumulation rate (MAR) for the entire core. The mean MAR value is 34.5 ± 2.2 mg/cm^2/yr and allows us to estimate a maximum age of 84 ± 5 yrs (1935 ± 5) at a 45.5 cm sediment depth. The age of the depth interval with the greatest ^{137}Cs activities (37–32 cm) is 1956–1965 according to the CFCS model, which is consistent with global fallout history. This comparison suggests that the lack of a Chernobyl peak is related to ^{137}Cs migration within the sediment column rather than physical sediment mixing.

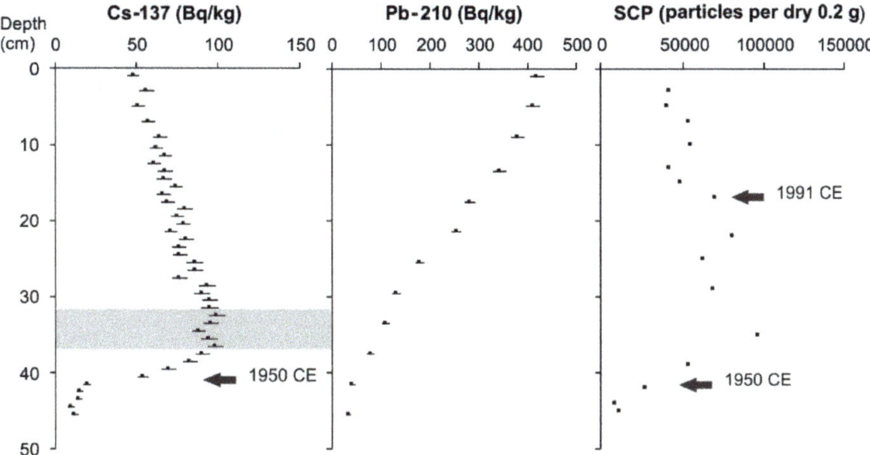

Figure 2. Downcore distribution of ^{137}Cs and ^{210}Pb activities as well as spheroidal carbonaceous particles (SCP) concentrations in the Lake Sekšu topmost sediments. The grey shaded area on the caesium diagram indicates time period 1956–1965 CE.

Based on the SCP results, it was possible to define the years 1950 and 1991 with an error of +/− 10 years at depths 42 and 17 cm, respectively. The locations of these peaks are, with relative certainty, consistent with the ^{137}Cs and ^{210}Pb data.

The final age-depth model, including the radionuclide and SCP data, shows relatively stable sedimentation rates with only a moderate increase in the topmost part of the profile (Figure 3). The maximum age of sediment at a depth of 46 cm is 84 ± 9 yrs (95%). The mean sedimentation rate is 0.6 cm per year.

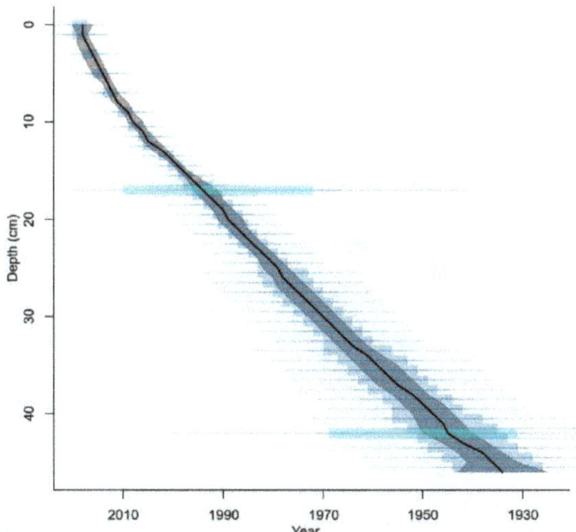

Figure 3. Lake Sekšu age depth model. The black solid line shows the weighted mean ages of all depths, whereas the grey area indicates a reconstructed 95% chronological uncertainty band. Dates of ^{210}Pb (pale blue boxes) and spheroidal carbonaceous particles (electric blue boxes) with their error margin and associated age-depth model uncertainties are displayed.

3.2. Sediment Composition

The Lake Sekšu sediment composition is dominated by organic matter (50%–61%) with a significant component of minerogenic matter (35%–46%, Figure 4). The sediments contain 2%–5% carbonate matter. Based on sediment composition and the variation in element concentrations, the sedimentary data are divided into two zones, Zones I and II, which correspond to the approximate time period before 1950 and between 1950 and 2018. Within Zone II, short-lived events can be distinguished.

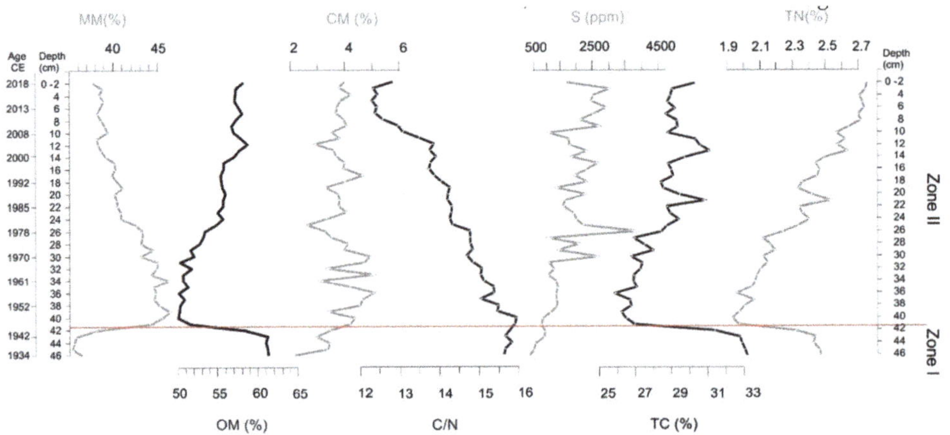

Figure 4. Lake Sekšu Sediment composition: minerogenic matter (MM), organic matter (OM), carbonate matter (CM), C/N expressed as atomic ratio, sulfur (S), carbon (TC), nitrogen (TN), and the division in two zones based on major changes in sedimentary composition.

Zone I

The relative organic matter (OM), total carbon (TC), and total nitrogen (TN) content show high values in this time period (Figure 4). The OM is around 61%, while TC is around 32%, and TN is around 2.5%. The C/N value, expressed as the atomic ratio, is around 15.7. The sulfur (S) content is approximately 500 ppm. The relative minerogenic matter (MM) (36%) displays the lowest values, while approximately 3.5% is carbonate matter (CM).

Zone II

At the boundaries of Zone I and Zone II, a large change occurs. The OM content shows a sudden drop from 61% to 51% accompanied by a fall in TC and TN content from 33% to 27% and from 2.45% to 1.95%, respectively. Simultaneously, the MM content increases from 36% to 45%. After this rapid change, a steady gradient emerges from 41 cm to the sediment surface, during which MM, OM, and TC gradually reset towards their Zone I value. However, TN increases over the same period to values far exceeding those of Zone I (approximately 2.8% at the sediment surface, Figure 4). Consequently, C/N displays a continuous gradient towards lower values throughout the entire core. Similarly, S content increases gradually throughout the entire core, with some pronounced variability within Zone II.

3.3. Pollen and Non-Pollen Palynomorphs

There is no significant change in surrounding vegetation before or after the pipe installation. The vegetation includes the stable dominance of pine (*Pinus*), birch (*Betula*), spruce (*Picea*), and alder (*Alnus*) over the studied period (Figure 5). The presence of conifer stomata in the lake sediment supports local abundance of spruce and pine. Although, the surrounding landscape is forested, there is continuous evidence of human-activities in the vicinity. For instance, the presence of pollen of flax (*Linum*), rye (*Secale cereale*), barley (*Hordeum*), and wheat/oat (*Triticum/Avena*) is a direct indication of agricultural practices. Our results also underline that there have not been agricultural fields directly at the Sekšu lake shores, but only further away, as evidenced by the high (nearly 85–90%) forest pollen component in the landscape. It is interesting that the rye pollen accumulation rates above 1000 cm^{-2} year^{-1} (Figure 6) point to cereal fields within a 2 km radius of Lake Sekšu [53,54]. Based on USSR topographic maps from the time period of 1941–1991, the cereal fields could be present in the northwest and east of the Sekšu Lake where pastoral activities have also been evident. It is, however, possible that the high concentration of rye pollen can be partially a result of soil erosion and water pumping (enlarging the pollen source area outside the watershed) into the Lake Sekšu. This assumption is further supported by a strong increase of fungi hyphae and corroded pollen grains.

Regarding in-lake vegetation, lake quillwort/Merlin's grass (*Isoëtes lacustris*) was recorded throughout the sediment sequence. The highest relative abundance of *I. lacustris* was recorded prior the pipe installation, after which values continuously declined and did not reach previous values. Sporadically, the pollen of water lilies (*Nymphaeaceae*), bulrush (*Typha*), spiked water-milfoil (*Myriophyllum spicatum*), pondweed (*Potamogeton*), and bur-reed (*Sparganium*) were found.

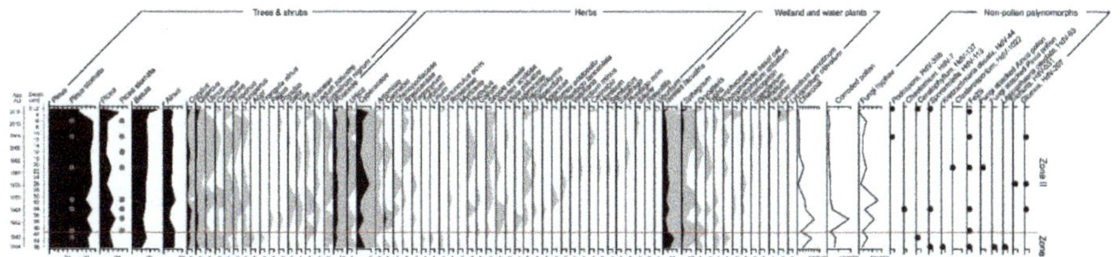

Figure 5. Pollen and non-pollen palynomorphs of Lake Sekšu. Trees, shrubs, herbs, crops, ruderal plants, bryophytes, and water plants expressed in percentages. Non-pollen palynomorphs and charcoal expressed as microscopic object accumulation rate cm^{-2} year^{-1}. Grey shaded areas indicate ×10 time exaggeration to underline the presence of microscopic remains.

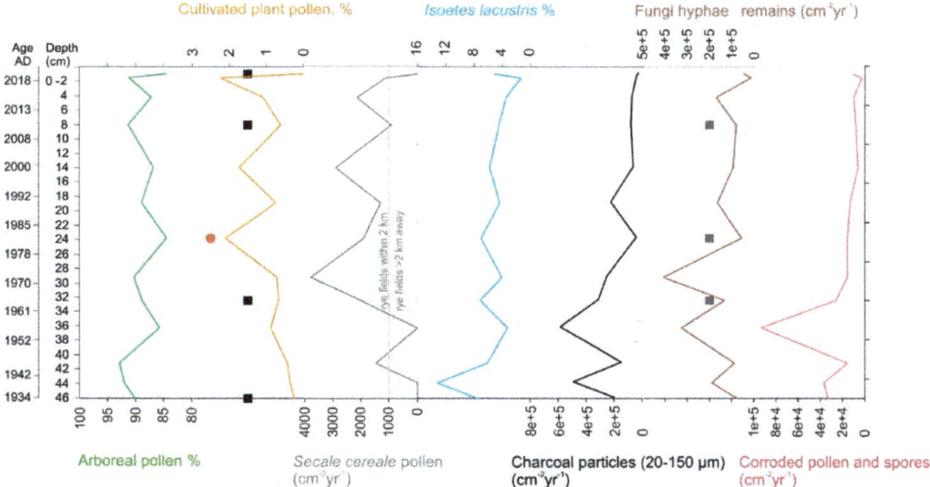

Figure 6. Selection of pollen and non-pollen palynomorphs, from the left: arboreal pollen (%), cultivated plant pollen (%), secale-cereale pollen (cm^{-2} yr^{-1}), *Isoëtes lacustris* (%), charcoal particles (cm^{-2}yr^{-1}), fungi hyphae remains (cm^{-2}yr^{-1}), and corroded pollen grains and spores (cm^{-2} yr^{-1}). The presence of whipworm in the sample is marked with a red circle, the presence of herbivores with black squares and the presence of *Glomus* spores indicating erosion with grey squares.

3.4. Diatom Analysis

The results show a medium to deplorable state for the diatom frustules, with numerous traces of destruction and dissolution. In total, 178 species of diatoms were identified. There were 47 dominant taxa, whose share was more than 2% of the relative abundance. Based on the changes in species composition and the proportions between the ecological diatom groups, the data are divided into two zones, Zones I and II, which closely correspond to the geochemical zones (Figures 7 and 8). Species were classified according to their habitat category, preference for pH, saprobic state, and trophic state (Figure 7). The total phosphorous (TP) reconstruction presented values between 21.43 and 102.14 µg L^{-1} (Figure 11).

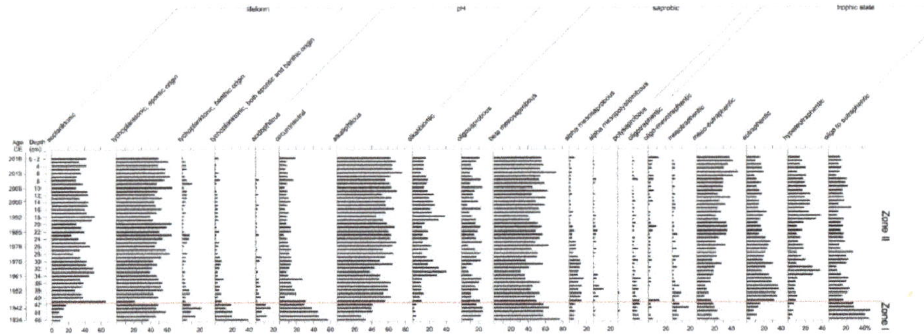

Figure 7. Relative abundance of the diatom group in Lake Sekšu according to habitat category (lifeform), pH, OM contamination (saprobic), and trophy preference.

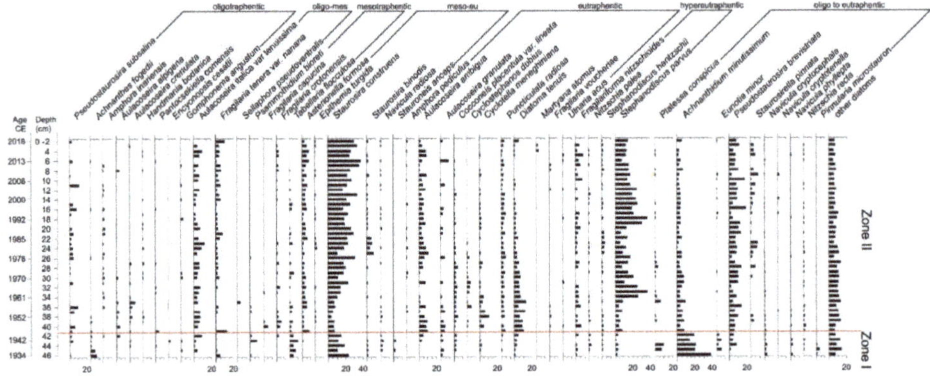

Figure 8. Relative abundance of dominant diatom species in Lake Sekšu (species with values >2%).

Zone I

This zone was characterized by the dominance of tychoplanktonic taxa, especially *Staurosira construens* and *Achnanthidium minutissimum* (Figure 8). In terms of pH, circumneutral taxa dominated. At the beginning of the phase, we observed an increase in alkaliphilous diatoms. The analysis of saprobic preferences revealed the domination of β-mesosaprobous diatoms. In terms of its trophic state, a high proportion of oligo to eutraphentic taxa was observed. The TP reconstruction remained low, between 21.43–57.66 µg L^{-1} (Figure 11).

Zone II

Several short-term fluctuations in species composition were observed. Around 1950, the planktonic taxa (euplanktonic) increased by up to 65%, represented by *Asterionella formosa*, *Aulacoseira ambigua*, *A. granulata*, *Diatoma tenuis*, *Fragilaria nanana*, *Cyclotella planktonica*, *Handmania bodanica*, and *Pantocsekiella comensis* (Figure 8). The alkaliphilous species dominated the pH group, with an abundance between 38.7 and 76.5 R. Around 1950, the α-mesosaprobous species increased by 0.3%–17.0%. An increase in eutrophic and hypereutrophic taxa was also observed (Figure 7). The TP reconstruction values increased up to 102.14 μg L^{-1} (Figure 11).

3.5. Cladocera Analysis

The remains of 44 cladoceran species were found (Figure 9). Throughout the core, the dominant planktonic species were *Bosmina (Eubosmina)* spp. and *Bosmina (Bosmina) longirostris*, while among the littoral species, *Chydorus* cf. *sphaericus*, *Alona quadrangularis* and *Alonella nana* dominated. The overall species composition changes, as well as the indicator species appearance and continuation, allowed us to distinguish between Zones I and II, which correspond to the time before and after the pipe started to operate (Figure 9).

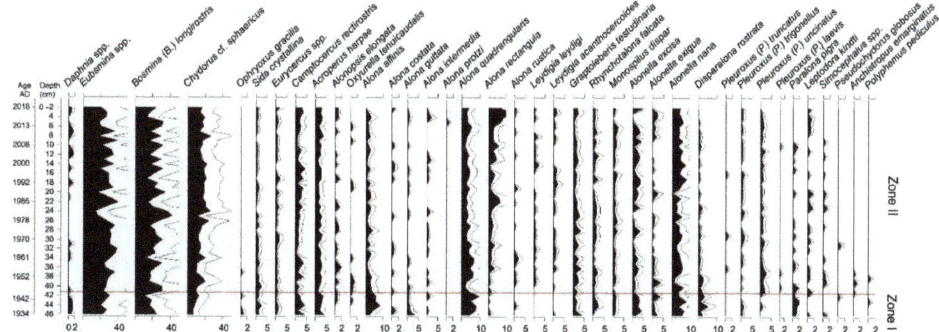

Figure 9. Relative abundance of the Cladocera species in Lake Sekšu.

Zone I

The littoral cladocerans comprised more than 50% of all species, among which *Alona affinis*, *A. quadrangularis*, and *Alonella nana* dominated. The Cladocera eutrophic indicator, which is based on the littoral species, showed the lowest values in the sequence. The rare species *Ophryoxus gracilis* was present almost continuously. Among the pelagic species, *Bosmina (E.)* spp. dominated, and *Daphnia* spp. remains were constantly present.

Zone II

During this time, several fluctuations in the species share and Cladocera indicator are observable. The pelagic species constitute more that 50% of all species until 2014 CE. However, an increase in the littoral taxa share was noted at the beginning of 1950, 1960, 1980, and ~1995. Next, *Chydorus* cf. *sphaericus* becomes the dominant littoral species. After 2014 CE, a clear decrease of pelagic species and an increase of littoral species occurred, and littoral *A. rectangula* started to dominate. The trophic indicator increased from around 1950 CE, but even more visibly from the ~1978 (Figure 9). The highest values of the indicator species were observed almost continuously from ~2010 until present.

3.6. Chironomidae Analysis

In the 45 sediment samples, 52 different chironomid taxa were observed. None of the taxa occurred in all the samples. The *Polypedilum nubeculosum*-type was the most frequent, as it was observed in 40 samples. The highest maximum abundance values belonged to *Procladius* (35.3%), *P. nubeculosum*-type (31.4%), and *Heterotanytarsus apicalis* (30.0%). The *P. nubeculosum*-type also had the highest mean abundance (9.3%), followed by the *Psectrocladius sordidellus*-type (7.9%) and the *Tanytarsus pallidicornis*-type (6.6%) (Figure 10). The hypolimnetic oxygen reconstruction showed values between 0.2 and 15.7 mg L^{-1} (Figure 11).

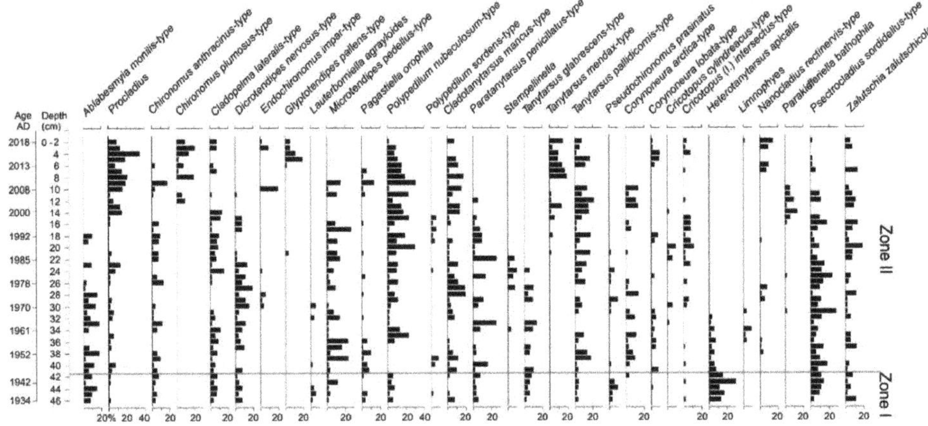

Figure 10. Relative abundance of Chironomidae in Lake Sekšu sediment.

Zone I

The initial part of the sediment profile (before ~1950 CE) was dominated by oligotrophic taxa, such as *H. apicalis* and the *Ablabesmyia monilis*-type. The hypolimnetic oxygen reconstructed values remained high at ~10 mg L^{-1}.

Zone II

In the middle portion of the sediment profile (1950–1990 CE), the *P. sordidellus*-type, which is a common species, was the most abundant. From 2000 CE onwards, *Procladius*, which prefers nutrient enriched waters, began to dominate. During the most recent years, the eutrophy indicating the *Chironomus plumosus*-type and *Glyptotendipes pallens*-type also significantly increased. From the 1950s until the 1980s, the oxygen values decreased to a level of ~4–8 mg L^{-1}. The hypolimnetic oxygen reconstructed values remained at ~4 mg L^{-1} from the late 1980s until 2010, after which the values further decreased to anoxic levels in recent years.

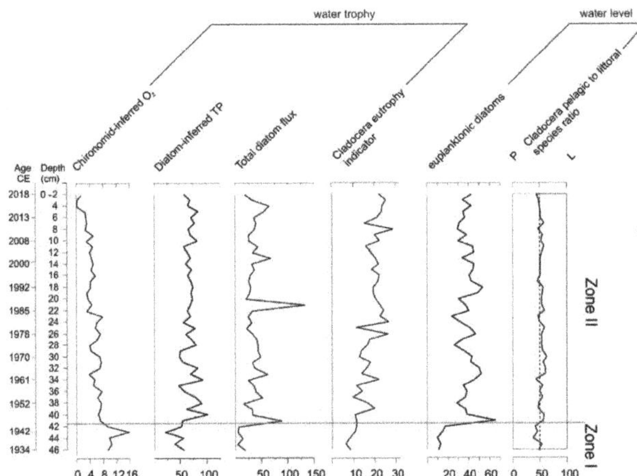

Figure 11. Comparison of indicators used to reconstruct trophic state changes: chironomid-inferred hypolimnetic oxygen reconstruction (O$_2$) mg L^{-1}, diatom-inferred total phosphorus reconstruction (TP) µg L^{-1}, total diatom flux (frustules × 10^6 /cm^2/yr^{-1}), Cladocera based eutrophy indicators, and water-level changes indicators, such as the euplanktonic diatoms (%) and pelagic (P) to littoral (L) Cladocera species ratio (%).

4. Discussion

Lake Sekšu exemplifies how environmental changes have occurred, in this case, as a result of artificial water pumping station activities, and also shows how such changes at the trophic level can have a lasting influence on lake functioning. In 1953 CE, the pipe was built and for 12 years (1953–1965) pumped water to Lake Sekšu from the large eutrophic lake, Mazais Baltezers. Our results clearly show that this action persistently changed the lake's ecosystem by causing increased eutrophication.

A major cause of eutrophication is an increase in the nutrient supply, particularly phosphorus [55]. Besides excessive plant production, algal blooms, and anoxia, the lake community structure changes through cascading trophic effects and benthic-pelagic coupling [56,57]. Another stressor controlling ecosystem functioning is water level fluctuations, which are essential for aquatic–terrestrial shared boundary processes. The physical and biological effects are especially pronounced in the littoral zone and in shallow water ecosystems in general (e.g., erosion, sedimentation, habitat alterations, and biota changes) [58]. Furthermore, erosion is a source of enhanced nutrient supply, and a greater sediment load decreases water quality. An increase of suspended solids reduces water transparency and increases turbidity, which negatively affects the photosynthesis processes but may also reduce the density of fish and invertebrates [56,59].

The Time Before ~1950

Before the pipe started to operate in 1953 CE, Lake Sekšu was a low productivity lake. However, the Cladocera community composition found in our research and in a study by Kuptsch [60] shows that already at the beginning of our core (~1935), the lake trophy had increased compared to 1924. This increase is indicated by a disappearance of the oligotrophy indicating *Holopedium gibberum* and the appearance of the increasing trophy indicator, *Bosmina (Bosmina) longirostris* [61–65].

The oldest available investigations from 1924 report a transparency of 5.8 m in the middle of June [60], i.e., almost throughout the lake. It was also reported that Lake Sekšu is one of the 14 lakes located in the vicinity of Latvia's capital city, Riga, that are *Lobelia–Isoëtes* population dominated lakes [66]. At the beginning of 20 century, there were at least three species typical for *Lobelia–Isoëtes* lakes found in Lake Sekšu—*L. dortmanna*, *Isoëtes lacustris*, and *Juncus bulbosus* [67]. Our research revealed the presence and high share of lake quillwort pollen *Isoëtes lacustris*, as well as Cladocera species *Ophryoxus gracilis*, around 1935 (Figure 9). *Ophyroxus gracilis* is a species characteristic of deep, transparent lakes, indicating low/moderate trophy [64,68]. Flössner [49] considers *O. gracilis* a species typical for *Lobelia–Isoëtes* lakes.

The diatom-based total phosphorus (TP) reconstruction for Lake Sekšu indicates values between 25–50 µg L^{-1} before the pipe started to operate (Figure 11), which, according to the OECD (Organisation for Economic Cooperation and Development) classification, indicates a meso- to eutrophic type lake [69]. However, this reconstruction seems to overestimate the TP values when considering the other paleoindicators. This might be due to weak diatom assemblage analogues [70–73]. Nevertheless, the diatom-inference model closely reflected a significant trend in the measured TP, namely, declining TP values before pipe installation and increasing TP values after pipe installation.

The low trophy status is confirmed by the high share of oligotraphentic diatoms and dominance of Chironomidae oligotrophic taxa [74,75]. The well oxygenated hypolimnetic water (values 9–16 mg L^{-1}, Figure 11) was inferred by chironomid-based reconstruction and supports a low trophic status. The chironomid-inferred oxygen values remained at a high level (>8 mg L^{-1}) from the beginning of the record until the mid-1940s, when a step change to a lower level (4–8 mg L^{-1} between the 1950s and early 1980s) occurred. The relative oxygen saturation exceeded 50% [56]. Lake Sekšu was slightly acidic to neutral at that time, indicated by the dominance of acidophilous and circumneutral diatoms [76–79].

The geochemical results support the conclusion of low trophy before the construction of the pipe. However, at that time, the sediment was dominated by organic matter (OM) (Figure 4) with a very high concentration of carbon compared to other lakes in the region [80–83]. The high carbon/nitrogen (C/N) values suggest a significant contribution of terrigenous OM. Atomic C/N values of phytoplankton vary between 4 and 10, while cellulose-rich vascular plants have C/N ratios of more than 20 [84,85].

Therefore, high C/N rates are expected for oligotrophic lakes with low autochthonous production, and the values observed here are consistent with those of the densely forested catchment of Lake Sekšu. The low water levels of this period due to water pumping, and the subsequent water level decreases as early as 1930, may also have led to the concentration of OM in the coring location due to erosion from exposed organic rich littoral sediments from the lake. The water level in 1930 was reported to be very shallow at 1 m a.s.l [86]. The dominance of littoral over planktonic Cladocera confirms this information [87]. In addition, we reconstructed the high hypolimnetic water's oxygen values (Figure 11), which suggest that the sediment was exposed for efficient bioturbation. Water mixing is also confirmed by the dominance of tychoplanktonic diatoms, i.e., random planktonic diatoms, which require turbulent waters to remain suspended within the photic zone [77,88,89].

The Time Between ~1950–2018

The Lake Sekšu water level was artificially increased by pumping water from Lake Mazais Baltezers starting in 1953 CE in order to elevate the ground water level. Apart from the physical changes caused by an increased water level, this action also caused significant changes to the lake's ecology and functioning. The lake became eutrophic according to the diatom-inferred TP values around 100 μg L^{-1} [69]. During the period of artificial water pumping, until 1965, a high P concentration existed at the water column. After the pipe operation was terminated, the TP decreased to about 70 μg L^{-1} until the present, indicating that elevated trophy continued. According to our results, the additional nutrient load increased the presence of planktonic eutrophic–hypereutrophic diatoms (Figure 7), which, in turn, decreased water transparency. For instance, in August 2013, the Secchi depth was 2.1 m [15], while in 2019, in April, June, and October, it was 1.6 m, 1.6 m, and 1 m, respectively [90].

The total diatom flux (Figure 11) shows an increase, suggesting higher primary production [91–93]. As a result, the *I. lacustris* population greatly decreased. In turn, eutrophic-indicating macrophyte species, such as pondweed (*Potamogeton*), reed (*Typha*), and water lily (*Nuphar*), started to flourish in the lake (Figure 6). *Lobelia–Isoëtes* lakes indicating species have disappeared from Lake Sekšu according to botanical observations [67].

Further evidence of the rapid shift towards higher trophy during artificial water pumping is provided by the other biological and geochemical results. The Cladocera eutrophic indicator, based on the species living in the littoral zone (*Alona rectangula* and *Chydorus* cf. *sphaericus*), increased, thereby indicating eutrophic conditions until the most recent sample (Figure 11). Simultaneously, the higher trophy in the pelagic zone is indicated by the increase of the *Bosmina (B.) longirostris* share. This species is associated with higher concentrations of TP [65,94].

The sudden change in the relative minerogenic matter and OM input coincides with the implementation of the pipeline and water pumping from Lake Mazais Baltezers to Lake Sekšu. The increased MM content of around 10% indicates substantial changes in the sediment sources. The pipe was installed tens of meters from today's lake shore, and flowing waters likely eroded the substrate—sandy soil that is prone to erosion. A comparable increase occurred in carbonate content, which also suggests intensified catchment erosion. The coeval decrease in OM accumulation, as well as carbon and nitrogen content, is likely a result of dilution by an excess MM supply (Figure 4). Despite the prominent relative changes in sediment composition, no evidence of significant increases in the total sedimentation rate are observable. This could be explained by a sudden change from the erosion of previous littoral sediments at a low-water level stage to channel erosion of the ditch, as well as erosion higher on the shore where no former lake sediments exist. Another indicator of the highly dynamic environment after the pipe started to operate is the increased share of corroded pollen grains [95]. The erosion of the shores is also indicated by the presence of herbivores (via coprophilous fungal spores), Fungi hyphae remains, and *Glomus* spores [54,96].

Pumping caused a significant elevation of the water level from 1 m a.s.l up to 4.5 m a.s.l [86], which is also reflected in the paleolimnological results. The higher water stand is expressed by the larger share of euplanktonic diatoms [97,98] and the dominance of pelagic over littoral Cladocera species [87] (Figures 9 and 11). However, the fluctuation of water level during the whole 12-year period can be deduced from the different share of the Cladocera pelagic/littoral taxa [87]. Both the increasing and decreasing water levels are related to increased catchment erosion [99–101]. The variable MM and carbonate matter (CM) content during the period of external water pumping indicate that fluctuating water levels enhanced catchment erosion in addition to channel erosion of the ditch from the pipe-end to the lake.

The C/N ratio (Figure 4) began to decrease soon after the major sedimentary change occurred, which suggests a steadily increasing proportion of phytoplankton in the total organic matter up to the present day. This is consistent with the gradual eutrophication of the lake, as indicated by the microfossil data. The sediment sulfur (S) profile also provides indirect evidence of eutrophication. Sulfur in lake sediments may be derived from natural processes, such as the weathering of sulfur-bearing rocks and the oxidation of organic sulfur in the catchment [102]. However, changes in S contents in recently deposited lake sediments generally indicate the balance between anthropogenic inputs and the rate of sulfate reduction leading to iron sulfide precipitation in sediments [103]. The steady increase in sedimentary S towards the present day suggests an additional input of sulfate from Lake Mazais Baltezers during the period of pumping, followed by enhanced iron sulfide precipitation in anoxic, OM-rich sediments during subsequent eutrophication. Eutrophication due to sewage water input from Lake Mazais Baltezers was suggested by Leinerte [10] (p.2) and is further supported by the high abundance of α-mesosaprobous diatoms [104,105]. This diatom group flourishes under the lower oxygen saturation in the water column [44].

In the middle part of the sediment record, representing the early 1980s, a threshold change can be observed in the chironomid-inferred hypolimnetic oxygen values (Figure 11). At this point, the values decreased from the previous level of 4–8 mg L^{-1} (between ~1945 and 1980) to a constant level of 4 mg L^{-1}, which prevailed until the 2010s. This is supported by the rapid decrease in hypolimnetic oxygen levels reconstructed from the chironomid data and recent measurements in April and June 2019, when the hypolimnetic oxygen concentration was 4.0 mg L^{-1} and 0.1 mg L^{-1}, respectively [90]. Jansons [9] (p. 27) and Zarina [15] (p.41) reported an amount of oxygen close to zero or zero in the hypolimnion of the deepest part of the basin, which agrees well with our reconstruction. Oxygen depletion was likely also favored by the water depth increase from <3 meters to 8 meters. Deeper waters reduced wind-driven mixing and enabled efficient thermal stratification.

From around 2000 CE, the water level lowered, reaching 2.5 m a.s.l. at present. This might have increased sedimentary nutrient release, further favoring eutrophication [58]. This higher productivity is indicated by the increase of the Chironomidae species *Procladius*, which prefers nutrient-rich waters [106] (Figure 10), but is also evidenced by the higher values of diatom-inferred TP (Figure 11). The lowering of the water level inferred from the Cladocera pelagic/littoral species ratio was also confirmed by the local authority [107]. The water level decrease and water trophy increase resulted in oxygen depletion in the hypolimnion, which was revealed by the chironomid reconstruction and recent measurements. Another threshold change based on the chironomid reconstruction occurred during the most recent decade, when the values decreased from a level of 4 mg L^{-1} (between the ~1940s and 2010s) to <2 mg L^{-1} (Figure 11). Such a decrease in oxygen concentration is ecologically highly significant for chironomid larvae [108].

The lake continues to be significantly overgrown with water plants, which is evidenced by the slight increase in the abundance of the Cladocera phytophilous species, *Alonella exigua*, and *A. excisa* [49,109,110]. More sand and mud associated species (*Rhynchotalona falcata*, *Leydigia* spp., and *Ilyocryptus* sp.) have joined the existing community since 1924 [58]. The lake surface pH in 2019 varied within a season from 7.9 to 8.4 [88], which is consistent with previous studies [9,15].

Due to inappropriate water resource management, many lakes in Riga and its vicinity have lost their high ecological quality. The pumping of water from the eutrophic Lake Mazais Baltezers into the low trophic Lake Sekšu increased nutrient inflow and pushed the lake over the threshold, causing eutrophication. Presently, Lake Sekšu continues to be overgrown with macrophytes such as *Phragmites australis*, *Nuphar* sp., and *Nymphaea* sp. [15]. Phytoplankton blooms can be observed in the lake. In 1995/1996, the majority of the cyanobacterial mass contained toxic species [111], while in 2002, Balode et al. [112] found no toxic species.

The "Baltezers" water pumping station is still in use. However, its contribution to the drinking water supply is not as significant as it was in previous years because other pumping stations were established, and additional resources for water supply were found [8].

5. Conclusions

The pumping of water from the eutrophic lake changed the water level and contributed to faster eutrophication in Lake Sekšu. Before the pipe began operations, the lake was shallow with low productivity and could be defined as a *Lobelia–Isoëtes* population-dominated biotope. The core studied represents the time period ~1935–2018. In 1953, the water level was artificially increased by pumping water from another lake. The sudden nutrient load increased the abundance of planktonic eutrophic–hypereutrophic diatoms, causing the lake water transparency to decrease and inducing hypolimnetic anoxia. The deterioration of the lake's ecological state, as indicated by the results of the biological analysis, was also confirmed by the geochemical results. Cladocera species indicating low trophy became suppressed or disappeared.

Reducing external nutrient loading is known as an important approach to improve the environmental state in small shallow lakes suffering from eutrophication. Nevertheless, in Lake Sekšu, after the water replenishment activities were terminated, the disturbance effect caused by the pumping continued to occur over decades, slowly changing assemblages of keystone species and the foodweb. This highlights the long legacy effects of eutrophication in lacustrine systems. Artificial restoration measures may, therefore, be considered as a strategy to accelerate the recovery of Lake Sekšu.

Author Contributions: Conceptualization, I.Z. and I.D.-D.; methodology, I.Z., T.P.L., S.S., N.S., W.T., T.J.; software NO; validation, NO; formal analysis, I.Z., T.P.L., M.R., S.S., N.S., W.T., T.J.; investigation, I.Z., I.D.-D., S.S., A.L., M.R.; resources, I.D.-D.; data curation, I.Z., T.P.L., M.R., S.S., N.S.; writing—original draft preparation, I.Z., I.D.-D., T.P.L., M.R., S.S., N.S., W.T., A.L., M.R., T.J.; writing—review and editing, I.Z., I.D.-D., M.R., S.S., N.S., W.T.; visualization, I.Z., N.S., W.T., M.R.; supervision, NO.; project administration, I.D.-D.; funding acquisition, I.D.-D., I.Z., N.S. All authors have read and agreed to the published version of the manuscript.

Funding: Research was funded by the European Regional Development Fund, 1.1.1.2 Post-doctoral project No.1.1.1.2/VIAA/2/18/359 and Latvian Institute of Aquatic Ecology. The effort of Izabela Zawiska was supported by a grant from Polish National Science Centre No.2016/23/D/ST10/03071. The work of Normunds Stivrins was supported by grants founded by Estonian Research Council grant PRG323 and the Latvian Council of Science Project No. LZP-2018/1-0171 and performance-based funding of the University of Latvia within the "Climate change and sustainable use of natural resources" program. Funding for Tomi P. Luoto was provided by the Kone Foundation (grant number 090140).

Acknowledgments: Our special thanks to lab assistants Juris Tunens, Nina Sunelika and all who participated in sampling. In particular, we would like to acknowledge Timo Saarinen for organizing sampling campaign, storyteller Pauls Timrots from Riga Water Supply Museum for providing historical background of the study site and expert in lake plants Uvis Susko from Latvian Fund for Nature for sharing his notes on historical observations.

Conflicts of Interest: The authors declare no conflict of interest.

References

1. Haas, M.; Baumann, F.; Castella, D.; Haghipour, N.; Reusch, A.; Strasser, M.; Eglinton, T.I.; Dubois, N. Roman-driven cultural eutrophication of Lake Murten, Switzerland. *Earth Planet Sci. Lett.* **2019**, *505*, 110–117. [CrossRef]
2. Wassmann, P. Cultural eutrophication: Perspectives and prospects. In *Drainage Basin Nutrient Inputs and Eutrophication: An Integrated Approach*; Wassmann, P., Olli, K., Eds.; University of Tromsø: Tromsø, Norway, 2005; pp. 224–234.
3. Hasler, A.D. Eutrophication of lakes by domestic drainage. *Ecology* **1947**, *28*, 383–395. [CrossRef]

4. Elmgren, R.; Larsson, U. Eutrophication in the Baltic Sea area: Integrated coastal management issues. In *Science and Integrated Coastal Management*; von Bodungen, B., Turner, R.K., Eds.; University Press: Berlin, Germany, 2001; pp. 15–35.

5. Smol, J.P. *Pollution of Lakes and Rivers. A Paleoenvironmental Perspective*, 2nd ed.; Blackwell Publishing Ltd.: Oxford, UK, 2008; pp. 1–396.

6. Yuksel, I.; Arman, H.; Ceribasi, G. Water Resources Management of Hydropower and Dams as Sustainable Development. In Proceedings of the International Conference on Sustainable Systems and the Environment (ISSE), Sharjah, UAE, 23–24 March 2011.

7. Chen, J.; Qian, H.; Gao, Y.; Wang, H.; Zhang, M. Insights into hydrological and hydrochemical processes in response to water replenishment for lakes in arid regions. *J. Hydrol.* **2020**, *581*. [CrossRef]

8. Dziluma, M. *Riga Water and the Riga Water Supply Museum*; Riga Water: Riga, Latvia, 2003; pp. 1–86.

9. Jansons, E. *Rīgas Dzeramais ūdens un tā Resursu Aizsardzība. Baltezeru un to Apkaimes Izpēte*; Rīgas Ūdens: Rīga, Latvija, 1997; pp. 1–53.

10. Krutofala, T.; Levins, I. *Pazemes ūdeņu Atradnes "Baltezers", Iecirkņa "Akoti" Pase. Pazemes ūdeņu Ekspluatācijas Krājumu Novērtējums. Atradne "Baltezers", iecirknis "Akoti". Gruntsūdens Horizonts*; Latvijas Vides, ģeoloģijas un Meteoroloģijas Aģentūra: Rīga, Latvija, 2006; pp. 1–38.

11. Pastors, A. Sekšu ezers. In *Latvijas Daba: Enciklopēdija*; Kavacs, G., Ed.; Latvijas Enciklopēdija: Rīga, Latvija, 1995; Volume 5, pp. 12–13.

12. Bauze-Krastiņš, M.G. *Garkalnes novada teritorijas plānojums 2009–2021. Gadam. Paskaidrojuma Raksts*; Garkalnes novads: Rīga, Latvija, 2009; pp. 1–169.

13. Latvijas Klimats. Latvijas Vides, ģeoloģijas un Meteoroloģijas Centrs. Available online: https://www.meteo. lv/lapas/laika-apstakli/klimatiska-informacija/latvijas-klimats/latvijas-klimats?id=1199&nid=562 (accessed on 10 March 2020).

14. Natural Data Management System OAK, Meža Valsts Reģistra Informācija. Available online: http://ozols. daba.gov.lv/pub/Life/ (accessed on 10 March 2020).

15. Zarina, D. Sekšu Ezera, Sudrabezera un Venču Ezera Ekoloģiskais Stāvoklis. Master's Thesis, Latvijas Universitāte, Riga, Latvija, 2014.

16. Tylmann, W.; Bonk, A.; Goslar, T.; Wulf, S.; Grosjean, M. Calibrating 210Pb dating results with varve chronology and independent chronostratigraphic markers: Problems and implications. *Quat. Geochronol.* **2016**, *32*, 1–10. [CrossRef]

17. Rose, N. A method for the selective removal of inorganic ash particles from lake sediments. *J. Paleolimnol.* **1990**, *4*, 61–68. [CrossRef]

18. Alliksaar, T. Spatial and temporal variability of the distribution of spherical fly-ash particles in sediments in Estonia. Ph.D. Thesis, Tallinn Pedagogical University, Tallinn, Estonia, 2000.

19. Hedges, J.I.; Eglinton, G.; Hatcher, P.G.; Kirchman, D.L.; Arnosti, C.; Derenne, S.; Evershed, R.P.; Kögel-Knabner, I.; de Leeuw, J.W.; Littke, R.; et al. The molecularly-uncharacterized component of non-living organic matter in natural environments. *Org. Geochem.* **2000**, *31*, 945–958. [CrossRef]

20. Masiello, C.A. New directions in black carbon organic geochemistry. *Mar. Chem.* **2004**, *92*, 201–213. [CrossRef]

21. Stivrins, N.; Wulf, S.; Wastegård, S.; Lind, E.M.; Allisaar, T.; Gałka, M.; Andersen, T.J.; Heinsalu, A.; Seppä, H.; Veski, S. Detection of the Askja AD 1875 cryptotephra in Latvia, Eastern Europe. *J. Quat. Sci.* **2016**, *31*, 437–441. [CrossRef]

22. Blaauw, M. Methods and code for 'classical' age-modelling of radiocarbon sequences. *Quat. Geochronol.* **2010**, *5*, 512–518. [CrossRef]

23. R Core Team. *R: A Language and Environment for Statistical Computing*; R Foundation: Vienna, Austria, 2018.

24. Heiri, O.; Lotter, A.F.; Lemcke, G. Loss on ignition as a method for estimating organic and carbonate content in sediments, reproducibility and comparability of results. *J. Paleolimnol.* **2001**, *25*, 101–110. [CrossRef]

25. Berglund, B.E.; Ralska-Jasiewiczowa, M. Pollen analysis and pollen diagrams. In *Handbook of Holocene Palaeoecology and Palaeohydrology*; Berglund, B., Ed.; John Wiley & Sons: Chichester, UK, 1986; pp. 455–484.

26. Stockmarr, J. Tablets with spores used in absolute pollen analysis. *Pollen Spores* **1971**, *13*, 615–621.

27. Fægri, K.; Iversen, J. *Textbook of Pollen Analysis*; John Wiley & Sons: Chichester, UK, 1986.

28. Miola, A. Tools for non-pollen palynomorphs (NPPs) analysis: A list of quaternary NPP types and reference literature in English language (1972–2011). *Rev. Palaeobot. Palynol.* **2012**, *186*, 142–161. [CrossRef]

29. Sweeney, C.A. A key for the identification of stomata of the native conifers of Scandinavia. *Rev. Palaeobot. Palynol.* **2004**, *128*, 281–290. [CrossRef]

30. Finsinger, W.; Tinner, W. New insights on stomata analysis of European conifers 65 years after the pioneering study of Werner Trautmann (1953). *Veg. Hist. Archaeobotany* **2019**. [CrossRef]

31. Grimm, E.C. *TILIA Software*; Illinois State Museum, Research and Collection Center: Springfield, IL, USA, 2012.

32. Battarbee, R.W. Diatom analysis. In *Handbook of Holocene Paleoecology and Paleohydrology*; Berglund, B.E., Ed.; John Wiley and Sons Ltd.: Chichester, UK, 1986; pp. 527–570.

33. Battarbee, R.W.; Kneen, M.J. The use of electronically counted microsphere sin absolute diatom analysis. *Limnol. Oceanogr.* **1982**, *27*, 184–188. [CrossRef]

34. Krammer, K.; Lange-Bertalot, H. Bacillariophyceae 4. Achnanthaceae. In *Süsswasserflora von Mitteleuropa*, 3rd ed.; Ettl, H., Gärtner, G., Gerloff, J., Heyning, H., Mollenhauer, D., Eds.; Fisher: Stuttgardt, Germany, 2011; Volume 2, p. 437.

35. Krammer, K.; Lange-Bertalot, H. Bacillariophyceae 1. Naviculaceae. In *Süsswasserflora von Mitteleuropa*, 4th ed.; Ettl, H., Gerloff, J., Heyning, H., Mollenhauer, D., Eds.; Fisher: Stuttgardt, Germany, 2010; Volume 2, p. 876.

36. Krammer, K.; Lange-Bertalot, H. Bacillariophyceae 2. Ephitemiaceae. Bacillariaceae. Surirellaceae. In *Süsswasserflora von Mitteleuropa*, 4th ed.; Ettl, H., Gerloff, J., Heyning, H., Mollenhauer, D., Eds.; Fisher: Stuttgardt, Germany, 2008; Volume 2, p. 596.

37. Krammer, K.; Lange-Bertalot, H. Bacillariophyceae 3. Centrales. Fragilariaceae. Eunotiaceae. In *Süsswasserflora von Mitteleuropa*, 3rd ed.; Ettl, H., Gerloff, J., Heyning, H., Mollenhauer, D., Eds.; Fisher: Stuttgardt, Germany, 2008; Volume 2, pp. 1–577.

38. Lange-Bertalot, H.; Hofmann, G.; Werum, M. *Diatomeen im Süßwasser-Benthos von Mitteleuropa*; A.R.G. Gantner Verlag, K.G.: Ruggell, Liechtenstein, 2011; pp. 1–908.

39. Lange-Bertalot, H.; Metzeltin, D. Indicators of Oligotrophy. 800 taxa representative of three ecologically distinct lake types. In *Iconographia Diatomologica: Annotated Diatom Micrographs*; Lange-Bertalot, H., Ed.; A.R.G. Gantner Verlag, K.G.: Ruggel, Liechtenstein, 1996; pp. 1–390.

40. Lange-Bertalot, H.; Bąk, M.; Witkowski, A.; Tagliaventi, N. *Eunotia* and related genera. In *Diatoms of Europe*; A.R.G. Gantner Verlag, K.G.: Ruggell, Liechtenstein, 2011; Volume 6, pp. 1–747.

41. AlgaeBase. Available online: https://www.algaebase.org/ (accessed on 19 February 2020).

42. Lecointe, C.; Coste, M.; Prygiel, J. "Omnidia": Software for taxonomy, calculation of diatom indices and inventories management. In *Twelfth International Diatom Symposium*; Springer: Dordrecht, The Netherlands, 1993; pp. 509–513.

43. Denys, L. *A Check-List of the Diatoms in the Holocene Deposits of the Western Belgian Coastal Plain with a Survey of their Apparent Ecological Requirements*; Service Geologique de Belgique: Brussels, Belgium, 1991; Volume 1, pp. 1–2.

44. Van Dam, H.; Mertens, A.; Sinkeldam, J. A coded checklist and ecological indicator values of freshwater diatoms from the Netherlands. *Neth. J. Aquat. Ecol.* **1994**, *28*, 117–133.

45. Juggins, S. *Version 1.0. User Guide. The European Diatom Database*; Newcastle University: Newcastle upon Tyne, UK, 2001.

46. Szeroczyńska, K.; Sarmaja-Korjonen, K. *Atlas of Subfossil Cladocera from Central and Northern Europe*; Friends of Lower Vistula Society: Świecie, Poland, 2007; pp. 1–84.

47. Kurek, J.; Korosi, J.B.; Jeziorski, A.; Smol, J.P. Establishing reliable minimum count sizes for cladoceran subfossils sampled from lake sediments. *J. Paleolimnol.* **2010**, *44*, 603–612. [CrossRef]

48. Juggins, S. *C2 Version 1.5. User Guide. Software for Ecological and Palaeoecological Data Analysis and Visualisation*; Newcastle University: Newcastle upon Tyne, UK, 2007.

49. Flössner, D. *Die Haplopoda und Cladocera (ohne Bosminidae) Mitteleuropas*; Backhuys Publishers: Leiden, The Netherlands, 2000; p. 428.

50. Brooks, S.J.; Langdon, P.G.; Heiri, O. *The Identification and Use of Palaearctic Chironomidae Larvae in Palaeoecology*; Quaternary Research Association: Edinburgh, UK, 2007.

51. Luoto, T.P.; Nevalainen, L. Inferring reference conditions of hypolimnetic oxygen for deteriorated Lake Mallusjärvi in the cultural landscape of Mallusjoki, southern Finland using fossil midge assemblages. *Water Air Soil Pollut.* **2011**, *217*, 663–675. [CrossRef]

52. Luoto, T.P.; Salonen, V.P. Fossil midge larvae (Diptera: Chironomidae) as quantitative indicators of late-winter hypolimnetic oxygen in southern Finland: A calibration model, case studies and potentialities. *Boreal Environ. Res.* **2010**, *15*, 1–18.

53. Meltsov, V.; Poska, A.; Odgaardd, B.V.; Sammula, M.; Kull, T. Palynological richness and pollen sample evenness in relation to local floristic diversity in southern Estonia. *Rev. Palaeobot. Palynol.* **2011**, *166*, 344–351. [CrossRef]

54. Stivrins, N.; Kołaczek, P.; Reitalu, T.; Seppä, H.; Veski, S. Phytoplankton response to the environmental and climatic variability in a temperate lake over the last 14,500 years in eastern Latvia. *J. Paleolimnol.* **2015**, *54*, 103–119. [CrossRef]

55. Schindler, D.W. The Dilemma of Controlling Cultural Eutrophication of Lakes. *Proc. R. Soc. B* **2012**, *279*, 4322–4333. [CrossRef] [PubMed]

56. Lampert, W.; Sommer, U. *Limnoecology*; Oxford University Press: Oxford, UK, 1997; pp. 1–382.

57. Schindler, D.W. Recent advances in the understanding and management of eutrophication. *Limnol. Oceanogr.* **2006**, *51*, 356–363. [CrossRef]

58. Leira, M.; Cantonati, M. Effects of water-level fluctuations on lakes: An annotated bibliography. *Hydrobiologia* **2008**, *613*, 171–184. [CrossRef]

59. Koralay, N.; Kara, O. Effects of Soil Erosion on Water Quality and Aquatic Ecosystem in a Watershed. In Proceedings of the 1st International Congress on Agricultural Structures and Irrigation, Antalya, Turkey, 26–28 September 2018; pp. 20–29.

60. Kuptsch, P. Die Cladoceren der Umgegend von Riga. *Arch. Hydrobiol.* **1927**, *18*, 273–315.

61. Hakkari, L. Zooplankton species as indicators of environment. *Aqua Fenn.* **1972**, *1*, 46–54.

62. Lyche, A. Cluster analysis of plankton community structure in 21 lakes along a gradient of trophy. *Verh. Internat. Verein Limnol.* **1990**, *24*, 586–591. [CrossRef]

63. Urtane, L. Cladocera as indicators of lake types and trophic state in Latvian lakes. Ph.D. Thesis, University of Latvia, Riga, Latvia, 1998.

64. Walseng, B.; Halvorsen, G. Littoral microcrustaceans as indices of trophy. *Verh. Internat. Verein Limnol.* **2005**, *29*, 827–829. [CrossRef]

65. Jensen, T.C.; Dimante-Deimantovica, I.; Schartau, A.K.; Walseng, B. Cladocerans respond to differences in trophic state in deeper nutrient poor lakes from Southern Norway. *Hydrobiologia* **2013**, *715*, 101–112. [CrossRef]

66. Umja Ezera Dabas Aizsardzības Plāns. Available online: https://www.daba.gov.lv/upload/File/DAPi_apstiprin/DL_Ummis-06.pdf (accessed on 20 March 2020).

67. Susko, U.; Latvian Fund for Nature, Riga, Latvia. Personal communication, 2020.

68. Sloka, N. *Latvijas PSR Dzīvnieku Noteicējs. Latvijas Kladoceru (Cladocera) Fauna un Noteicējs*; P. Stučkas Latvijas valsts universitāte: Rīga, Latvia, 1981; p. 146.

69. *Eutrophication of Waters, Monitoring, Assessment and Control*; OECD (Organisation for Economic Cooperation and Development): Paris, France, 1982.

70. Andrén, E. Changes in the Composition of the Diatom Flora During the Last Century Indicate Increased Eutrophication of the Oder Estuary, South-western Baltic Sea. *Estuar. Coast. Shelf Sci.* **1999**, *48*, 665–676. [CrossRef]

71. Andrén, E.; Clarke, A.L.; Telford, R.J.; Wesckström, K.; Vilbaste, S.; Aigars, J.; Conley, D.; Johnsen, T.; Juggins, S.; Korhola, A. *Defining Reference Conditions for Coastal Areas in the Baltic Sea*; TemaNord Series; Nordic Council of Ministers: Copenhagen, Denmark, 2007.

72. Andrén, E.; Shimmield, G.; Brand, T. Environmental changes of the last three centuries indicated by siliceous microfossil records from the southwestern Baltic Sea. *Holocene* **1999**, *9*, 25–38. [CrossRef]

73. Weckström, K. Assessing recent eutrophication in coastal waters of the Gulf of Finland (Baltic Sea) using subfossil diatoms. *J. Paleolimnol.* **2006**, *35*, 571–592. [CrossRef]

74. Milecka, K.; Bogaczewicz-Adamczak, B. Zmiany żyzności trofii w ekosystemach miękkowodnych jezior Borów Tucholskich. *Przegl. Geol* **2006**, *54*, 81–86.

75. Christensen, K.K.; Sand-Jensen, K. Precipitated iron and manganese plaques restrict root uptake of phosphorus in Lobelia dortmanna. *Can. J. Bot.* **1998**, *76*, 2158–2163.

76. Battarbee, R.W.; Simpson, G.L.; Shilland, E.M.; Flower, R.J.; Kreiser, A.; Yang, H.; Clarke, G. Recovery of UK lakes from acidification: An assessment using combined palaeoecological and contemporary diatom assemblage data. *Ecol. Indic.* **2014**, *37*, 365–380. [CrossRef]

77. Battarbee, R.W.; Jones, V.J.; Flower, R.J.; Cameron, N.G.; Bennion, H.; Carvalho, L.; Juggins, S. Diatoms. In *Tracking Environmental Change Using Lake Sediments. Terrestrial, Algal, and Siliceous Indicators*; Smol, J.P., Birks, H.J.B., Last, W.M., Eds.; Kluwier Academic Publishers: Norwell, MA, USA, 2001; Volume 3, pp. 155–202.

78. Battarbee, R.W.; Charles, D.F.; Dixit, S.S.; Renberg, I. Diatoms as indicators of surface water acidity. In *The Diatoms: Applications for the Environmental and Earth Sciences*; Stoermer, E.F., Smol, J.P., Eds.; Cambridge University Press: Cambridge, UK, 1999; pp. 85–127.

79. Battarbee, R.W.; Charles, D.F. Diatom-based pH reconstruction studies of acid lakes in Europe and North America: A synthesis. *Water Air Soil Pollut.* **1986**, *30*, 347–354. [CrossRef]

80. Klavins, M.; Kokorite, I.; Jankevica, M.; Mazeika, J.; Rodinov, V. Trace elements in sediments of lakes in Latvia. In Proceedings of the 4th WSEAS International Conference on Energy and Development—Environment—Biomedicine, Corfu Island, Greece, 14–16 July 2011.

81. Kļaviņš, M.; Briede, A.; Kļaviņa, I.; Rodinov, V. Metals in sediments of lakes in Latvia. *Environ. Int.* **1995**, *21*, 451–458. [CrossRef]

82. Liiv, M.; Alliksaar, T.; Amon, L.; Freiberg, R.; Heinsalu, A.; Reitalu, T.; Saarse, L.; Seppä, H.; Stivrins, N.; Tõnno, I.; et al. Late glacial and early Holocene climate and environmental changes in the eastern Baltic area inferred from sediment C/N ratio. *J. Paleolimnol.* **2019**, *61*, 1–16. [CrossRef]

83. Terasmaa, J.; Puusepp, L.; Marzecová, A.; Vandel, E.; Vaasma, T.; Koff, T. Natural and human-induced environmental changes in Eastern Europe during the Holocene: A multi-proxy palaeolimnological study of a small Latvian lake in a humid temperate zone. *J. Paleolimnol.* **2013**, *49*, 663–678. [CrossRef]

84. Meyers, P.A.; Ishiwatari, R. Lacustrine organic geochemistry—An overview of indicators of organic matter sources and diagenesis in lake sediments. *Org. Geochem.* **1993**, *20*, 867–900. [CrossRef]

85. Meyers, P.A.; Teranes, J.L. Sediment Organic Matter. In *Tracking Environmental Change Using Lake Sediments: Physical and Geochemical Methods, Developments in Paleoenvironmental Research*; Last, W.M., Smol, J.P., Eds.; Springer: Dordrecht, The Netherlands, 2001; pp. 239–269.

86. Buzajevs, V.; Levina, N.; Levins, I. *Pazemes ūdeņu bilance un kvalitāte Baltezera ūdensgūtnēs*; Valsts Ģeoloģijas Dienests: Riga, Latvia, 1997; pp. 1–56.

87. Sarmaja-Korjonen, K. Correlation of fluctuations in cladoceran planktonic: Littoral ratio between three cores from a small lake in southern Finland: Holocene water-level changes. *Holocene* **2001**, *11*, 53–63. [CrossRef]

88. Cooper, J.T.; Steinitz-Kannan, M.; Kallmeyer, D.E. Bacillariophyta: The Diatoms. In *Algae: Source of Treatment*; Ripley, M.G., Ed.; American Water Works Association: Denver, CO, USA, 2010; pp. 207–215.

89. Zelnik, I.; Balanč, T.; Toman, M. Diversity and Structure of the Tychoplankton Diatom Community in the Limnocrene Spring Zelenci (Slovenia) in Relation to Environmental Factors. *Water* **2018**, *10*, 361. [CrossRef]

90. Dimante-Deimantovica, I.; Latvian Institute of Aquatic Ecology, Riga, Latvia. Personal communication, 2020.

91. Ardiles, V.; Alcocer, J.; Vilaclara, G.; Oseguera, L.A.; Velasco, L. Diatom fluxes in a tropical, oligotrophic lake dominated by large-sized phytoplankton. *Hydrobiologia* **2012**, *679*, 77–90. [CrossRef]

92. Romero, O.E.; Thunell, R.C.; Astor, Y.; Varela, R. Seasonal and interannual dynamics in diatom production in the Cariaco Basin, Venezuela. *Deep-Sea Res.* **2009**, *56*, 571–581. [CrossRef]

93. Ryves, D.B.; Jewson, D.H.; Sturm, M.; Battarbee, R.W.; Flower, R.J.; Mackay, A.W.; Granin, N.G. Quantitative and qualitative relationships between planktonic diatom communities and diatom assemblages in sedimenting material and surface sediments in Lake Baikal, Siberia. *Limnol. Oceanogr.* **2003**, *48*, 1643–1661. [CrossRef]

94. Ceirans, A. Zooplankton indicators of trophy in Latvian lakes. In *Acta Universitatis Latviensis. Biology*; University of Latvia: Riga, Latvia, 2007; Volume 723, pp. 61–69.

95. Veski, S.; Amon, L.; Heinsalu, A.; Reitalu, T.; Saarse, L.; Stivrins, N.; Vassiljev, J. Lateglacial vegetation dynamics in the eastern Baltic region between 14,500 and 11,400 cal yr BP: A complete record since the Bolling (GI-1e) to the Holocene. *Quat. Sci. Rev.* **2012**, *40*, 39–53. [CrossRef]

96. Kołaczek, P.; Zubek, S.; Błaszkowski, J.; Mleczko, P.; Włodzimierz, M. Erosion or plant succession—How to interpret the presence of arbuscular mycorrhizal fungi (Glomeromycota) spores in pollen profiles collected from mires. *Rev. Palaeobot. Palynol.* **2013**, *189*, 29–37. [CrossRef]

97. Wang, L.; Mackay, A.W.; Leng, M.J.; Rioual, P.; Panizzo, V.N.; Lu, H.; Gu, Z.; Chu, G.; Han, J.; Kendrick, C.P. Influence of the ratio of planktonic to benthic diatoms on lacustrine organic matter $\delta13C$ from Erlongwan maar lake, northeast China. *Org. Geochem.* **2013**, *54*, 62–68. [CrossRef]

98. Wolin, J.A.; Stone, J.R. Diatoms as Indicators of Water-Level Change in Freshwater Lakes. In *The Diatoms Applications to the Environmental and Earth Sciences*; Stoermer, E.F., Smol, J.P., Eds.; Cambridge University Press: Cambridge, UK, 2010; pp. 174–185.

99. Carmignani, J.R.; Roy, A.H. Ecological impacts of winter water level drawdowns on lake littoral zones: A review. *Aquat. Sci.* **2017**, *79*, 803–824. [CrossRef]

100. Magny, M.; Arnaud, F.; Billaud, Y.; Marguet, A. Lake-level fluctuations at Lake Bourget (eastern France) around 4500–3500 cal. a BP and their palaeoclimatic and archaeological implications. *J. Quat. Sci.* **2012**, *27*, 494–502. [CrossRef]

101. Theuerkauf, E.J.; Braun, K.N.; Nelson, D.M.; Kaplan, M.; Vivirito, S.; Williams, J.D. Coastal geomorphic response to seasonal water-level rise in the Laurentian Great Lakes: An example from Illinois Beach State Park, USA. *J. Great Lakes Res.* **2019**, *45*, 1055–1068. [CrossRef]

102. Holmer, M.; Storkholm, P. Sulphate reduction and sulphur cycling in lake sediments: A review. *Freshw. Biol.* **2001**, *46*, 431–451. [CrossRef]

103. Couture, R.M.; Fischer, R.; Van Cappellen, P.; Gobeil, C. Non-steady state diagenesis of organic and inorganic sulfur in lake sediments. *Geochim. Cosmochim. Acta* **2016**, *194*, 15–33. [CrossRef]

104. Aasif, A.; Nilli, A.; Aasif, L.; Hema, A.; Rayees, S. Diatom Diversity and Organic Matter Sources in Water Bodies around Chennai, Tamil Nadu, India. *MOJ Ecol. Environ. Sci.* **2017**, *2*, 141–146.

105. Rosińska, A.; Rakocz, K. Rola biodegradowalnej materii organicznej w procesie dezynfekcji wody. *Inż. Ochr. Środ.* **2013**, *16*, 511–521.

106. Brodersen, K.P.; Quinlan, R. Midges as palaeoindicators of lake productivity, eutrophication and hypolimnetic oxygen. *Quat. Sci.* **2006**, *25*, 1995–2012. [CrossRef]

107. Timrots, P.; Riga Water Supply Museum, Riga, Latvia. Personal communication, 2019.

108. Brodersen, K.P.; Pedersen, O.; Lindegaard, C.; Hamburger, K. Chironomids (Diptera) and oxy-regulatory capacity: An experimental approach to paleolimnological interpretation. *Limnol. Oceanogr.* **2004**, *49*, 1549–1559.

109. Kacalova, O.; Laganovsla, R. *Zivju Barības Bāze Latvijas PSR Ezeros*; Latvijas PSR Zinātņu akadēmijas izdevniecība: Rīga, Latvija, 1961; pp. 17–23.

110. Bledzki, L.A.; Rybak, J.I. *Freshwater Crustacean Zooplankton of Europe Cladocera and Copepoda (Calanoida, Cyclopoida), Key to Species Identification, with Notes on Ecology, Distribution, Methods and Introduction to Data Analysis*; Springer: New York, NY, USA, 2016; pp. 1–918.

111. Lielā un Mazā Baltezera Eitroficēšanās Izpēte. Available online: https://www.google.com/url?sa=t&rct=j&q=&esrc=s&source=web&cd=1&ved=2ahUKEwiPufXQp7_oAhXpxIsKHfMmDOcQFjAAegQIBRAB&url=https%3A%2F%2Fwww.ezeri.lv%2Fblog%2FDownloadAttachment%3Fid%3D347&usg=AOvVaw1qjWh6fVLl6-JGxc-_EbzG (accessed on 10 March 2020).

112. Balode, M.; Purina, I.; Strake, S.; Purvina, S.; Pfeifere, M.; Barda, I.; Povidisa, K. Toxic cyanobacteria in the lakes located in Rīga (the capital of Latvia) and its surroundings: Present state of knowledge. *Afr. J. Mar. Sci.* **2006**, *28*, 225–230. [CrossRef]

 *water*

Article

The Effect of Human Impact on the Water Quality and Biocoenoses of the Soft Water Lake with Isoetids: Lake Jeleń, NW Poland

Piotr Klimaszyk [1,*], **Dariusz Borowiak** [2,3], **Ryszard Piotrowicz** [1], **Joanna Rosińska** [4], **Elżbieta Szeląg-Wasielewska** [1] **and Marek Kraska** [1]

[1] Deparment of Water Protection, Adam Mickiewicz University, Uniwersytetu Poznańskiego 6, 61-642 Poznań, Poland; ryszardp@amu.edu.pl (R.P.); eszelag@amu.edu.pl (E.S.-W.); makra@amu.edu.pl (M.K.)

[2] Department of Limnology, University of Gdańsk, Bażyńskiego 4, 80-309 Gdańsk, Poland; geodb@univ.gda.pl

[3] Limnological Station University of Gdańsk, 83-323 Kamienica Szlachecka, Poland

[4] Department of Environmental Medicine, Poznań University of Medical Sciences, Rokietnicka 8, 60-806 Poznań, Poland; joaros@ump.edu.pl

* Correspondence: pklim@amu.edu.pl

Received: 11 February 2020; Accepted: 25 March 2020; Published: 26 March 2020

Abstract: Soft water lakes with isoetids (SLI) are ecosystems prone to degradation due to the low buffer capacity of their waters. One of the main threats resulting from human impact is eutrophication due to agriculture, catchment urbanization and recreational use. In this paper, changes in the water chemistry and transformation of biocoenoses of one of the largest Polish SLI, Lake Jeleń, over the past 30 years are presented. The lake is located within the borders of a city, and a significant part of its catchment is under agriculture and recreation use. The physicochemical (concentration of nutrients, organic matter, electrical conductivity, oxygen saturation and water pH) and biological parameters (macrophytes and phytoplankton) were measured in summer 1991, 2004, 2013 and 2018. Since the beginning of the 1990s, a gradual increase in the trophy of the lake has been observed as indicated by increased nutrient availability, deterioration of oxygen conditions and a decrease in water transparency. The alterations of water chemistry induce biological transformations, in particular, an increase in phytoplankton abundance (4-fold increase of biomass in epilimnion) as well as a gradual reduction in the range of the phytolittoral (from 10 to 6 m), a decrease in the frequency of isoetids, *Lobelia dortmanna* and *Isoetes lacustris*, and expansion of plant species characteristic for eutrophy.

Keywords: soft water lake; water quality; *Lobelia dortmanna*; *Littorella uniflora*; *Isoëtes lacustris*; eutrophication; catchment; human impact

---

1. Introduction

Soft water lakes with isoetids (SLI) are unique ecosystems that occur mainly in the temperate and boreal zones of the northern hemisphere. In Europe, such lakes are associated with low calcareous, sandy soils or develop near high or transitional peatlands [1–3]. The SLI are considered as valuable natural habitats and are included in the Natura 2000 protected area network as habitat 3110—"Oligotrophic waters containing very few minerals of sandy plains (Littorelletalia uniflorae)" [4,5]. In Poland, this habitat is usually known by the term "lobelia lakes", which is derived from one of the representative isoetids (plants that are associated with such a habitat) [6]. The SLI are usually small in surface, closed water bodies, fed exclusively by precipitation or surface runoff from catchment areas poor in nutrients and calcium [7]. Low availability of carbon dioxide in the water column and sediments and low concentration of nitrogen and phosphorus shape the specific vegetation structure of SLI. The isoetids *Lobelia dortmanna* L., *Isoetes lacustris* L., and *Littorella uniflora* (L.) Ascherson constitute an essential feature of SLI vegetation [3,8]. These small perennial plants are usually accompanied by several other plant species like *Myriophyllum alterniflorum* DC., *Luronium natans* (L.) Raf. or charophyte *Chara delicatula* Ag. [9]. Due to the low concentration of calcium, manganese and hence low buffer capacity, the SLI are sensitive to external impact [1]. Natural aging processes—eutrophication of the SLI—occurs very slowly only when the catchment areas are natural, or human impact is scarce. However, not only changes in the direct catchment but also global pollution may affect the functioning of SLI. Increased atmospheric deposition of nitrogen and sulfur oxides leads to a decrease in water pH (acidification) of SLI ecosystems causing subsequent disappearance of characteristic plant species [2]. Acidification may also be caused by an increased inflow of humic acids from the adjacent land ecosystem. Extremely high loads of humic compounds reach lakes during clear-cutting of nearby forests and the drainage of high bogs [9]. Moreover, changes in the character of rainfall (an increase of heavy rains) driven by climate changes may result in increased surface runoff and transport of humic substances to the lakes [10,11]. Humic substances not only acidify SLI but also affect underwater light conditions since they increase the color of water, decrease light penetration and thus affect the survival of macrophytes [12]. The accelerated eutrophication of SLI is usually the result of a complex of processes. However, it is always triggered by intense human activity and pressure exerted directly on the lakes or their catchment. Increasing the share of agricultural areas and human settlements leads to changes in the water and chemical balance [1]. Increased loads of nutrients and other chemical elements reach the SLI and promote phytoplankton development which gradually reduces light penetration. Changes in the light regime and competition from plant species adapted to high turbidity and calcium content lead to the progressive loss of isoetids. Due to their natural values, many SLIs are also under intense recreational pressure. Clearwater, sandy bottom and forested shores attract people for bathing, resting and water sports. The main consequences of high recreational pressure are: (i) changes in transport of water and chemical elements from the catchment area to the lake—trampled ducts increase the surface runoff and accelerate the transfer of chemical substances, (ii) destruction of land–water ecotonal zones that play a significant role in maintaining the homeostasis of adjacent ecosystems [13,14], (iii) mechanical devastation of plants within the bathing areas. Eventually, intense recreational impact may trigger the eutrophication process and degradation of the ecosystem [15]. The effects are particularly visible in the case of ecosystems in the vicinity of which human settlements are located.

Lake Jeleń is one of the largest and, until recently, one of the best-preserved lobelia lakes in Poland, with a very abundant population of isoetids. However, this lake is subjected to intense human pressure. A significant part of the lake catchment area is used for agriculture and recreational purposes. In the vicinity of the shoreline, there is a campsite, summer houses, a restaurant, a playground and various types of infrastructure (roads and parking). On every sunny day, hundreds of tourists can be found resting on the shores of the lake. The disturbances and changes in the structure of the catchment and the functioning of the entire ecosystem can be linked to recreational activities, which can, in turn, pose a threat to the population of isoetids [16,17]. For these reasons, this lake is subject to periodic monitoring, during which it has been possible to repeat the research several times over for almost 30 years and observe the changes occurring during this time. Based on biological (phytoplankton and macrophytes) and physico-chemical parameters (nutrient concentration) of water over the years, it was possible to determine the direction of lake changes. We hypothesized that intense human pressure has caused an increase in nutrient concentration and phytoplankton abundance, a decrease in water transparency and rebuilding in the phytolittoral (disappearance of isoetids and domination of eutrophic elodeids).

2. Materials and Methods

2.1. Study Site

Lake Jeleń is located in northern Poland (54°12′04,4″ N, 17°31′31,5″ E) near the southeastern borders of the city of Bytów (population 17 k). It is a closed lake with a surface area of 81.6 ha.

Its shape is irregular, elongated in the E-W direction (Figure 1). The central part of the reservoir is deep, with several depressions: 33 m (max. depth), 26 m, 24 m and 23 m, separated by shallowings. The mean depth is 9.2 m and volume about 7.54 hm^3. At the northern shore, there are three shallow bays with a depth lower than 5 m. The bay located at the southeastern end of the lake is separated from the main lake basin by a narrow and shallow channel (Figure 1). In periods of lower water level, contact between the lake and bay is severely limited. The lake catchment covers an area of 281.2 ha. The largest (53%) part of the catchment (over 150 ha) is used for agriculture. Only 35% of the catchment is covered by forest (100 ha (mainly beech and oaks)). There are no compact built-up areas in the direct catchment. However, a recreational resort is located on the southern bank. The resort consists of a restaurant, several summer houses, a BBQ area, campsite, swimming platforms and a parking lot [16]. A fishing club belonging to the Polish Angling Society is located close to the resort.

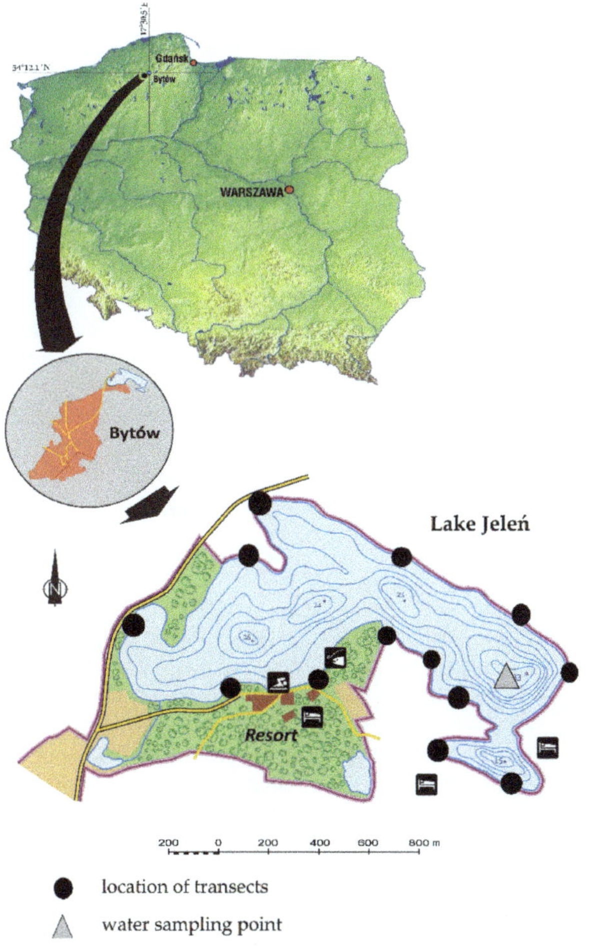

Figure 1. Location of Lake Jeleń and sampling points.

2.2. Methods

The studies were carried out between 1991 and 2018, in the years 1991, 1995, 1998, 2000, 2004, 2007, 2013, 2015 and 2018 during the middle of the stagnation period (August). At the deepest point of the lake field, measurements of oxygen saturation, temperature, pH, and electrical conductivity (EC) were conducted using Elmetron S410 and YSI R3650 probes. Water transparency was measured with a Secchi disc. In the years 1991, 2014, 2013, and 2018, also during the stagnation period (August), a wide range of physico-chemical and biological analyses ware made. According to field measurements, three specific layers were determined (epi-, meta-, and hypolimnion) and subsamples from each meter were mixed to obtain one sample of water representative for the layer. The water samples were transported in PVC containers to the laboratory and filtered through GF/C filters. The following parameters were analyzed: ammonium (N-NH$_4{}^+$, using the Nessler method), nitrites (N-NO$_2{}^-$, using the sulphonic acid method), nitrates (N-NO$_3{}^-$, using the sodium salicylate method), Norg (using the Kjeldahl method), total phosphorus (P tot., using the molybdate method after mineralization) and orthophosphates (TRP, using the molybdate method) [18]. Biological oxygen demand (BOD) was analyzed using the dark bottle method, as a difference in the concentration of oxygen after five days. A set of equations developed by Carlson [19] was used to determine the trophic state. Trophic indices were determined for three individual components separately (chlorophyll "a", total phosphorus, Secchi depth), and then summarized to express the overall trophic state of the lake (TSI). The simplified Carlsons' trophic state equations used in this study are as follows:

$$\text{TSI(SD)} = 60 - 14.41\ \ln(\text{SD}) \tag{1}$$

$$\text{TSI(CHL)} = 9.81\ \ln(\text{CHL}) + 30.6 \tag{2}$$

$$\text{TSI(TP)} = 14.42\ \ln(\text{TP}) + 4.15 \tag{3}$$

where SD is Secchi disc depth (in meters), CHL and TP are concentrations of chlorophyll "a" pigments and total phosphorus, respectively (both in μg L^{-1}). The TSI was given as a sum of above-mentioned calculations. According to obtained value, the trophy state of the lake was determined, where: <110—oligotrophy, 110–140—mesotrophy, 140–160—meso-eutrophy, 160–210 eutrophy [19].

Empirical equations linking the euphotic zone depth (zEU) with water transparency (Secchi disc depth) were derived from field measurements [20,21]. Downwelling irradiance profiles were obtained using the profiling radiometers: PER-700 (Biospherical Instruments Inc.) and HyperPro (Satlantic Inc.), respectively. Using values of diffuse attenuation coefficients for photosynthetically active radiation (Kd,PAR) calculated from the light profiles and relation Kd,PAR = 4.6/zEU, as well as Secchi disc depths (SD), the following formula was obtained:

$$\text{zEU} = 2.88\ \text{SD}^{0.79} \tag{4}$$

The samples for phytoplankton studies were taken in the same period as samples for the physico-chemical analyses. Within the thermal layers, the subsamples from each meter were mixed to obtain one sample of water representative for the layer. The samples were preserved at the site with Lugol's solution. Organisms, if possible, were identified to the species level or assigned to a genus only. Phytoplankton larger than 2 μm were analyzed under an inverted microscope after sedimentation in 14-mL or 24-mL chambers, according to the method by Wetzel and Likens [22]. Abundance was expressed as numbers of cells per 1 mL. The biovolume of each species was calculated on the basis of cell shape, size, and number, while their biomass was expressed as wet weight. Biomass was estimated assuming that the volume of 10^9 μm^3 is equivalent to 1 mg. The trophic state was assessed based on indicator taxa [23] and expressed on a scale 1–3, where 1 denotes oligotrophic, 2—mesotrophic, and 3—eutrophic. The trophic index of the community was calculated on the basis of the biomass of trophic state indicators according to the Hörnström formula [24]:

$$Ist = \frac{\sum(fs \times Is)}{\sum fs} \tag{5}$$

where Ist = trophic index, Is = trophic index of individual species, fs = frequency of the given species (biomass of specimens).

The composition, depth and frequency of macrophyte occurrence were analyzed using the transects method at the peak of the growing season in the same periods as phytoplankton. The investigations were conducted by applying the phytosociological approach [25]. The research was conducted from a boat. The number of transects was calculated using Jensen's formula (which takes into account the area of the lake and shoreline length) [26]. Every year the study was carried out at the same 13 transects, which were evenly distributed. During the first study in 1991, 13 transects perpendicular to the shoreline were determined, with a width of about 30 m and length depending on the maximum depth of occurrence of plants (Figure 1). The location of the beginning of each transect was noted on a detailed bathymetric plan of the lake on a scale of 1:500. The macrophyte species were determined in each transect with their bottom coverage in percentage. The depth of occurrence, changes in the density and presence of individual taxa were also noted, starting from the shoreline up to the maximum depth of plant occurrence. At greater depths, sea scope for underwater observations or an anchor were used to identify plants. In 2004, the location of these transects was verified and refined by GPS. Therefore, it was possible to repeat the studies with the same methods in the following years, prepare the distribution maps of individual plant taxon and analyze the observed alterations in the structure of the phytolittoral. Particular attention was focused on changes in the occurrence of the isoetids *L. dortmanna*, *L. uniflora* and *I. lacustris*. Changes in the presence of elodeids, typical for eutrophic reservoirs (*Ceratophyllum demersum* L. and *Elodea canadensis* Michx.), as species that could potentially displace isoetids under an increase of water trophy conditions were also analyzed. Maps of the occurrence of vegetation in 1991 and 2018 were prepared based on the data from neighboring transects supported by additional field observations made during the field survey and extrapolated on the bathymetric plan. The maps were drawn with ArcGIS for Desktop 10.5.1 program. The transects consisted of phytosociological relevés, in which each uniform patch of vegetation constituted one relevé. The frequency of species was calculated as a percentage share of relevés with species occurrence to all relevés in the lake. Statistical analyses were performed using Statistica 8.0 (StatSoft, Tulsa, OK, USA). The differences in the concentration of elements and values of parameters between years were analyzed with ANOVA and post-hoc Tukey's HSD test. In all analyses, $p < 0.05$ was considered as statistically significant.

3. Results

3.1. Physical and Chemical Parameters of the Lake Water

Changes in the values of most of the analyzed physical and chemical parameters of Lake Jeleń water in the period 1991–2018 indicate a deterioration of its quality. At the beginning of the studies, almost all the water column was well oxygenated, and even in the sub-bottom layer, the saturation of oxygen was over 50% (Figure 2). Over time, gradual deterioration of oxygen conditions, especially in the deeper parts of the lake, was observed. In 1998 a total oxygen deficit appeared in the sub-bottom zone of the lake. The anoxic layer became more and more extensive. In August 2018, the volume of deoxygenated water was almost 500 000 m^3 and the total O_2 deficit extended from a depth of 22 m to the bottom of the lake (Figure 2). Between 1991 and 2018, a gradual increase in the concentration of nutrients in Lake Jeleń could be observed.

The average concentration of P tot. increased almost 5-fold while the concentration of orthophosphates increased more than 30 times (Table 1).

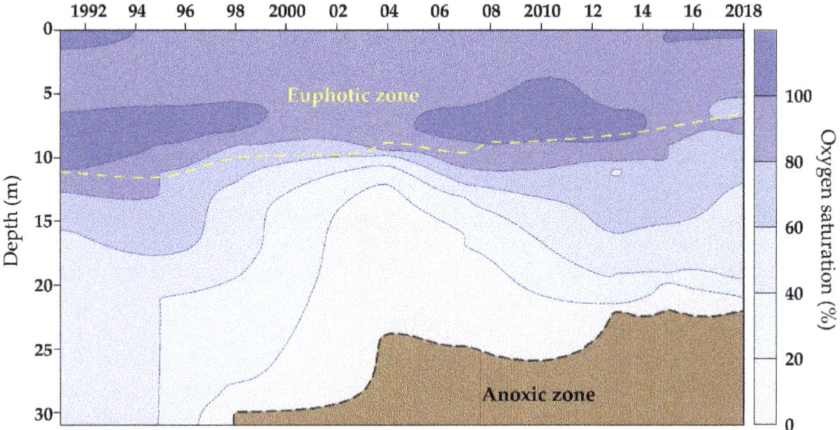

Figure 2. Long-term changes of the oxygen saturation (%O_2) and the euphotic zone depth in Lake Jeleń during summer stagnation. Isopleths of the oxygen saturation were interpolated and smoothed by means of the kriging method using Surfer software (ver. 8).

Table 1. Physical and chemical features of Lake Jeleń (average value for water column, n = 9) and significance of the statistical difference.

Parameter		1991	2004	2013	2018	ANOVA
color	mgPt/L	9.2 ± 1.09	10.2 ± 2.1	13 ± 3.16	15.1 ± 5.1	**
pH		6.79 ± 0.527	7.02 ± 0.31	7.18 ± 0.58	7.7 ± 0.93	*
BOD	mgO$_2$/L	2.3 ± 1.27	4.3 ± 1.89	4.8 ± 1.43	5.54 ± 1.1	***
N-NH$_4$	mgN/L	0.08 ± 0.03	0.2 ± 0.03	0.23 ± 0.043	0.3 ± 0.074	***
N-NO$_3$	mgN/L	0.02 ± 0.044	0.13 ± 0.08	0.2 ± 0.034	0.24 ± 0.066	***
N-NO$_2$	mgN/L	0	0.001± 0.0002	0	0.001± 0.0004	
N min.	mgN/L	0.1 ± 0.03	0.33 ± 0.07	0.42 ± 0.058	0.53 ± 0.11	***
N org.	mgN/L	0.74 ± 0.14	1.68 ± 0.6	1.76 ± 0.6	2.36 ± 1.33	**
P tot.	mgP/L	0.027 ± 0.004	0.06 ± 0.012	0.088 ± 0.011	0.159 ± 0.02	***
PO$_4$	mgPO$_4$/L	0.001 ± 0.003	0.02 ± 0.011	0.027 ± 0.013	0.037 ± 0.01	***
Ca	mgCa/L	7.02 ± 0.3	8.9 ± 0.9	7.2 ± 0.7	7.7 ± 0.9	
EC	µSm/cm	71.9 ± 13.2	79.4 ± 10.4	77.9 ± 9.98	85.4 ± 6.2	
Chl	µg/L	0.7 ± 0.49	2.36 ± 1.8	3.46 ± 2.5	4.15 ± 2.9	*
TSI		111	146	155	169	

EC—electrical conductivity, Chl—chlorophyll "a", TSI—Trophy State Index by Carlsson, *—$p < 0.05$, **—$p < 0.005$, ***—$p < 0.001$.

In 1991 orthophosphates were found only in the hypolimnion (with a concentration of about 0.005 mg PO$_4$/L). By 2004 this form of phosphorus was found in the entire water column, and gradually over time, the concentration increased (Figure 3). For the period 1991–2018, for all thermal layers, the difference in PO$_4^{3-}$ concentration was statistically significant (ANOVA, $p < 0.001$). Moreover, the concentration of almost all forms of nitrogen significantly increased over the 27 years of surveys. Between 1991 and 2018, the average concentration of mineral nitrogen rose over 5-fold and organic nitrogen over 3-fold (Table 1). The most significant increase was observed for organic N in the hypolimnetic zone of the lake (ANOVA, $p < 0.005$). In the studied period, an increase of easily degradable organic matter (expressed as BOD) was also found (Table 1, Figure 3).

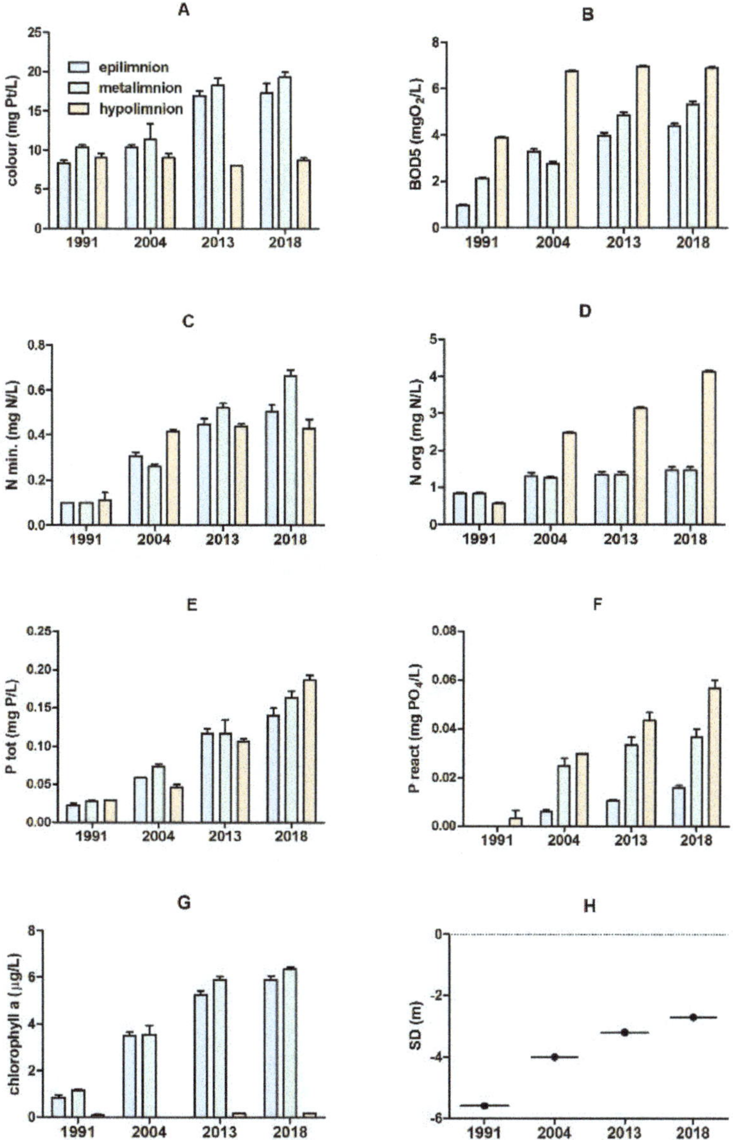

Figure 3. Changes in selected physical and chemical parameters of Lake Jeleń water. **A**—color, **B**—biochemical oxygen demand, **C**—mineral nitrogen, **D**—organic nitrogen, **E**—total phosphorus, **F**—orthophosphates, **G**—chlorophyll "a" and **H**—Secchi disc depth (box—mean, whisker—standard error, n = 3).

BOD increased gradually in all layers of the lake. However, the highest values were noted in the hypolimnion. Furthermore, the content of chlorophyll "a" increased in the studied period. The most profound and statistically significant changes in chlorophyll were found in the epilimnion and metalimnion (Figure 3) (ANOVA, $p < 0.005$). With the increase in the content of chlorophyll and watercolor the water transparency decreased significantly (Table 1, Figure 3). In 1991, in the middle of summer, SD visibility exceeded 5.5 m, while in 2004 it was about 4 m, and in 2018 only 2.6 m. The depth of the euphotic zone (1% of incident light) decreased gradually from 11.1–11.3 in 1991 to 5.6–6.3 m in 2018 (Figure 2). Thus, the euphotic zone reduced by half. There was no statistically significant difference between years in the electrical conductivity of lake water. Despite this, comparing 1991 and 2018, an approximately 20% increase in the average EC of lake water was found (Table 1). Carlson's trophic state index (TSI) shows the deterioration of water quality and the increase of trophy in the lake. The TSI value of 111 calculated for 1991 is characteristic of oligo-mesotrophy, approximately 150 for 2004 and 2013 indicate mesotrophy, while almost 170 found in 2018 points to a meso-eutrophic state. Over the last 30 years, a deterioration in the water quality of Lake Jeleń was found. It should be noted, however, that the most significant changes took place in the years 1991–2004. The fastest increase in the concentration of nutrients, organic matter and the simultaneous deterioration of water transparency occurred during this period.

3.2. Phytoplankton

In 1991, 42 taxa were found in the phytoplankton of Lake Jeleń. Green algae had the largest share in the species structure (33.3%), followed by cyanobacteria and cryptophytes (each of these groups with a 14.3% share). The highest phytoplankton biomass was in the metalimnion, due to the presence of the microplankton species *Gymnodinium uberrimum*. This dinophyte had as much as a 72% share in the total biomass, which in this layer amounted to 1.24 mg/L. In the hypolimnion, biomass was almost seven times less than in the metalimnion. In turn, the epilimnion with biomass of 0.8 mg/L was also co-dominated by the dinophytes *G. uberrimum* and *G. oligoplacatum*, and algae headed by *Coenococcus planktonicus* (Tables 2 and 3, Figure 4).

Table 2. Phytoplankton taxa with the highest share in biomass.

No.	Taxon	Biomass (mg/L)
	1991	
1	*Gymnodinium uberrimum* (G.J.Allman) Kofoid & Swezy	0.898 (m)
2	*Gymnodinium oligoplacatum* Skuja	0.138 (e)
3	*Coenococcus planktonicus* Korshikov	0.104 (e)
	2004	
1	*Planktothrix agardhii* (Gomont) Anagnostidis & Komárek	0.707 (e)
2	*Gymnodinium uberrimum* (G.J.Allman) Kofoid & Swezy	0.090 (e)
3	*Eudorina elegans* Ehrenberg	0.053 (m)
	2013	
1	*Gymnodinium* sp.	0.212 (e)
2	*Ceratium hirundinella* (O.F. Müller) Dujardin	0.167 (m)
3	*Peridinium inconspicuum* Lemmermann	0.139 (e)
	2018	
1	*Dolichospermum spiroides* (Klebhan) Wacklin, Hoffmann & Komárek	1.312 (e)
2	*Planktothrix agardhii* (Gomont) Anagnostidis & Komárek	0.157 (h)
3	*Dolichospermum affine* (Lemmermann) Wacklin, Hoffmann & Komárek	0.116 (e)

(e-epilimnion, m-metalimnion, h-hypolimnion).

Table 3. The number of species of phytoplankton of Lake Jeleń in studied periods.

Taxon/Year	1991	2004	2013	2018
Cyanobacteria	6	13	9	13
Euglenophyceae	3	1	0	1
Cryptophyceae	6	4	5	5
Dinophyceae	5	4	4	4
Chrysophyceae	5	3	4	3
Bacillariophyceae	2	5	2	4
Xanthophyceae	1	1	0	0
Chlorophyceae	13	13	17	13
Conjugatophyceae	1	8	7	8
Total	42	52	48	51

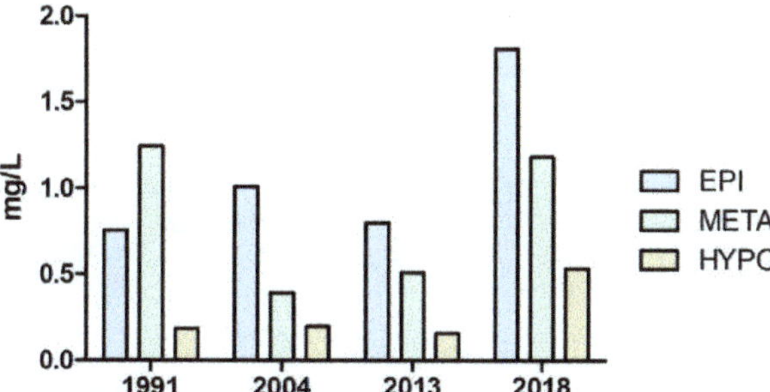

Figure 4. The changes in the phytoplankton biomass in water layers of Lake Jeleń in subsequent years of surveys.

In the following years, i.e. 2004, 2013 and 2018, the number of phytoplankton taxa was quite similar, changing in the range from 48 to 52 (Table 4). Green algae (40.4–50.0% share in the species composition) and cyanobacteria (18.8–25.5% share) had the greatest species richness. The increase in the number of cyanobacteria and green algae taxa, as compared to 1991, was accompanied by a decrease in the number of taxa of other algae groups, e.g. chrysophytes, euglenophytes and dinophytes.

Table 4. Trophic state index of Lake Jeleń based on phytoplankton indicator species biomass.

Year	Epilimnion	Metalimnion	Hypolimnion	Average Trophic State Index
1991	1.32	1.17	1.75	1.41
2004	2.10	2.16	1.84	2.03
2013	1.57	2.11	2.17	1.95
2018	2.62	2.45	2.53	2.58

Total phytoplankton biomass in 2004, 2013 and 2018, unlike 1991, was the highest in the epilimnion (Figure 4).

It gradually decreased with depth in all studied years. In 2004 and 2013, total biomass was low, and it slightly exceeded the value of 1 mg/L only in the epilimnion of 2004, but the course of its changes with depth in both years was similar. In 2018, the total biomass of phytoplankton in the epi- and metalimnion was 2-fold higher than in the hypolimnion (Figure 4).

In 2004, the filamentous cyanobacteria *Planktothrix agardhii* had the highest share in all biomass. The next species in terms of biomass was *G. uberimmum* in the epilimnion, while it was the green alga *Eudorina elegans* in the metalimnion (Table 2). In 2013, in all layers, dinophytes had the greatest biomass, with a fairly large share in the biomass of cyanobacteria only in the metalimnion. *Gymnodinium* sp., *Ceratium hirundinella* and *Peridinium inconspicuum* belonged to the species achieving the highest biomass in this year (Table 2). The last analyzed year of research differed the most from the others because cyanobacteria predominated in the whole water column and the share of other groups in the total biomass of phytoplankton did not exceed 10% at any depth. Among cyanobacteria, *Dolichospermum spiroides*, *D. affine* and *P. agardhii* had successively the largest biomass (Table 2).

Trophic state index values calculated based on the biomass of indicator species did not show the same trends in the water column. However, higher values were generally observed in the hypo- or metalimnion. The exception was 2018 when the index value for the epilimnion was the highest (2.62). The average index values for the water column during the studied years ranged from 1.41 in 1991 to 2.58 in 2018 (Table 4). This indicates that in 2018, Lake Jeleń was eutrophic (calculated trophic index values were within limits indicating eutrophy: 2.25–2.74), in 1991 the lake was mesotrophic (1.25–1.74), while in all other years Lake Jeleń was characterized by an intermediate state between mesotrophy and eutrophy (1.75–2.24).

3.3. Macrophytes

During the research, 46 taxa of plants representing various ecological groups of macrophytes were found—from emerged plants (helophytes) to charophytes and bryophytes building "underwater meadows" in the deeper parts of the phytolittoral (Figure 5, Table 5). The number of plant taxa noted in each year of research has changed slightly, from 32 to 39 taxa.

The typical belt system of littoral vegetation for lobelia lakes was observed. There was a very narrow strip of emerged plants at the bank. However, in some transects, they were not even noted. In addition, the belt with floating-leaf plants was very poorly developed. In contrast, the isoetid belt was well developed. In shallower places, it was dominated by *L. dortmanna* and *L. uniflora*, and in deeper areas by *I. lacustris*. Isoetids were accompanied by other submerged plants, mainly *M. alterniflorum* (a species considered as an indicator of moderately eutrophic soft-water lakes) and *C. demersum* and *E. canadensis* (species typical of eutrophic reservoirs). In 1991, at greater depths (up to 10 m) there were abundant macroscopic algae (*Charophyta*) and bryophytes (Figure 5).

Comparing the occurrence of submerged plants in 4 periods, a gradual 2-fold decrease of mean depth (± stand. dev.), from 3.31 ± 2.43 m in 1991 to 1.49 ± 1.57 m in 2018, was observed. A limitation of the zone of deeper phytolittoral built mainly by bryophytes and charophytes was noticed at the turn of the 20th and 21st centuries. The maximum depth of occurrence decreased from 10.0 m in 1991 to 6.0 m in 2013 and about 8.0 m in 2018 (Figure 6). Recently, in 2018, the deep phytolittoral was poorly developed. The charophyte zone was visible, although at a shallower depth (up to 6.5 m compared to about 10 m in 1991). In addition, it was entirely covered by one species, *Nitella translucens* (Persoon) C.Agardh, which had replaced the species found earlier in this lake of the genera *Nitella* and *Chara*.

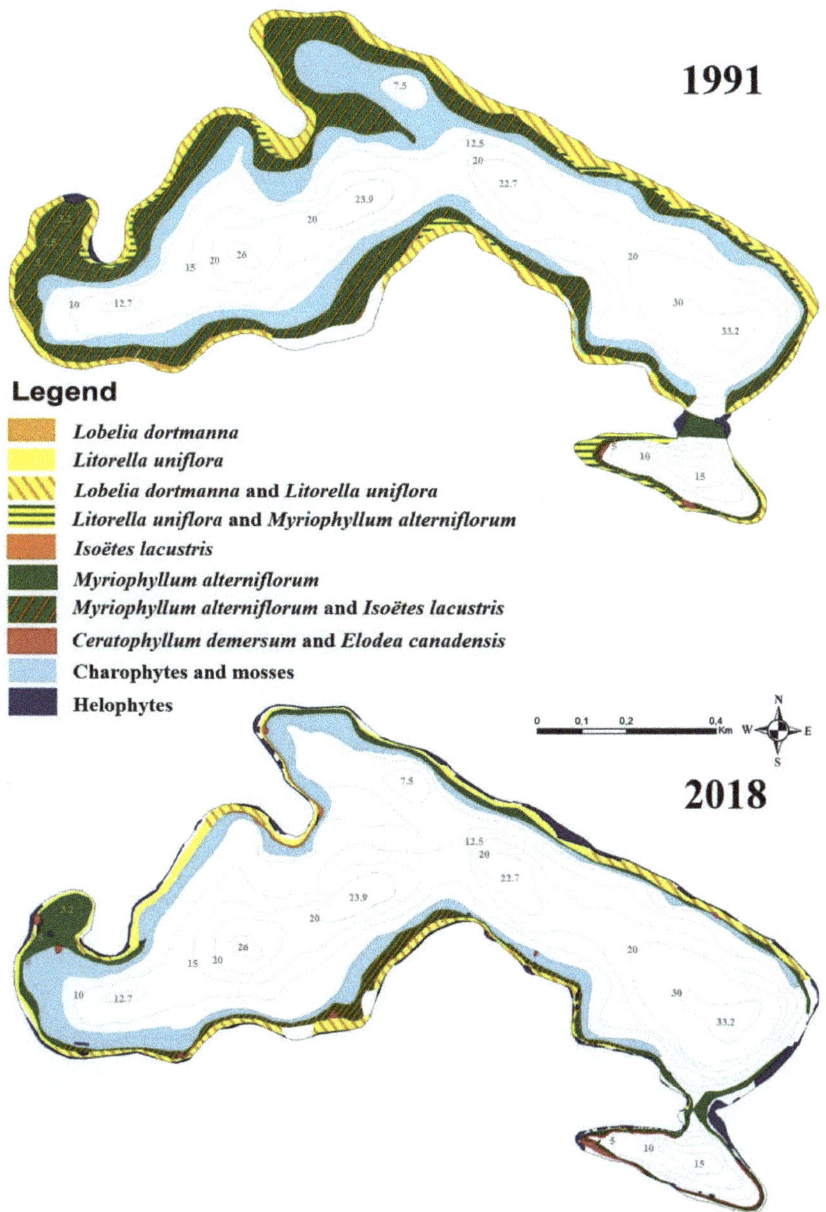

Figure 5. The changes in the range of the patches with a dominance of characteristic macrophytes in the studied period.

Table 5. The occurrence of plant species in the Lake Jeleń in the studied years.

Name of Species	1991	2004	2013	2018
Emergent plants (amphiphytes and helophytes)				
Acorus calamus L.	−	+	+	+
Alisma plantago-aquatica L.	+	+	+	+
Calla palustris L.	−	+	−	−
Carex acutiformis L.	−	+	+	+
Carex paniculata L.	−	−	+	+
Carex rostrata Stokes	+	+	+	−
Eleocharis palustris (L.) R. et Schr.	+	+	+	+
Epilobium palustre L.	+	+	−	−
Equisetum limosum L.	+	+	−	+
Glyceria plicata Fries	+	+	+	+
Hydrocotyle vulgaris L.	+	+	+	+
Iris pseudoacorus L.	+	+	−	+
Juncus bulbosus L.	−	−	+	−
Juncus effusus L.	−	+	+	+
Lycopus europaeus L.	+	+	−	+
Lysimachia thyrsiflora L.	+	+	−	−
Lysimachia vulgaris L.	−	+	+	+
Lythrum salicaria L.	+	+	+	+
Mentha aquatica L.	+	+	+	−
Mentha verticillata L.	+	+	−	−
Myosostis palustris (L.) Nathorst	+	+	+	+
Phalaris arundinacea L.	−	+	+	+
Phragmites australis (Cav.) Trin. ex Steud	+	+	+	+
Scutellaria galericulata L.	+	+	−	+
Sparganium erectum Huds.	−	+	+	+
Sparganium emersum Huds.	+	+	+	−
Typha angustifolia L.	−	+	+	+
Floating-leaved macrophytes (pleustophytes and nymphaeids)				
Lemna minor L.	+	+	+	+
Luronium natans (L.) Raf.	+	+	+	−
Polygonum amphibium L.	+	+	+	+
Sparganium angustifolium Michx	−	+	+	+

Table 5. *Cont.*

Name of Species	1991	2004	2013	2018
Submerged macrophytes (elodeids)				
Ceratophyllum demersum L.	+	+	+	+
Elodea canadensis Rich.	+	+	+	+
Myriophyllum alterniflorum DC.	+	+	+	+
Potamogeton obtusifolius Mert.et Koch.	+	+	−	−
Potamogeton crispus L.	−	−	+	−
Stuckenia pectinata (L.) Börner	−	−	+	+
Isoetids				
Isoëtes lacustris L.	+	+	+	+
Littorella uniflora (L.) Aschers.	+	+	+	+
Lobelia dortmanna L.	+	+	+	+
"Underwater meadows" (Charophytes and Bryophytes)				
Chara delicatula Agardh	+	+	+	−
Nitella flexilis (L.) Agardh	+	+	+	−
Nitella translucens (*Persoon*) Agardh	−	−	+	+
Nitellopsis obtusa (Desvaux) J. Groves	+	−	−	+
Drepanocladus sendtneri (Schimp. ex H.Müll.) Warnst.	+	+	+	+
Fontinalis antipyretica Hedw.	+	−	+	−
Total number of species	32	39	35	32

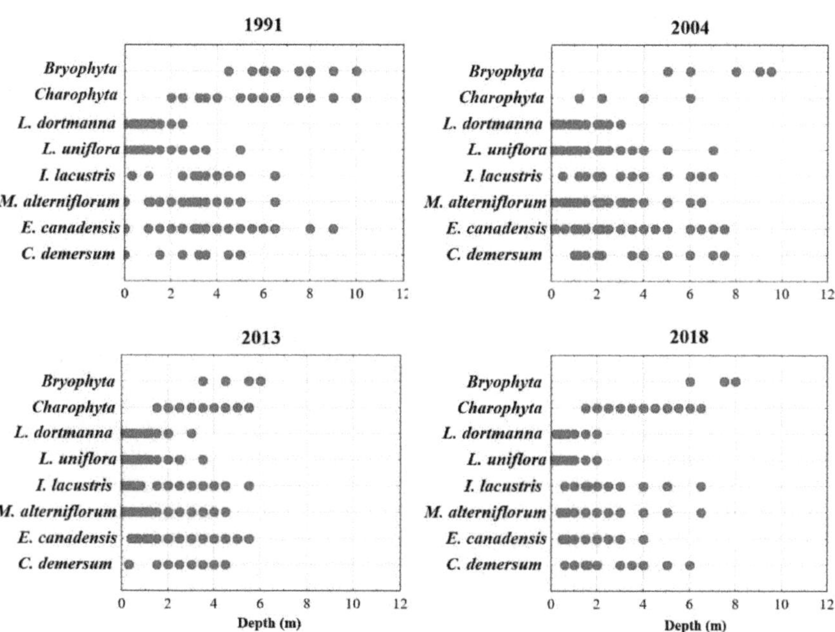

Figure 6. The occurrence depth of characteristic macrophytes in subsequent years of surveys.

In the last year of the study, isoetids occurring in the shallower littoral, i.e., *L. dortmanna* and *L. uniflora*, were recorded only to a depth of 2.0 m, while previously they were found at 4.0 m, and *L. uniflora* sporadically even deeper (Figures 5 and 6).

The frequency of the most important taxa—isoetids and eutrophic elodeids—was analyzed. Comparing 1991 and 2018, each isoetid species showed different trends. The frequency of *L. dortmanna* decreased 2-fold, and the frequency of *L. uniflora* increased, whereas the frequency of *I. lacustris* at first increased and between 2013 and 2018 decreased. The 3-fold increase in the frequency of *C. demersum* (from 6 to ca. 20%) and a slight increase of *E. canadensis* indicated changes in the lake (Figure 7).

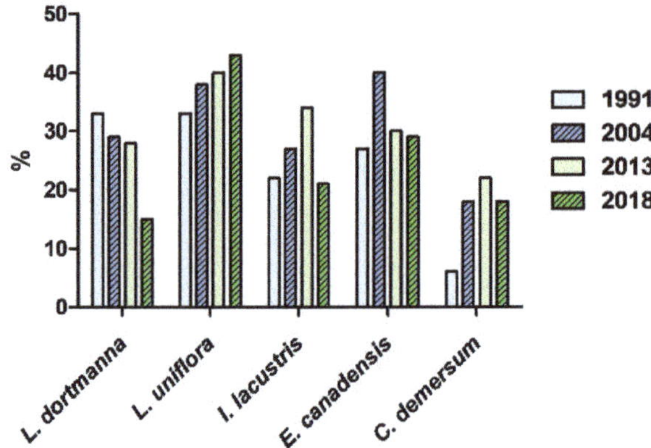

Figure 7. Frequency of the occurrence of characteristic plant species in subsequent years of surveys.

These species had been noted in Lake Jeleń in 1991. However, their share in the structure of vegetation clearly increased.

4. Discussion

SLI are valuable and relatively rare water ecosystems. Low concentrations and bioavailability of carbon (bicarbonates), nitrogen, and phosphorus limit the growth of plants and result in low productivity and stable functioning [1,7,9,27,28]. On the other hand, a low ion content (mainly calcium and manganese), and thus a low water buffer capacity makes these lakes sensitive to human-driven impacts and highly prone to degradation [1,2]. Deterioration of SLI leading to the disappearance of isoetids, in a different pace in various geographical locations, has been continuously observed [1]. According to Arts [1], since the early 20th century, the percentage reduction of isoetids sites in the Netherlands and Germany has amounted to approx. 90% and 95%, respectively. The main threats to the soft water lakes are related to acidification, humification, alkalinisation and eutrophication [15,29]. In Poland, the eutrophication has been identified as a primary process responsible for the degradation of SLI [9,30]. At the beginning of the 20th century, there were over 190 soft water lakes with isoetids [7], while surveys conducted at the turn of the 20th and 21st centuries have shown that 14% of these ecosystems have undergone substantial deterioration resulting in the loss of all characteristic plant species [6]. An estimated 35% of the remaining Polish SLI lost at least one characteristic plant species and approx. 10% lost two plant species. In most of the lakes the coverage of isoetids has significantly decreased [31].

Our study also showed that one of the largest SLI in Poland, Lake Jeleń, underwent fast man driven eutrophication. The first mention on isoetids (*L. dortmanna*, *I. lacustris*) in Lake Jeleń dates from the mid-nineteenth century [32]. According to surveys conducted by Szmal [33] at the end of the 1930s, Lake Jeleń was an oligo-mesotrophic ecosystem with high water transparency, good oxygenation of hypolimnetic water and had dense populations of isoetids. A similar trophic state (oligo-meso), high water transparency, good oxygenation in all water zones and low nutrient concentration was found almost 50 years later in 1991. This indicated a slow evolution of the lake until the beginning of the 1990s. Over the years 1991–2018, there was a gradual increase in the concentration of biogenic elements in the lake water. In the first period, this increase was caused only by an allochthonous supply, but since the formation of the anaerobic zone in the hypolimnion, the nutrients previously deposited in the bottom sediments were easily resuspended into the water column. The steadily increasing concentration of soluble reactive phosphorus in the hypolimnion has confirmed the significant share of internal enrichment in the lake's nutrient balance in recent years. Kraska et al. [9] also pointed out that internal supply plays a significant role in the acceleration of SLI eutrophication.

Increasing loads of nutrients in the water column trigger primary production [34,35]. A gradual increase in phytoplankton biomass and the chlorophyll "a" concentration in Lake Jeleń are clear indicators of the growing fertility of the lake. According to the surveys of Polish SLI made in the early1990s, Lake Jeleń with a range of 33–42 taxa was classified as medium-rich [23,36]. The results of phytoplankton surveys from 2004 to 2018 show that species richness increased to over 50 taxa—a high value for SLI in Poland. The dominance of green algae with the codominance of cyanobacteria observed in Lake Jeleń is characteristic of the array of summer phytoplankton of many other Polish SLI [37–43]. Considering phytoplankton biomass, in the light of the criteria of Heinonen [44], an increase of trophy of the epilimnetic zone from olig-meso to mesoptrophic was found. According to biomass, the cyanobacteria were the group with the highest increase in the studied period and in 2018 prevailed over other phytoplankton groups. The codominance of *P. agardhii*, a species that produces toxic, bioactive metabolites [45,46], may be alarming in the context of its intensive recreational lake use.

Deterioration of the underwater light regime caused by increasing phytoplankton biomass together with changes in chemistry alters the abundance and composition of an aquatic plant community [9,11,47,48]; this was also observed in Lake Jeleń. Undoubtedly, the deterioration of isoetids during the SLI eutrophication arose from several overlapping processes. Changes in water chemistry (availability of nutrients, carbon, water acidity) may lead to the competitive exclusion of isoetids by other plants adapted to more fertile habitats [1,49]; other deleterious effects may arise from an increase in the rate of sedimentation and decomposition of organic matter [50,51] and oxygen depletion in the root zone [52]. A characteristic feature in the 1990s was the occurrence of dense mats of macrophytes to considerable depths, charophytes to 8 m and mosses to over 10 m, and the covering of significant areas of the lake bottom. These stabilized the lake functioning due to the ability to capture and immobilize nutrients from the water column and counteract the resuspension of N and P from the sediments [53,54]. Changes in the physical and chemical parameters of water, especially light reduction [1,3,55], triggered substantial alterations in Lake Jeleń vegetation: reduction of the maximum depth of plant occurrence, and thus of the phytolittoral area, decrease in the frequency and density of isoetids, with a simultaneous expansion of *C. demersum* and *E. canadensis* characteristic of eutrophy and progressing alkalinization [1,56]. Such a reaction of SLI ecosystems on progressing eutrophication have already been described many times [2,9,11,30]. Our findings confirmed the highest tolerance of *L. uniflora* to deterioration of water quality during eutrophication [1,57,58]. The appearance in 2013 [17] of the extremely rare and critically endangered in Poland stonewort *N. translucens* confirms the unstable character of Lake Jeleń. The colonization of SLI by this stonewort during the change of the chemical properties of water were also noted by Brouwer and Roloefs [56] in lake Banen.

Due to the unique character of the lake and its natural value, as well as the continuous ongoing deterioration of its ecological condition, it is necessary to introduce protective measures. The restoration of the SLI is a very complex issue and the measures to be taken depend on many factors such as the character of the catchment area, the primary cause and the degree of the ecosystem degradation, the morphology of the lake and many others [5,9,29,59]. Undoubtedly the restoration of characteristic plant species in SLI depends on the restoration of the physical and chemical conditions of the water and sediments that were present before ecosystem deterioration [9,56,60]. The biotope restoration should have been preceded by the identification and elimination of the main causes of degradation. In the case of surveyed Lake Jeleń, the main threats come from nonpoint sources of pollution: agriculture, urban areas and recreation [16,49], so primarily these sources should be excluded or significantly reduced. There are many examples of natural recovery of water ecosystems, including soft water lakes, after the reduction of local nutrient inputs [9,61,62]. Although isoetids regenerate well thanks to their ability to germinate seeds even 20 years after the species has disappeared [56], the natural recovery of the soft water lakes occurs either in the case of minor degeneration or in the case of a drastic reduction of the pollution [61]. There are many measures supporting natural lakes recovery and reducing the inflow of allochthonous nutrients. Ensuring the proper sewage system, introducing sustainable agriculture, establishing wide buffer zones with wetland and forest vegetation around the lakes are most often used [16,63]. There are also a number of methods of SLI restoration consisting of measures within the water ecosystems. One of the non-invasive methods that could be applied is top-down biomanipulation leading to the eventual reduction of phytoplankton biomass and increase in water transparency. However, biomanipulation may be successfully implemented in lakes in the early stages of degradation [64,65]. Other methods of SLI restoration depend on the main degradation factor. The liming is the main strategy to counteract acidification of boreal SLI. However, this measure is implemented to benefit fish populations, rather than to recover soft water vegetation [66,67]. Experimental data shows that liming affects the shoot:root ratio of isoetids. Shoot biomass and length of *L. dortmanna*, *L. uniflora* and *I. lacustris* significantly increase after liming [67]. It may be a factor causing the isoetids to uproot during heavy storms and water waving [29,67]. Boreal SLI also lost their characteristic vegetation after liming due to overshadowing by rapidly-developing caulescent plants (e.g. *Juncus bulbosus*) and alkalinization and fertilization of sediments [67–69]. A similar reaction of characteristic plant species was found in limed Polish SLI [9]. Several cases of the restoration of acidified Atlantic SLI also indicate the lack of effectiveness of this method [29]. One-time liming is usually followed by rapid reacidification while regular liming leads do the alkalinization of sediments and nutrient mobilization [29,70]. In acidified Atlantic SLI, better results of restoration were observed after the removal of recently accumulated organic matter from both the riparian and aquatic zone. However, the ecosystem restoration and soft water macrophyte recolonization was usually short term. Reacidification or eutrophication leads to the decline of isoetids in a few years [56]. Organic matter removal has had a much better effect in Atlantic eutrophicated SLI. Mud removal in several such lakes resulted in a decrease of the concentration of soluble reactive phosphorus, a decrease of algal blooms, an increase of transparency and the reestablishment of isoetids and other plant species characteristic for SLI. Such effects persisted for over a decade after the restoration [29,56]. However, the case of Lake Banen in the Netherlands shows that such measures must cover all of the lake areas. Otherwise, the unrestored parts are the source of re-eutrophication and degradation [29,56]. The implementation of such measures in Lake Jeleń, due to the significant area and development of shoreline inaccessibility of a part of the shore, would be very expensive. Improvement of underwater light regimes and oxygenation conditions during the restoration of lakes is achieved by the flocculation of algae blooms with aluminum or iron-based coagulants. However, in the case of SLI, these coagulants should not be used, due to reports of adverse effects of these measures on aquatic macrophytes [71,72]. Due to the significant volume of hypolimnion and an anoxic zone in Lake Jeleń, such measures as the removal of hypolimnetic water, or supplementation of sub-bottom water with oxygen is inefficient in relation to costs [16].

5. Conclusions

Until the last decade of the 20th century, Lake Jeleń was a stable soft water ecosystem with dense populations of isoetids. Human impact exerted on the catchment and directly on the lake has led to ongoing deterioration of water quality and an increase in lake fertility. Altogether these changes have caused a biological reaction: increase in the biomass of phytoplankton assemblages, the dominance of cyanobacteria, reduction of the range of the phytolittoral and decreased bottom coverage by *L. dortmanna* and *I. lacustris*. In view of the progressive eutrophication and degradation of the valuable ecosystem, protective measures should be implemented. Reduction of allochthonous nutrient loads and a change in the structure of ichthyofauna (an increase of predatory species) are suggested to achieve a significant reduction in algae biomass, an increase in water transparency and rehabilitation of soft water macrophytes.

Author Contributions: Conceptualization, P.K., R.P. and D.B.; methodology, P.K.; software, P.K., J.R., D.B., R.P.; validation, P.K., D.B., J.R. formal analysis, P.K.; investigation, P.K., M.K., D.B., R.P., J.R., E.S.-W.; resources, P.K.; writing—original draft preparation, P.K., D.B., R.P., J.R. and E.S.-W.; writing—review and editing, P.K., D.B., J.R. and E.S.-W.; visualization, P.K., D.B. and J.R.; supervision, P.K. and M.K.; finding acquisition, P.K., D.B. All authors read and approved the final manuscript.

Funding: This research received no external funding.

Conflicts of Interest: The authors declare no conflict of interest.

References

1. Arts, G.H.P. Deterioration of atlantic soft water macrophyte communities by acidification, eutrophication and alkalinisation. *Aquat. Bot.* **2002**, *73*, 373–393. [CrossRef]
2. Smolders, A.J.P.; Lucassen, E.C.H.E.T.; Roelofs, J.G.M. The isoetid environment: Biogeochemistry and threats. *Aquat. Bot.* **2002**, *73*, 325–350. [CrossRef]
3. Spierenburg, P.; Lucassen, E.C.H.E.T.; Lotter, A.F.; Roelofs, J.G.M. Competition between isoetids and invading elodeids at different concentrations of aquatic carbon dioxide. *Freshw. Biol.* **2010**, *55*, 1274–1287. [CrossRef]
4. European Commission, DG Environment, Nature ENV B.3, B. *European Commission, Interpretation Manual of European Union Habitats–EUR28*; European Commission: Brussels, Belgium, 2013.
5. Kolada, A.; Piotrowicz, R.; Wilk-Woźniak, E.; Dynowski, P.; Klimaszyk, P. Conservation status of the Natura 2000 habitat 3110 in Poland: Monitoring, classification and trends. *Limnol. Rev.* **2017**, *17*, 215–222. [CrossRef]
6. Kraska, M. Jeziora lobeliowe (Lobelia lakes). In *Siedliska Przyrodnicze. Poradniki Ochrony Siedlisk i Gatunków Natura 2000–Podręcznik Metodyczny. Tom 2: Wody Słodkie i Torfowiska (Natural Habitats. Guides for the Protection of Habitats and Natura 2000 Species–A Methodological Handbook*; Herbich, J., Ed.; Ministerstwo Środowiska: Warsaw, Poland, 2004; Volume 2, pp. 29–36.
7. Borowiak, D.; Piotrowicz, R.; Nowiński, K.; Klimaszyk, P. Soft-water lobelia lakes in Poland. In *Polish River Basins and Lakes–Part I. The Handbook of Environmental Chemistry*; Krzeniewska, E., Harnisz, M., Eds.; Springer: Berlin/Heidelberg, Germany, 2020; pp. 89–118.
8. Free, G.; Bowman, J.; McGarrigle, M.; Caroni, R.; Donnelly, K.; Tierney, D.; Trodd, W.; Little, R. The identification, characterization and conservation value of isoetid lakes in Ireland. *Aquat. Conserv. Mar. Freshw. Ecosyst.* **2009**, *19*, 264–273. [CrossRef]
9. Kraska, M.; Klimaszyk, P.; Piotrowicz, R. Anthropogenic changes in properties of the water and spatial structure of the vegetation of the lobelia lake Lake Modre in the Bytów Lakeland. *Oceanol. Hydrobiol. Stud.* **2013**, *42*, 302–313. [CrossRef]
10. Klimaszyk, P.; Rzymski, P. Catchment vegetation can trigger lake dystrophy through changes in runoff water quality. *Ann. Limnol.-Int. J. Limnol.* **2013**, *49*, 191–197. [CrossRef]
11. Klimaszyk, P.; Piotrowicz, R.; Rzymski, P. Changes in physico-chemical conditions and macrophyte abundance in a shallow soft-water lake mediated by a Great Cormorant roosting colony. *J. Limnol.* **2015**, *74*, 114–122. [CrossRef]
12. Joniak, T.; Klimaszyk, P.; Kraska, M. Diel dynamics of vertical changes of chlorophyll and bacteriochlorophyll in small humic lakes. *Oceanol. Hydrobiol. Stud.* **2010**, *39*, 103–111. [CrossRef]
13. Naiman, R.J.; Decamps, H. *The Ecology and Management of Aquatic-Terestrial Ecotones*; UNESCO/MAB(05)/M2/v.4: Paris, French, 1990.
14. Serafin, A.; Sender, J.; Bronowicka-Mielniczuk, U. Potential of shrubs, shore vegetation and macrophytes of a lake to function as a phytogeochemical barrier against biogenic substances of various origin. *Water* **2019**, *11*, 290. [CrossRef]
15. Murphy, K.J. Plant communities and plant diversity in softwater lakes of northern Europe. *Aquat. Bot.* **2002**, *73*, 287–324. [CrossRef]

16. Szyper, H.; Gołdyn, R.; Romanowicz, W.; Kraska, M. Możliwości ochrony oligotroficznego jeziora Jeleń przed czynnikami antropogennymi (Possibilities of protecting the oligotrophic Lake Jeleń against influence of man). In *Jeziora Lobeliowe. Charakterystyka, Funkcjonowanie i Ochrona. Cz. I.*; Kraska, M., Ed.; Sorus: Poznań, Poland, 1994; pp. 105–115.

17. Rosińska, J.; Piotrowicz, R.; Celiński, K.; Dabert, M.; Rzymski, P.; Klimaszyk, P. The reappearance of an extremely rare and critically endangered Nitella translucens (Charophyceae) in Poland. *J. Phycol.* **2019**, *55*, 1412–1415. [CrossRef] [PubMed]

18. American Public Health Association. *Standard Methods for the Examination of Water and Wastewater*, 17th ed.; American Public Health Association: Washington, DC, USA, 1991.

19. Carlson, R.E. A trophic state index for lakes. *Limnol. Ocean.* **1977**, *22*, 361–369. [CrossRef]

20. Borowiak, D. *Właściwości Optyczne wód Jeziornych Pomorza (Optical properties of the Pomeranian Lake Waters)*; Wydawnictwo Uniwersytetu Gdańskiego: Gdańsk, Poland, 2011.

21. Ficek, D. *Właściwości Biooptyczne wód jezior Pomorza oraz ich poróWnanie z Właściwościami wód Innych Jezior i Morza Bałtyckiego (Bio-Optical Properties of Waters in Pomeranian Lakes and Their Comparison to Optical Properties of Other Lake Waters and the Baltic Sea)*; Rozpr. Mon.; PAN IO: Sopot, Poland, 2013.

22. Wetzel, R.G.; Likens, G.E. *Limnological Analyses*, 2nd ed.; Springer: Berlin/Heidelberg, Germany, 1991.

23. Szeląg-Wasielewska, E.; Gołdyn, R.; Bernaciak, A. Fitoplankton a stan trofii wód jeziora Jeleń na Pojezierzu Bytowskim (Phytoplankton versus the trophic state of lake Jeleń in the Pomeranian Lakeland). In *Badania Fizjograficzne nad Polską Zachodnią, Seria B–Botanika*; PTPN: Poznań, Poland, 1999; pp. 203–223.

24. Hörnström, E. Trophic characterization of lakes by means of qualitative phytoplankton analysis. *Limnologica* **1981**, *13*, 249–261.

25. Braun-Blanquet, J. *Pflanzensoziologie*; Springer: Berlin/Heidelberg, Germany, 1964.

26. Jensén, S. An objective method for sampling the macrophyte vegetation in lakes. *Vegetatio* **1977**, *33*, 107–118. [CrossRef]

27. Kraska, M.; Piotrowicz, R. Roślinność wybranych jezior lobeliowych na tle warunków fizyczno-chemicznych ich wód. In *Jeziora Lobeliowe. Charakterystyka, Funkcjonowanie i Ochrona. Cz. I.*; Kraska, M., Ed.; Sorus: Poznań, Poland, 1994; pp. 37–83.

28. Madsen, T.V.; Olesen, B.; Bagger, J. Carbon acquisition and carbon dynamics by aquatic isoetids. *Aquat. Bot.* **2002**, *73*, 351–372. [CrossRef]

29. Brouwer, E.; Bobbink, R.; Roelofs, J.G.M. Restoration of aquatic macrophyte vegetation in acidified and eutrophied softwater lakes: An overview. *Aquat. Bot.* **2002**, *73*, 405–431. [CrossRef]

30. Szmeja, J. Evolution and conservation of lobelia lakes in Poland. *Fragm. Flor. Geobot.* **1997**, *40*, 89–94.

31. Kraska, M.; Piotrowicz, R. Lobelia lakes: Specificity, trophy, vegetation and protection. In *Protection of Lakes and Wetlands of Pomerania Region*; Malinowski, B., Ed.; Towarzystwo Ekologiczno-Kulturalne: Bobolice, Poland, 2000; pp. 48–52.

32. Doms, F.A. Ein neuer Standort von Isoetes lacustris (L.) Dur. (A new location for Isoetes lacustris (L.) Dur.). *Werh. Bot. Ver. Pr. Brand.* **1862**, *3–4*, 387–388.

33. Szmal, Z. Badania hydrochemiczne jezior lobeliowych Pojezierza Zachodniego. *PTPN* **1959**, *XIX*, 106.

34. Roelofs, J.G.M. Soft-water macrophytes and ecosystems: Why are they so vulnerable to environmental changes? Introduction. *Aquat. Bot.* **2002**, *73*, 285–286. [CrossRef]

35. Piotrowicz, R.; Kraska, M.; Klimaszyk, P.; Szyper, H.; Joniak, T. Vegetation richness and nutrient loads in 16 lakes of Drawieński National Park (northern Poland). *Polish J. Environ. Stud.* **2006**, *15*, 467–478.

36. Szeląg-Wasielewska, E.; Gołdyn, R. Zbiorowiska glonów w pelagialu jezior lobeliowych (Algal communities in the pelagial zone of lobelian lakes. In *Jeziora Lobeliowe. Charakterystyka, Funkcjonowanie i ochrona. Cz. I.*; Kraska, M., Ed.; Sorus: Poznań, Poland, 1994; pp. 37–65.

37. Szeląg-Wasielewska, E. Vertical distribution of phototrophs in the pelagic zone of three small Lobelia lakes. *Oceanol. Hydrobiol. Stud.* **2010**, *39*, 121–133. [CrossRef]

38. Oleksowicz, A. Phytoplankton communities of Lobelia–type lakes in the Kashubian Lake District (Pomerania, northern Poland). *Acta Hydrobiol.* **1989**, *31*, 259–271.

39. Luścińska, M.; Oleksowicz, A.S. The structure and production of algal communities in the lakes of the Bory Tucholskie region. In *Some Ecological Processes of the Biological Systems in North Poland*; Bohr, R., Nienartowicz, A., Witkoń-Michalska, J., Eds.; N. Copernicus University Press: Toruń, Poland, 1992; pp. 283–297.

40. Szeląg-Wasielewska, E. Phytoplankton community structure in non-stratified lakes of Pomerania (NW Poland). *Hydrobiologia* **2003**, *506–509*, 229–236. [CrossRef]

41. Szeląg-Wasielewska, E. Phytoplankton structure of two small lakes–changes after a decade. *Oceanol. Hydrobiol. Stud.* **2007**, *36*, 113–120.

42. Pełechata, A. Zbiorowiska fitoplanktonu jezior lobeliowych Pojezierza Pomorskiego (Phytoplankton communities of lobelia lakes of the Pomeranian Lake District). In *Jeziora Lobeliowe w Drugiej Dekadzie XXI Wieku*; Bociąg, K., Borowiak, D., Eds.; Fundacja Rozwoju Uniwersytetu Gdańskiego: Gdańsk, Poland, 2016; pp. 80–96.

43. Szeląg-Wasielewska, E. Phytoplankton structure in lakes with a low trophic level in north-west Poland. In *Physicochemical Problems of Natural Waters Ecology*; Gurgul, H., Ed.; Wydawnictwo Naukowe Uniwersytetu Szczecińskiego: Szczecin, Poland, 2005; pp. 281–296.

44. Heinonen, P. Quantity and composition of phytoplankton in Finnish inland waters [eutrophication, water quality, diversity, indicator species, odour index]. *Water Res. Inst.* **1980**, *37*, 1–91.

45. Komárek, J.; Komárkova, J. Taxonomic review of the cyanoprokaryotic genera Planktothrix and Planktothricoides. *Czech. Phycol.* **2004**, *4*, 1–18.

46. Welker, M.; Christiansen, G.; von Döhren, H. Diversity of coexisting Planktothrix (Cyanobacteria) chemotypes deduced by mass spectral analysis of microystins and other oligopeptides. *Arch. Microbiol.* **2004**, *182*, 288–298. [CrossRef]

47. Rosińska, J.; Gołdyn, R. Response of vegetation to growing recreational pressure in the shallow Raczyńskie Lake. *Knowl. Manag. Aquat. Ecosyst* **2018**, *419*, 1. [CrossRef]

48. O'Hare, M.T.; Aguiar, F.C.; Asaeda, T.; Bakker, E.S.; Chambers, P.A.; Clayton, J.S.; Elger, A.; Ferreira, T.M.; Gross, E.M.; Gunn, I.D.M.; et al. Plants in aquatic ecosystems: Current trends and future directions. *Hydrobiologia* **2018**, *812*, 1–11. [CrossRef]

49. Klimaszyk, P.; Kraska, M.; Piotrowicz, R. Human impact on the functioning lobelia lake: Lake Jeleń (Bytów Lakeland). In *Diagnosing the State of the Environment, Methods and Prognoses*; Garbacz, J.K., Ed.; Bydgoskie Towarzystwo Naukowe: Bydgoszcz, Poland, 2014; pp. 131–138.

50. Chappuis, E.; Lumbreras, A.; Ballesteros, E.; Gacia, E. Deleterious interaction of light impairment and organic matter enrichment on Isoetes lacustris (Lycopodiophyta, Isoetales). *Hydrobiologia* **2015**, *760*, 145–158. [CrossRef]

51. Ribaudo, C.; Tison-Rosebery, J.; Buquet, D.; Gwilherm, J.; Jamoneau, A.; Abril, G.; Anschutz, P.; Bertrin, V. Invasive aquatic plants as ecosystem engineers in an oligo-mesotrophic shallow lake. *Front. Plant. Sci.* **2018**, *9*, 1781. [CrossRef] [PubMed]

52. Sand-Jensen, K.; Claus, L.; Borum, J. High resistance of oligotrophic isoetid to oxic and anoxic dark exposure. *Freshwat. Biol.* **2015**, *60*, 1044–1051. [CrossRef]

53. Srivastava, J.; Gupta, A.; Chandra, H. Managing water quality with aquatic macrophytes. *Rev. Environ. Sci. Biotechnol.* **2008**, *7*, 255. [CrossRef]

54. Kuczyńska-Kippen, N.; Klimaszyk, P. Diel microdistribution of physical and chemical parameters within the dense Chara bed and their impact on zooplankton. *Biologia* **2007**, *4*, 432–437. [CrossRef]

55. Borowiak, D.; Bociąg, K.; Nowiński, K.; Borowiak, M. Light requirements of water lobelia (Lobelia dortmanna L.). *Limnol. Rev.* **2017**, *17*, 171–182. [CrossRef]

56. Brouwer, E.; Roelofs, J.G.M. Degraded softwater lakes: Possibilities for restoration. *Restor. Ecol.* **2001**, *9*, 155–166. [CrossRef]

57. Kłosowski, S.; Szańkowski, M. Habitat differentiation of the Myriophyllum alterniflorum and Littorella uniflora phytocoenoses in Poland. *Acta Soc. Bot. Pol.* **2011**, *73*, 79–86. [CrossRef]

58. Dierssen, K. Littorelletea uniflorae. In *Prodromus der Europäischen Pflanzengesellschaften*; Tuxen, R., Ed.; Vaduz, Liechtenstein, 1975.

59. Roelofs, J.G.M. Restoration of eutrophied shallow softwater lakes based upon carbon and phosphorus limitation. *Neth. J. Aquat. Ecol.* **1996**, *30*, 197–202. [CrossRef]

60. Pulido, C.; Keijsers, D.J.H.; Lucassen, E.C.H.E.T.; Pedersen, O.; Roelofs, J.G.M. Elevated alkalinity and sulfate adversely affect the aquatic macrophyte Lobelia dortmanna. *Aquat. Ecol.* **2012**, *46*, 283–295. [CrossRef]

61. Kochanowski, J.; Tobolski, K. A new locality of lobelia in Lake Krzywce Wielkie: Bory Tucholskie National Park. *Stud. Lim. Tel.* **2010**, *4*, 61–64.

62. Riera, J.L.; Ballesteros, E.; Pulido, C.; Chappuis, E.; Gacia, E. Recovery of submersed vegetation in a high mountain oligotrophic soft-water lake over two decades after impoundment. *Hydrobiologia* **2017**, *794*, 139–151. [CrossRef]

63. Zalewski, M.; Robarts, R. Ecohydrology–a new paradigm for integrated water resources management. *SILnews* **2003**, *40*, 1–5.

64. Romo, S.; Van Donk, E.; Gylstra, R.; Gulati, R. A multivariate analysis of phytoplankton and foodweb changes in a shallow biomanipulated lake. *Freshw. Biol.* **1996**, *36*, 683–696. [CrossRef]

65. Rosińska, J.; Romanowicz-Brzozowska, W.; Kozak, A.; Gołdyn, R. Zooplankton changes during bottom-up and top-down control due to sustainable restoration in a shallow urban lake. *Environ. Sci. Pollut. Res.* **2019**, *26*, 19575–19587. [CrossRef]

66. Larssen, T.; Cosby, B.J.; Lund, E.; Wright, R.F. Modeling future acidification and fish populations in Norwegian surface waters. *Environ. Sci. Technol.* **2010**, *44*, 5345–5351. [CrossRef]

67. Lucassen, C.H.E.T.; Smolders, A.J.P.; Roelofs, J.G.M. Liming induces changes in the macrophyte vegetation of Norwegian softwater lakes by mitigating carbon limitation: Results from a field experiment. *App. Veget. Sci.* **2012**, *15*, 166–174. [CrossRef]

68. Lindmark, G.K. Acidified lakes sediment treatment with sodium carbonate, a remedy? *Hydrobiologia* **1987**, *92*, 537–547.

69. Sand-Jensen, K.; Borum, J.; Binzer, T. Oxygen stress and reduced growth of Lobelia dortmanna in sandy lake sediments subject to organic enrichment. *Freshw. Biol.* **2005**, *9*, 1–11. [CrossRef]

70. Bellemakers, M.J.S.; Maessen, M.; Verheggen, G.M.; Roelofs, J.G.M. Effect of liming on shallow acidified moorland pools: A culture and seed bank experiment. *Aquat Bot.* **1996**, *54*, 37–50. [CrossRef]

71. Rybak, M.; Kołodziejczyk, A.; Joniak, T.; Ratajczak, I.; Gąbka, M. Bioaccumulation and toxicity studies of macroalgae (Charophyceae) treated with aluminium: Experimental studies in the context of lake restoration. *Ecotoxicol. Environ. Saf.* **2017**, *145*, 359–366. [CrossRef] [PubMed]

72. Rybak, M.; Gąbka, M.; Ratajczak, I.; Woźniak, M.; Sobczyński, T.; Joniak, T. In-situ behavioural response and ecological stoichiometry adjustment of macroalgae (Characeae, Charophyceae) to iron overload: Implications for lake restoration. *Water Res.* **2020**, *173*, 115602. [CrossRef] [PubMed]

Article

Oxygen Gradients and Structure of the Ciliate Assemblages in Floodplain Lake

Roman Babko [1,*], Tetiana Kuzmina [2], Yaroslav Danko [3], Joanna Szulżyk-Cieplak [4] and Grzegorz Łagód [5]

[1] Department of Invertebrate Fauna and Systematics, Schmalhausen Institute of Zoology NAS of Ukraine, 01030 Kyiv, Ukraine
[2] Faculty of Technical Systems and Energy Efficient Technologies, Sumy State University, 40007 Sumy, Ukraine; t.kuzmina@ecolog.sumdu.edu.ua
[3] Faculty of Natural Sciences and Geography, Sumy Makarenko State Pedagogical University, 40002 Sumy, Ukraine; yaroslavdanko@gmail.com
[4] Fundamentals of Technology Faculty, Lublin University of Technology, 20-618 Lublin, Poland; j.szulzyk-cieplak@pollub.pl
[5] Faculty of Environmental Engineering, Lublin University of Technology, 20-618 Lublin, Poland; g.lagod@pollub.pl
* Correspondence: rbabko@ukr.net

Received: 23 May 2020; Accepted: 17 July 2020; Published: 23 July 2020

Abstract: This paper presents the results of studies on the structure of the ciliate population in a freshwater lake. The classification of the ciliated communities based on the analysis of the distribution of ciliate population density in the lake along the oxygen gradients, taking into account their oxygen preferences, was proposed. It was shown that the distribution of ciliated protozoa in the space of a reservoir is determined not by such spatial units as the water column, bottom, and periphytal, but by the oxygen gradients. Four types of habitats with different oxygen regimes were distinguished: With stably high oxygen concentration, stably low oxygen concentration, stably oxygen-free conditions, and conditions with a high amplitude of diurnal oxygen variations. The location of these habitats in the space of the lake and their seasonal changes were determined. On the basis of the quantitative development of ciliate populations, zones of optima and tolerance ranges of some ciliate species in the oxygen gradient were established. The oxygen preferences were established for the species from four distinguished assemblages: Microoxyphilic, oxyphilic, euryoxyphilic, and anoxyphilic (anaerobic). The presence or the absence of a certain type of assemblage in the reservoirs depends solely on the parameters of the oxygen gradients. The diversity of the ciliated protozoa in water bodies also depends on the stability and diversity of the oxygen gradients.

Keywords: ciliate assemblages; oxygen concentration; ecological optimum; lake; freshwater habitats

---

1. Introduction

Ciliates are one of the diverse groups of protozoa; today about 1500 morphospecies of freshwater ciliates have been described [1]. However, according to forecasts, the total diversity of these unicellular organisms can reach from 12 to 40 thousand species [2,3]. The diversity of ciliated protozoa on a global scale is fairly uniform. Nevertheless, numerous studies show that in some territories, and in water bodies of different types, the diversity of ciliated protozoa differ significantly [4–12]. Compared to the potential diversity, the realized diversity is dynamic and reflects the specific conditions of each water body, leading to the dissimilarity of the fauna. This local diversity remains unpredictable and varies in space and time. Nevertheless, it has been successfully used to assess the status of aquatic ecosystems. In assessing the species diversity in a separate time-limited study, only a small fraction of the available genetic potential ("seedbank of protozoan") is usually detected [1]. Assessing the local diversity of even a small, one-hectare pond can never be complete [13]. Obtaining reliable information about the local diversity of ciliates remains a problem. It depends on the duration of the studies, season coverage, the geographical location of the reservoir, and so on. The number of identified species is also affected by the number of samples taken, the frequency of sampling, the number and variety of the examined habitats, etc. Obviously, an effective study of the diversity of protozoa also depends on our ideas about their structural organization in the hyperspace of water bodies.

Despite a significant number of studies, they are not enough to predict the local diversity of ciliated protozoa, its changes in time and its relation to the type of reservoir. The assessment of diversity remains problematic due to the conservatism of ideas about the nature of species distribution. Namely, most studies proceed from the idea that the ciliate population is structured according to the traditional biotopes: Pelagial, benthal, periphytal [4,14–23].

At the same time, it has been shown quite convincingly that the ciliated protozoa are very sensitive to the concentration of oxygen dissolved in water [24,25]. Nevertheless, the oxygen gradients are far from being always taken into account when studying the diversity of protozoa and assessing the state of the ecosystem of a reservoir. In most cases, the traditional approach is prevailing in dividing the space of a reservoir into listed biotopes, although the conditions, including the concentration of oxygen, within each biotope may be heterogeneous [24].

This makes it relevant to analyze the spatial distribution of the ciliated protozoa without reference to the classical biotopes, particularly, to study the structure of the population of ciliates in the context of their reaction to the oxygen gradients.

In this regard, the aim of our work was to generalize the results of two-year studies on the ciliated protozoa under the conditions of a floodplain lake in the context of their spatial organization and the relationship of this organization with the oxygen gradients that existed in different parts of the water body, regardless of its traditional separation into the classical biotopes.

2. Materials and Methods

2.1. Study Area

The authors studied the distribution of ciliated protozoa in water in a floodplain eutrophic lake. The floodplain lake is located on the right bank of the Vorskla River in the Sumy Oblast (Ukraine). The lake stretches from northwest to southeast approximately parallel to the modern channel of the Vorskla River. Table 1 shows the location, morphometric data of the lake, and selected water quality indicators.

Table 1. Geographical location and characteristics of the lake.

Parameters	Values
Location	50.313942, 34.839761
Area, m^2	21,300
Length, m	400.0
Maximum width, m	59.5
Maximum depth, m	8.5
Secchi depth, m	0.7–1.0
pH	8.15–8.30
Permanganate index, mg O_2/L	7.8–15.0

The lake is located 100 m from the main channel of the river and during periods of high water—mainly in the spring during floods—is connected to the main channel by a narrow channel, which dries up in summer. The lake never completely dries up. In the low-water period (summer), the maximum depth is 8–8.5 m. Macrophyte thickets were formed by *Phragmites australis* (Cav.) Trin. ex Steudel, *Ceratophyllum demersum* L., *Lemna minor* L., *Spirodela polyrrhiza* (L.) Schleid., *Nuphar lutea* (L.)

Sibth. & Sm. Samples from each type of habitat were taken at five points (places). The lake and the localization of sampling stations are shown in Figure 1.

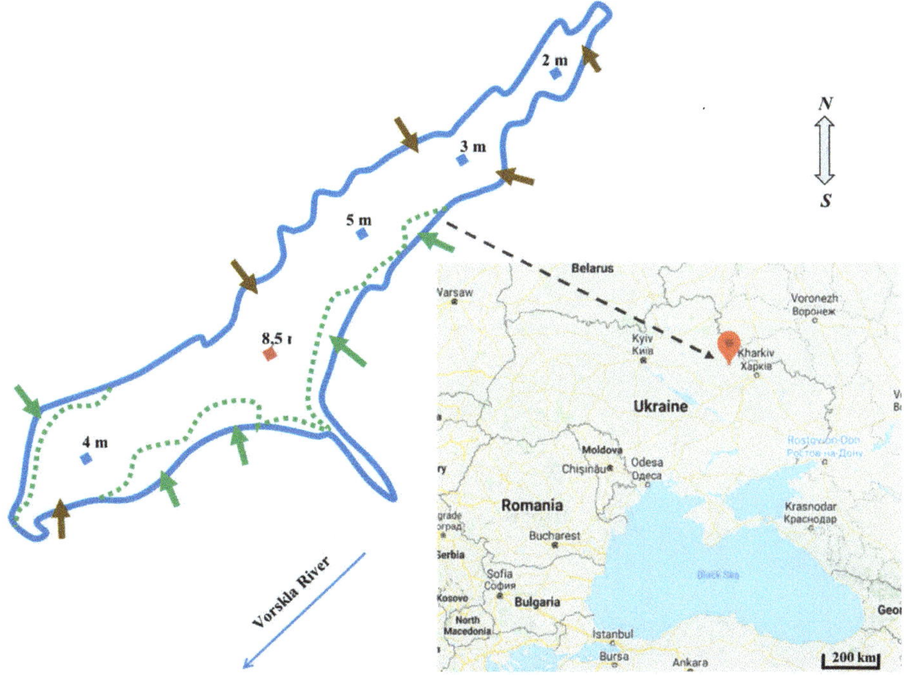

Figure 1. Location of the study area and sampling points. The green arrows indicate the places of sampling of plants, the brown arrows indicate the places of sampling of bottom sediments in the littoral, the rhombuses indicate the places of sampling of water and bottom sediments in the pelagial and profundal, the red rhombus indicates the place of the maximum depth of the lake. The dashed green line shows littoral sections of macrophyte thickets.

In the littoral, samples of bottom sediments were taken at a depth of 0.5 to 1.5 m depending on the slope of the bottom, usually at a distance of 1 m from the water edge. Plant samples were taken from the central part of the plant spot at a depth of about 0.5 m. Metaphyton samples were taken at a distance of 0.5–1.0 cm from the plant shoots at a depth of 0.5 m.

2.2. Habitats

In spring and autumn, water turnover was observed in the lake. In winter, during the ice cover, anaerobic conditions were observed in the lake in almost the entire water column. From June to September, a distinct temperature stratification is observed in the lake. In the water column during this period, three habitats that differ in the oxygen content can be distinguished. The epilimnion has a high oxygen content of up to 13 mg/L and a layer of water from the surface to a depth of 1–1.5 m. The metalimnion has an oxygen content of from 0 to 3 mg/L and a layer from 1.5 to 4.0 m. In the upper part of the metalimnion at a depth of 1.5 to 2.5 m, an accumulation of detritus is observed. There is a sharp decrease in the oxygen concentration here to 0–1 mg/L. The hypolimnion was anoxic.

According to the bottom profile, two different habitats are distinguished: The bottom in the littoral zone, represented by silted sand with detritus, with a low oxygen content of 0 to 3 mg/L; and the bottom in the profundal zone with sapropel deposits and anoxic conditions. The difference between these two habitats is that in the profundal zone oxygen is always absent, and in the silted sand the conditions range from microaerophilic to anaerobic.

Periphytal, the surface of the plants and metaphytal, the water among the thickets of plants, were considered as separate habitats. Both in the layer of water near the plants' surface and—to a lesser extent—in the metaphytal, the oxygen content varied drastically over the course of the day: From 0 to 250% of saturation. Oxygen surges (high amplitude of diurnal fluctuations in the oxygen content) were the main difference between these habitats and all others, where the oxygen content did not change significantly during the day.

2.3. Sampling Methods and Calculations

The samples for counting of ciliates were taken in all of the habitats. Samples from the water column (1.0 L) were taken by a Rutner bathometer. At each of the five stations, samples were taken every 0.5 m vertically. The sampling of the bottom sediments was carried out with a microbentometer [26]. Using it, the sediment cores of a cylindrical shape with an area of 7 cm^2 and a length of up to 6 cm were taken. For analysis of bottom samples from littoral, a 1-cm layer of bottom sediments and a 3-cm layer of water above the bottom, where oxygen was recorded, were treated as a single sample. The layer of bottom sediments below 3 cm depth, where there was no oxygen, was detached and examined as a separate sample. The water samples from the thickets of plants were taken with a syringe (volume 200 mL, hole diameter 4 mm) onto which instead of a needle, a plastic tube of the corresponding diameter was installed. The plant samples were taken using a glass tube with a diameter of 3.5 cm, which was carefully draw on the shoot of the plant; the shoot was cut off, afterwards the tube was closed up on both sides (Figure 2a–c) [10]. Semi-submerged plants were previously cut off above the surface of the water, and a sample was taken, truncating part of the underwater leaf or stem and enclosing it in a glass tube. Samples of floating plants (duckweed) were taken by passing a glass tube along the surface covered with duckweed plants; after the tube was closed on both sides (Figure 2d). Samples of anaerobic sediments (sapropel) and samples from oxygen free hypolimnion were taken and sealed in containers without contact with oxygen. Oxygen-free samples were opened only at the time of sub-sampling.

The studies were carried out in full in the first and last decade of each month. In each habitat, temperature and O_2 measurements were carried out using a HACH HQ40d portable multi-meter with a sensor attached to a 10 m cable. Oxygen measurements in all environments were performed before each sampling. Each time vertical sampling using the HACH HQ40d multi-meter, oxygen and temperature were measured from the surface to the bottom. The distribution of these parameters by depths were established. Seasonal averages are shown in graphs.

To establish the permanganate index, a part of the sample taken by a bathometer from each depth was used. When sampling plants, metaphyton, and bottom, a sample for the permanganate index was taken in the same place with a 100 mL syringe.

The samples were transported to the laboratory, where they were immediately processed. During the analysis of the sub-samples, the main samples were kept in the refrigerator at a temperature of +5 °C, which ensured the maximum preservation of quantitative ratios of populations.

In the quantitative treatment of samples, the abundance of ciliates was determined in their actual living space (ALS). By actual living space, we mean part of the volume of the water body in which the population of a given species is recorded. This can be the entire pelagic zone (or only part of it), the water in bottom sediments, or a layer of water around the submerged surface of plants. The population may be limited in its distribution to one, or be distributed among several habitats. Moreover, as a rule, there is a habitat in which conditions correspond to their ecological optimum, while their presence in neighboring habitats is a reflection of their environmental tolerance expanding their ecological niche. Different species of plants vary significantly in the spatial organization of the surface, which creates unequal conditions for the existence of protozoa [27]. Therefore, various approaches were used for calculating of the density of protozoa to determine the volume of their ALS.

The ALS of the protozoa on the *Ceratophyllum* is equal to the volume of the cylinder with the shoot diameter minus the volume of the plant itself (Figure 2). The ALS of protozoa on the plants with simple leaves such as *Phragmites australis* was calculated based on the surface area of the sample and the water layer 1 cm above the surface. For the floating plants, such as *Lemna minor* and *Spirodela polyrrhhiza*, ALS was calculated as the volume corresponding to the area of leaf blades and the length of the dropping roots. The volume of living space for the inhabitants of bottom sediments was determined by measuring the volume of water between the particles of bottom sediments. The bottom sediments were drying and re-filling it with water.

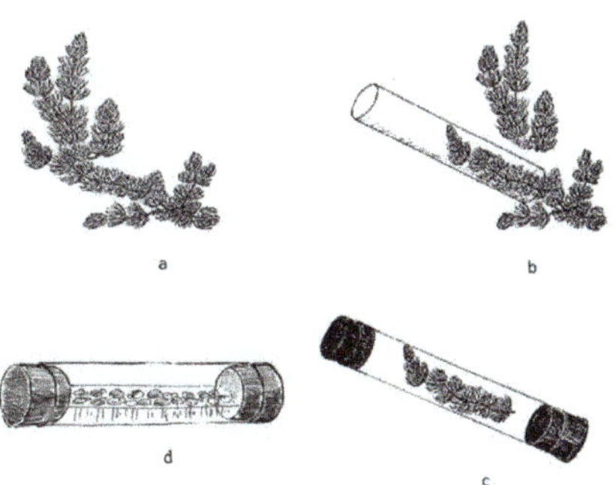

Figure 2. Sampling of periphyton on submerged macrophyte *Ceratophyllum demersum*: (**a**) The shoot of a plant; (**b**) the tube is put on the shoot so as to minimally disrupt its position in the water; (**c**) a separated shoot in the tube, ready for transfer to the laboratory; (**d**) sample of *Lemna minor* and *Spirodela polyrrhiza*.

From a pre-mixed sample of the water from pelagial and metaphytal, 5–7 sub-samples were taken with a sampler. The sub-sample volume was 1 mL. Each sub-sample was transferred to a 1.5 mL chamber and the protozoa were counted under a binocular microscope with magnification ×28 and ×56. In cases of a small number of organisms—less than 3 specimens per sub-sample—the number of sub-samples was increased to 9–11. Sub-samples data were averaged.

The procedure for processing bottom sediment samples was similar. However, in the case of a high abundance of ciliates, they were counted in a drop of 25 µL on a microscope slide. Subsequently, the population density was calculated per unit volume of their living space.

Laboratory processing of periphyton samples was carried out in two stages [10]. Three fragments (each about 1 cm^2) were separated from a reed sample under water with a scalpel in a Petri dish. Each fragment was placed in a small Petri dish (diameter 5 cm) and fragmented into strips with a width of about 3 mm. On each fragment, under the magnifying glass, sessile ciliates were counted. The data were averaged over 1 cm^2 and the abundance was calculated on the volume of living space for periphyton. The rest of the sample (leaf fragment) was thoroughly cleaned with a stiff brush on both sides in a large Petri dish. The fragment was removed, and its area was measured. The sample obtained by this procedure was analyzed in the same way as in the case of the water column. At this stage, crawling and swimming forms were counted. The data were averaged based on the area of the processed fragment and counted on the volume of periphyton living space—1 cm^3. In the case of duckweed, sessile forms were counted by placing individual plants in small diameter Petri dishes with a small amount of water. Usually 9 to 11 plants were treated. The duckweed plants remaining in the sample were shredded with scissors, and the sample was mixed and analyzed in the same way as the bottom samples. The data obtained were recounted on the volume of living space based on the number of duckweed plants involved in the calculations, the sample volume, and the average root length. When calculating the population density on the hornwort, three plant fragments 1 cm long were placed in separate Petri dishes. The whorl of the plant with scissors or a scalpel was divided into fragments on which the sessile forms were counted. The remaining fragment of the hornwort sample was shredded and analyzed for crawling and swimming forms according to the previously described procedure. The obtained data were counted on the volume of living space.

Quantitative processing of samples was carried out by three specialists and took from 10 to 12 h. As a standard, the sampling and calculation of population density was performed three times, and the results obtained over three days were averaged. Due to the fact that the quantitative processing of a significant number of samples is difficult to combine with species identification, before the start of research, during the year, the lake was studied for species diversity in various habitats.

Species identification of ciliates was carried out in vivo with an Olympus CX41 microscope in transmitted light as well as using dark field and phase contrast methods. When necessary, ciliates were stained with 1% methyl green or silver nitrate [28]. Species identification was based on Kahl [29–32], Foissner and Berger [33], Foissner et al. [34–37], Jankowski [38], Warren [39,40], and others.

2.4. Statistical Analysis

Data were processed using R [41]. Hierarchical clustering was performed with hclust from core R package stats, fuzzy clustering with function fanny from package cluster [42], and plotted on the plain of principal components with fviz_cluster from factoextra [43] and principal components analysis (PCA) from FactoMineR [44]. The data on abundances of each of ciliate species before PCA and clustering were transformed by dividing abundances in each sample by the total abundance of a given species in all samples. Plots were produced using ggplot2 [45].

3. Results

The studied eutrophic floodplain lake remains stratified from the June to the September. In winter, during the freezing period, oxygen practically disappears in the water column (Figure 3). As mentioned in the methods section, a pronounced oxygen gradient was observed in the pelagic zone of the lake, within which three zones of different O_2 content were distinguished. In the profundal zone, where the bottom sediments are represented by sapropel, stable oxygen-free conditions are maintained.

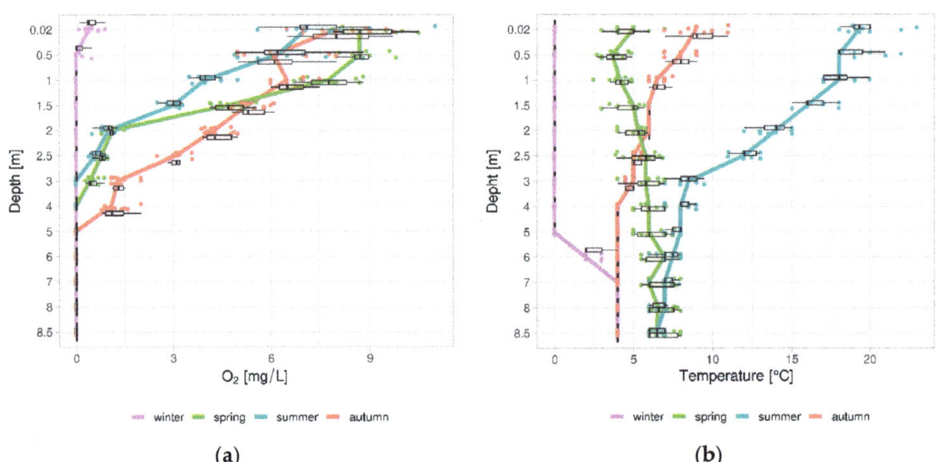

Figure 3. Vertical distribution of (**a**) oxygen and (**b**) temperature in the water column of the floodplain lake in different seasons.

The bottom sediments in the littoral zone are represented by silted sand with a low oxygen content. The oxygen content in silted sand was kept in the range from 1 to 3 mg per liter. Deeper than 3–5 cm from the bottom surface, a stable oxygen-free zone was formed (Figure 4).

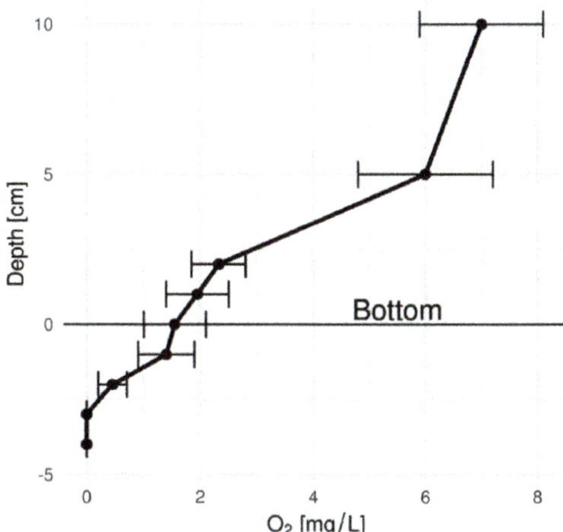

Figure 4. The oxygen content in the upper layer of the littoral sand.

In the littoral of the lake, the coastal part was overgrown with aquatic vegetation with the dominance of *Ceratophyllum* and *Phragmites*. The surface of the water between the thickets of submerged and semi-submerged vegetation is occupied by duckweeds (*Lemna* and *Spirodela*). In the thickets of aquatic vegetation, the oxygen regime was significantly different from that in the pelagic and benthic zones. As a result of fluctuations in the photosynthetic activity, the oxygen content in the thickets of plants cyclically changed within a day (Figure 5). At night, in thickets and water layers adjacent to plants (metaphytal), the respiration of plants and periphyton organisms, including bacteria, led to a decrease in the oxygen content to almost zero. At the same time, by noon, the oxygen content could reach 100–300% of the saturated concentration.

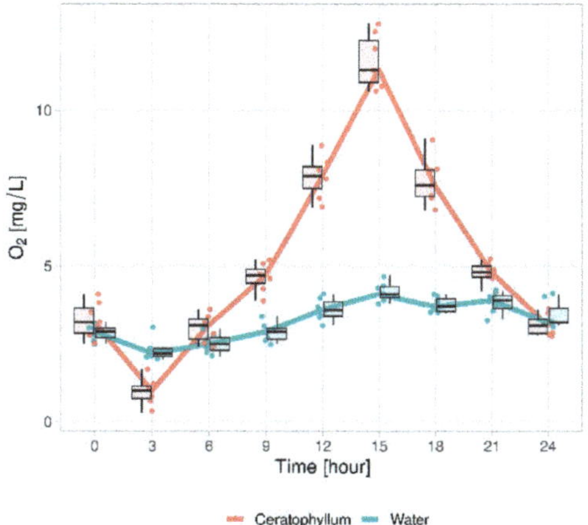

Figure 5. Daily changes in the oxygen content in the thickets of *Ceratophyllum* and in the epilimnion.

During the period of research in the lake, 209 taxa of ciliated protozoa (of them 183 identified to species) were observed (Table A1). Populations of many species are distributed within the range of several habitats identified, under the conditions of different oxygen contents. The presence of part of the population in the neighboring habitats, as a rule, with a lower density than in the optimum zone, indicated a range of tolerance of the species with respect to oxygen. The effect of oxygen on the functioning of ciliates and their spatial distribution is a known fact [24]. There are experimental data on the oxygen optima of a number of species of ciliates [46,47]. In our long-term field studies in different seasons, we observed maximum population densities of these species at the same values of oxygen contents. Such a coincidence of experimental and field data on the relationship between the densities of species populations and the concentration of oxygen in water allowed us to simplify reality and recognize that oxygen concentration is crucial, although, of course by no means the only factor in the ecological niche of ciliated species.

An analysis of principal components (PCA) was performed to determine how locations with different oxygen contents are grouped based on the abundances of 154 species of ciliates present in them. (In PCA analysis species occurring only sporadically were excluded.) During the research, 135 samples taken in all habitats were grouped by the oxygen concentrations rounded to integers. As a result, 15 locations with oxygen concentrations from 0 to 19 were obtained (Figure 6). The abundances of 154 species of ciliates found in our studies were averaged over these locations. Four groups can be seen in the plane of the first two principal components: (1) Oxygen-free conditions (0 mg/L), (2) locations with a high oxygen content (10–19 mg/L), (3) a group with a fairly narrow concentration range from 3 to 5 mg/L, and (4) locations in which the oxygen concentration is in the ranges of 1–2 and 6–8 mg/L (Figure 6).

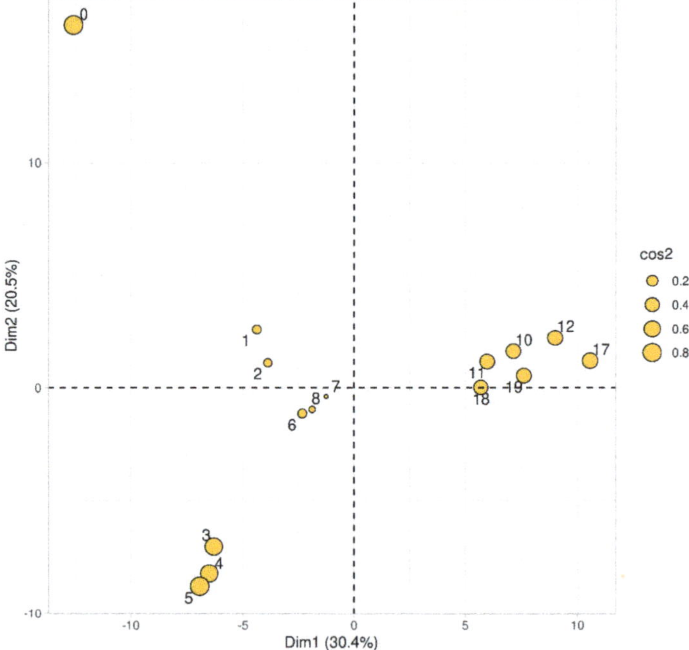

Figure 6. Principal component analysis (PCA) of locations with different oxygen contents based on the species abundances of the ciliates observed under these conditions. The numbers from 0 to 19 correspond to the rounded oxygen concentrations in mg/L. The area of the points is proportional to the quality of their representation on the plane of the first two principal components.

In order to verify our conclusions about the presence of four groups of locations by the oxygen content, the results of fuzzy clustering with a given number of clusters of 4 were superimposed on the

PCA plane (Figure 7). It can be seen that the results of fuzzy clustering are in good agreement with the results obtained by means of the PCA method.

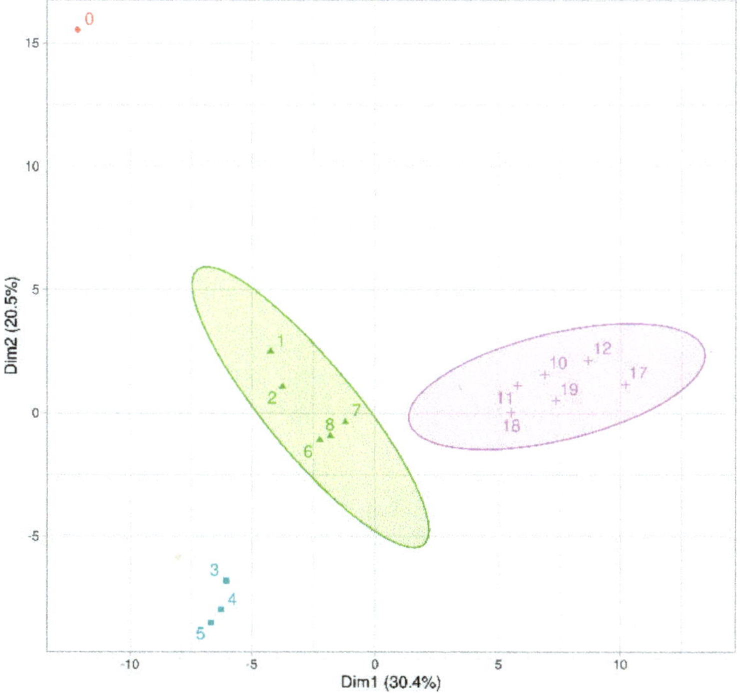

Figure 7. Fuzzy clustering superimposed on the plain of the first two principal components. The numbers from 0 to 19 correspond to rounded oxygen concentrations in mg/L.

The results of hierarchical clustering performed with the Ward method, pertaining to the locations ordered by depth and proximity of plants, are presented in Figure 8. Here, four clusters that correspond well to the groups isolated on the basis of oxygen concentration can also be seen. This is logical, since the concentration of oxygen in the investigated lake naturally varies with depth and in the vicinity of plants. Conditions in metaphytal—in the water between plants—are characterized by large daily fluctuations in oxygen contents. Moreover, this zone is practically devoid of ciliates. Single species from the decaying parts of plants found in this zone predetermined the affinity with the microaerophilic zone. Statistically, this position of the metaphyton is justified, but does not have a real biological meaning. The metaphytal can be considered as a dead or transit zone in which you can meet solitary species from adjacent assemblages. Most often, floating microoxyphilic species were recorded here.

On the basis of the analysis of the occurrence of 154 ciliates species in an oxygen gradient, four assemblages were identified: Microooxyphilic, oxyphilic, euryoxyphilic, and anoxyphilic [48]. The microoxyphilic assemblage includes the species that display preferences for the conditions with a stably low oxygen content. The upper limit of the tolerance range of the microoxyphilic species is usually limited to 3 mg of oxygen per liter, but they reach a maximum concentration in an environment with an oxygen content below 2 mg/L (Figure 9a). Such conditions existed in the water column at a depth of 1.5–4 m from the end of spring to the beginning of autumn homogenization, as well as in the littoral zone; constantly in the bottom sediments, and on the surfaces of decaying dead parts of plants. The oxyphilic assemblage includes species that reach the largest abundances under the conditions of relatively high (5–7 mg/L) and stable oxygen content, and the amplitude of diurnal variations is small (1–2 mg/L). Distribution of abundances of the oxyphilic species is shown in Figure 9b. The euryoxyphilic assemblage includes species living at the plants where the oxygen concentration can vary in a very wide range: From 0 to 15 mg/L, periodically reaching extremely high values—up to 300% of saturation (Figure 9c). The presence of many species exclusively on plants is determined by their tolerance to high oxygen content. The anoxyphilic assemblage unites species of habitats with anoxic conditions. Among anoxyphilic species, some species demonstrate endurance to the oxygen content in the environment (Figure 9d).

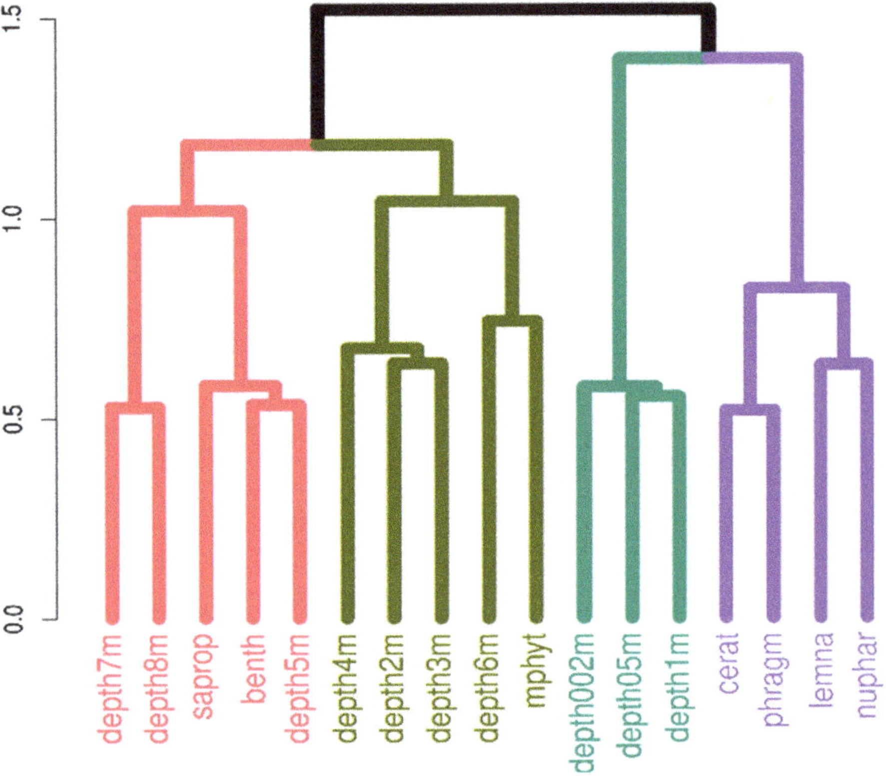

Figure 8. Ward's clustering of locations by depth and proximity of plants. Designation of habitats: benth—benthal in littoral, saprop—sapropel, depth 0.02, 0.5, 1–8 m—water column at depths of 0.02, 0.5, from 1 to 8 m, mphyt—metaphytal, cerat—*Ceratophyllum*, phragm—*Phragmites*, lemna—*Lemna* and *Spirodela*, nuphar—*Nuphar*.

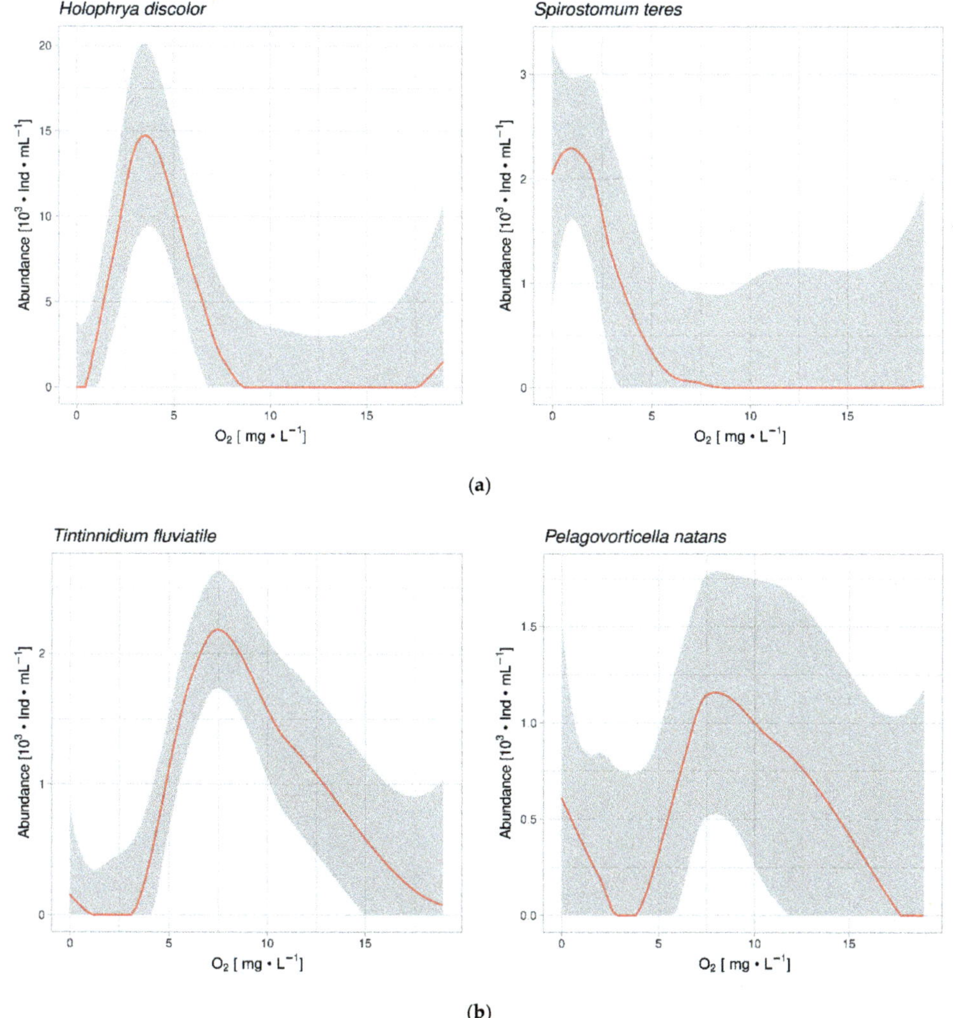

(**a**)

(**b**)

Figure 9. *Cont.*

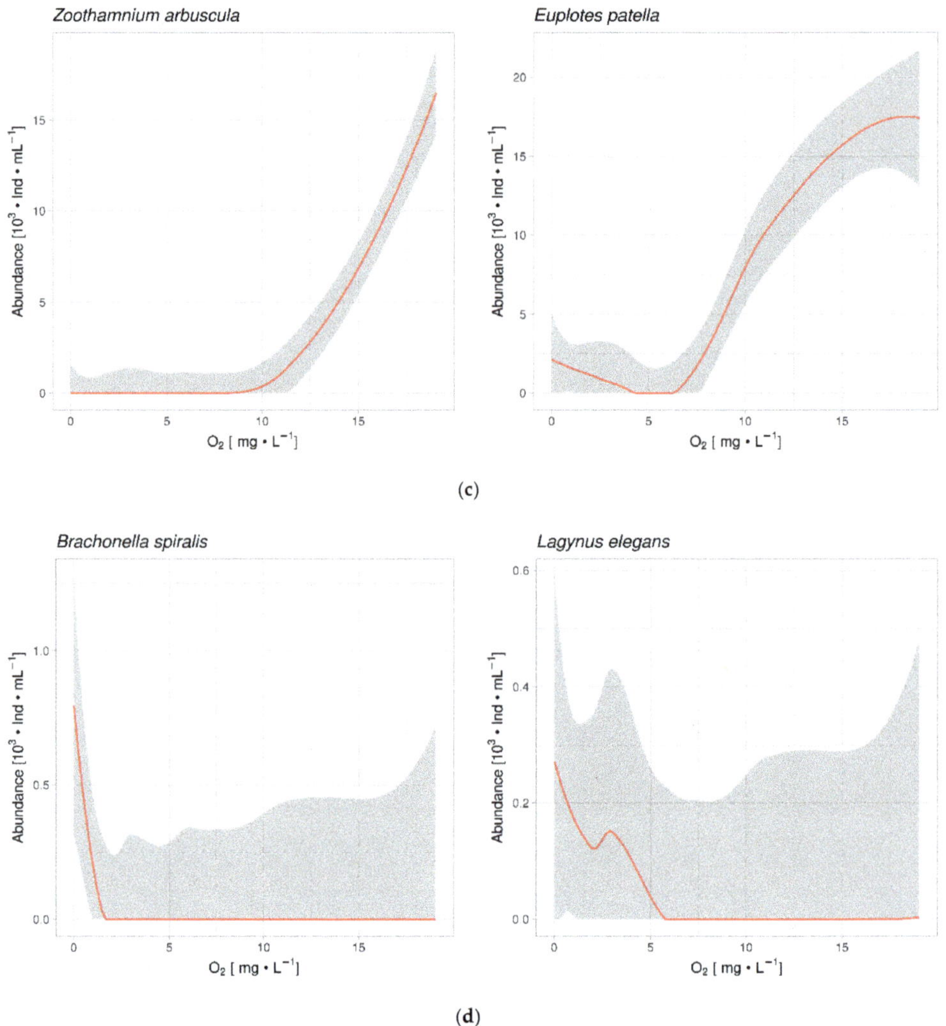

Figure 9. Distribution of the abundances of the species in the oxygen gradient: (**a**) Microoxyphilic species—*Holophrya discolor* and *Spirostomum teres*; (**b**) oxyphilic species—*Tintinnidium fluviatile* and *Pelagovorticella natans*; (**c**) euryoxyphilic species—*Zoothamnium arbuscula* and *Euplotes patella*; (**d**) anoxyphilic species—*Brachonella spiralis* and *Lagynus elegans*.

4. Discussion

4.1. Assemblages Structure Analysis

Clarification of the patterns of formation and functioning of biological diversity at the community level is one of the fundamental problems of biology and hydrobiology. The difficulties associated with the identification of local communities have been reported by many researchers [49,50]. The localization of the ciliate assemblages is usually based on the a priori division of the water body into biotopes: Benthal, pelagial, and periphytal. Further, on the basis of this classification, the organisms found within these a priori identified spatial boundaries are automatically assigned to the corresponding community: Benthic, periphytic, or pelagic [51,52].

The specialists working on a scale of one biotope developed the idea of its structure [52–54]. The consequences of this are the difficulties in distinguishing communities, in particular, the presence of the same species in the different communities. Thus, the researchers of plankton register the regular presence of the species that are considered to be benthic in the pelagic zone [51,52]. Conversely, the researchers of benthos note the planktonic species in the bottom sediments, for example, *T. fluviatile*, *T. cylindrata*, *S. humile*, *S. minimum*, *S. vernalis*, and some species of the Urotrichidae [55]. Significant difficulties are associated with the separation of the periphyton and benthic communities. The ciliates living in mountain rivers with a rocky bottom can equally be attributed to both benthos and periphyton, which shows the conventionality of dividing the fauna of such streams into the mentioned ecological groups [56].

There is an approach to the classification of communities based on the dominant species or on a group of co-occurring abundant species [33]. On the basis of this approach, the authors give an approximate species composition of several communities identified on the basis of the species co-occurring under conditions with different levels of organic pollution, in the pelagic zone, in swamps, and in capillary water of the soil [57]. For example, the Metopetum community is named after a group of indicator species from the genus *Metopus*. Most members of this community are anaerobic. Foissner and Berger [33] named a number of communities after the dominant species, for example, the Colpidietum community with the indicator species *Colpidium colpoda*, Trithigmostometum with the indicator species *Trithigmostoma cucullulus*, Carchesietosum with the indicator species *Carchesium polypinum*, and the Stentoretum community with indicator genus *Stentor*. They also distinguished the pelagic community: Oligotrichetea. However, the authors themselves noted that, for example, the euplanktonic species are common in most of those groups [33]. It was also proposed to distinguish the communities by their relation to environmental factors, while remaining within the framework of the biotopic approach paradigm and recognizing the existence of the "euplanktonic" species. The attempts to describe the spatial structure of the assemblages of ciliates in the body of water as a whole, rather than within a single biotope, have been made earlier, but the interpretation of the results remained within the boundaries of the biotopic paradigm [5,9,27,58].

Since there are no physical obstacles that would prevent the population from spreading over the entire volume of water, the authors argue that the entire volume of the lake is a potential biotope for each species. But, despite the absence of physical barriers to the spread of species, their populations were concentrated mainly in places with oxygen conditions that meet their physiological needs. Oxygen is the most important factor determining the spatial distribution of protozoa [25]. The reasons for this dependence lie in the presence of a connection between the oxygen content and the food objects of protozoa, the needs of their symbionts, and, most importantly, the oxygen toxicity for some of them [24].

Therefore, at least three assemblages represented the pelagic zone: Anoxyphilic (in hypolimnion), microooxyphilic (in metalimnion), and oxyphilic, represented by the species traditionally attributed to plankton. However, the oxyphilic assemblage itself was restricted to the zone from the surface to a depth of 1.5–2 m. Microooxyphils predominated in the metalimnion, whereas the anoxyphilic species were almost exclusively present in the water column from a depth of four meters.

The microxyphilic group existed in the oxycline zone from late spring to the beginning of autumn homogenization. At the same time, the microooxyphilic assemblage also occupied the bottom sediments in the littoral. The same composition of species was recorded on the dead parts of plants, where the processes of decay took place, and the oxygen content did not exceed 3 mg/L. Thus, the microooxyphilic assemblage was present simultaneously in three classical biotopes: Pelagial, periphytal, and benthal.

This illustrates the Ricklefs' claim that when solving the problem of local species diversity, we either ask the wrong questions or the right questions but on the wrong scale. In his opinion, a new understanding of the ecology of communities will come in the study of factors that affect the distribution of species in the whole space, in accordance with the gradients of environmental conditions [49]. This is also confirmed by our data obtained during the freezing period, when the amount of oxygen in the surface layer was below 1 mg/L, and the entire volume of the lake was practically oxygen-free. During this period, the anoxyphilic species prevailed from the bottom to the surface. Obviously, the pelagic zone was occupied by a single assemblage: Anoxyphilic. A small amount of anoxia-resistant microooxyphils was found in vegetation and in the surface water layer due to the local presence of a small amount of oxygen (1–2 mg/L) caused by photosynthesis and artificial holes in the ice sheet.

There are also two assemblages in the periphytal, which contradicts its status as a single biotope. Thus, on dying plants and inanimate objects, where bacteria are actively developing, only the species from the microooxyphilic assemblage are present. Hence, on the subjects immersed to a depth of only 0.5 m, we recorded the species living usually at a depth of three meters in the littoral. Periphytal on dying plants, as well as surfaces raised above the bottom, such as stones, sticks, and piles, is in every sense a continuation of the bottom. On these surfaces, due to the accumulation of sediments, the bacteria developing there provide stable microooxyphilic conditions. Most of the species found on the surface of vegetating plants are adapted to high oxygen content and, at the same time, to its significant fluctuations, which allows us to consider them as euryoxyphils. Such conditions are realized exclusively on the surfaces of vegetating plants. On vegetating plants, the composition of ciliates is predictable and excludes the presence of microooxyphils living on the rotting plant debris and on the bottom. These two groups can be slightly separated spatially, for example, occupying different tiers of one shoot of the hornwort, in which the lower dying parts are inhabited by microooxyphils, and the upper, actively vegetating, by euryoxyphils.

In view of the foregoing, in the absence of the information on the spatial distribution of oxygen, the samples taken at the boundary of two assemblages localized in the same biotope can pose significant problems for interpretation, giving grounds for erroneous conclusions about the unpredictable distribution of protozoa.

In undisturbed water bodies, the oxygen gradients in many respects coincide with visually distinct biotopes, which gave reason to consider the latter as habitats of communities. At present, when the water quality deteriorates due to the pollution of water bodies, a decrease in the diversity of protozoa is observed, largely due to the disappearance of suitable oxygen conditions for the oxyphilic and euryoxyphilic assemblages.

Summing up, it can be argued that protozoa are quite strictly determined in the space of water bodies by the oxygen gradients, rather than by according to common biotopes.

4.2. Population Characteristics

The formation of groups or assemblages is based on the similarity of the species requirements to the environmental conditions. Obviously, each species at the same time exists in its own niche to which it is adapted. Our results provide information on two important parameters of the ecological niche: The oxygen optimum and the range of the ecological tolerance of the species in the oxygen gradient. The species belonging to a particular assemblage can occur in a wide range of oxygen concentrations, but their optima mostly fit into a rather narrow range. Those optima are determined by the maximum population density in the space of the reservoir along the oxygen gradient. Within one assemblage, various types of distribution of populations in the oxygen gradient are observed.

For example, for *Holophrya discolor*, attributed by us to microooxyphils, various authors report a very wide range of oxygen concentrations at which this species was encountered. According to Bereczky [59], this species was found at the oxygen concentration of 10.2 mg/L. According to Detcheva [60,61], for this species the interval was between 3.0 and 9.3 mg/L. Patrick et al. [62] indicated an interval of 7.0 to 9.0 mg/L. The interval between 2.8 and 11.2 mg/L was pointed out by Foissner et al. [63]. The data presented give an idea of the range of tolerance of this species to oxygen, but do not allow judging which conditions are preferred or optimal for the species. According to our data, the concentration close to 3 ± 0.5 mg/L can be considered optimal for this species. The lower limit of tolerance of this species is 0.5 mg/L.

For microoxyphil *Spirostomum teres*, various authors mentioned the following oxygen concentrations at which it was found: 0.6–11 mg/L [64], from 0 to 11 mg/L [65], 8.9–22 [59], 2.9–10.4 [61,66], 2.3–3.9 [67], and 0–7.2 [68,69]. In general, a very wide amplitude is indicated for this species—from 0 to 22 mg/L. On the basis of our observations of its occurrence in the oxygen gradient, this species is tolerant to the anoxic conditions, and its optimum is in the range of 1–2 mg/L O_2.

The representatives of the oxyphilic assemblage are the species, most of which belong to the plankton ecological group. The data obtained on the tolerance ranges and optima of the representatives of this assemblage were illustrated on the example of the frequently occurring species. A typical representative of lake plankton and, according to our classification, the oxyphilic assembly, is *Tintinnidium fluviatile*. A number of authors noted that this species also occurs at fairly low values of dissolved oxygen. Reck [69] indicated a range from 0.6 to 15.9 mg/L, other authors considered the oxygen concentrations above 4 mg/L as a lower limit: 9–14 mg/L [59], 4.7–6.5 mg/L [70]. Reck [69] indicated that a maximum population density of 500 ind/L was observed in the oxygen range 1.2–10.7 mg/L. According to our data, the population of *Tintinnidium fluviatile* in the lake reached a density of 500 ind/L with the oxygen concentrations of 4 mg/L (below the optimum) and 15 mg/L (above the optimum), and under optimal conditions, in the range of 7–8 mg/L, the density increased to 2000 ind/L.

Another planktonic species from the oxyphilic group is *Pelagovorticella natans*. The species is usual for standing and low-flowing bodies of water. There is little information about its oxygen priorities. According to Sládeček and Sládecková [71], the species was found at the oxygen concentration of 11.6 mg/L. According to our observations, *Pelagovorticella natans* has a fairly wide range of tolerance to the oxygen content and is able to withstand its fairly low concentrations. The species was noted during the blooming of water. A wide range of tolerance allows the species to be retained during the blooming periods, when the decrease in oxygen at night is sharp. In this case, the optimum for the species is in the range: 7–8 mg/L.

A rather specific group, euryoxyphils, is represented mainly by the attached and crawling forms. As an example of attached euryoxyphils, we note *Zoothamnium arbuscula*. This species is rarely encountered today due to the pollution of water bodies. According to our studies, the species is found exclusively under the conditions of high oxygen concentrations: From 10 mg/L and higher, which is consistent with data of other authors. Schönborn [72] found this species under oxygen concentrations of 8–10 mg/L. The population density of this species increased along with the oxygen content (the graphs in Figure 9c show the oxygen concentrations in daylight).

A similar reaction to an increase in the oxygen content was demonstrated by the crawling representative of the euryoxyphilic assemblage—*Euplotes patella*. According to the published data, the species was found at oxygen concentrations of 7–9 mg/L [62], 0–12 mg/L [73], and 0.4–12 [60,61,66,74–77], 4.7–11.3 [78], 3.5–10.9 [63]. All authors indicate that the upper limit of the oxygen content for the species is above 10 mg/L. At the same time, the its lower limit of 0 mg/L confirms our assertion that the species of this group are the species that are adapted to exist in a wide range of fluctuations of this factor and are able to survive anoxia as well as high oxygen concentrations, which are avoided by representatives of other assemblies. According to our data, *Euplotes patella* reached the maximum population densities under the oxygen content above 15 mg/L.

The anoxyphilic assemblage in the lake was quite diverse, and its distribution was limited mainly by the sapropel-filled profundal, where the oxygen-free conditions were maintained all the time, and by hypolimnion during the periods of summer and winter stagnation. Most species of the anoxyphilic group were not observed even under the conditions of minimal oxygen concentrations. Some representatives of the anoxyphilic assemblage showed oxygen tolerance. These include the omnivore *Lagynus elegans*. According to the published data, this species occurred at the concentrations of O_2 up to 6 mg/L. Our studies have confirmed such an amplitude of its tolerance. Nevertheless, based on the quantitative representation of *Lagynus elegans*, the optimum of this species was localized in the oxygen-free region, and with an increase in the oxygen content in the medium, the population density rapidly decreased.

5. Conclusions

A study on the distribution of the ciliate populations in the oxygen gradient, regardless of the differentiation of the space of the reservoir into the water column, bottom, and periphytal, enabled us to identify the structuring of the ciliate population in the volume of the reservoir. The oxyphilic assemblage of species exists under conditions with a stably high oxygen content, which are maintained for most of the year in the epilimnion. It consists of species whose oxygen optima are about 5–7 mg/L. The microooxyphilic assemblage of species exists under the conditions with a consistently low oxygen content. These conditions are stably preserved on the surface of the bottom sediments in the littoral and are periodically formed in the metalimnion, as well as on the surface of decaying macrophytes. The oxygen optimum for it is about 2 mg/L. The euryoxyphilic assemblage is formed under conditions with a large amplitude of diurnal oxygen variations. The euryoxyphilic assemblage consists of species that are tolerant of both extremely high and low oxygen concentrations. The maximum oxygen concentrations are formed on the surface of photosynthetic macrophytes. Anoxyphilic assemblage is formed under conditions of a stable anoxia—in the profundal zone and, periodically, in the hypolimnion. The species composition of these assemblages is predictable and relatively stable.

Changing the configuration of the distribution of zones with relatively stable oxygen conditions leads to a change in the spatial localization of assemblages. The diversity of the ciliated protozoa in water bodies also depends on the stability and diversity of the oxygen gradients.

Author Contributions: Conceptualization, R.B.; methodology, R.B. and T.K.; software, Y.D. and R.B.; validation, R.B.; formal analysis, R.B.; investigation, R.B. and T.K.; resources, R.B. and T.K.; data curation, R.B., T.K., J.S.-C., and G.Ł.; writing—original draft preparation, R.B., T.K., and Y.D.; writing—review and editing, R.B., T.K., and G.Ł.; visualization, R.B. and T.K.; supervision, R.B. All authors have read and agreed to the published version of the manuscript.

Funding: This research received no external funding.

Acknowledgments: The authors are much obliged to the two anonymous reviewers for very valuable comments.

Conflicts of Interest: The authors declare no conflict of interest.

Appendix A

Table A1. Ciliate species found in the different sites in the floodplain lake (basin of the Vorskla River).

Sites	Pelagial			Bottom		Metaphytal	Periphytal of Macrophytes **
				Littoral	Profundal		
Oxygen condition	oxy	microoxy	anoxy	microoxy	anoxy	euryoxy	euryoxy
Acaryophrya sphaerica (Gelei, 1934)	x						
Acineria incurvata Dujardin, 1841							x
Actinobolina vorax (Wenrich, 1929)	x		*			x	
Amphileptus agilis Penard, 1922							x
Amphileptus pleurosigma (Stokes, 1884)				x			x
Amphileptus procerus (Penard, 1922)							x
Amphileptus sp.							x
Anteholosticha monilata (Kahl, 1928)							x
Apsiktrata gracilis (Penard, 1922)			xx		xx		
Askenasia sp.	x	x					
Askenasia volvox (Eichwald, 1852)	xx	x				x	x
Aspidisca cicada (Müller, 1786)	x				x		xx
Aspidisca lynceus (Müller, 1773)				x	x		xx
Astylozoon faurei (Kahl, 1935)	x	x	*	x			
Blepharisma sp.	x						
Bothrostoma sp.			xx				
Bothrostoma undulans (Stokes, 1887)			xx				*
Brachonella campanula (Kahl, 1932)					xx		*
Brachonella spiralis (Smith, 1897)				*	xx		
Bursellopsis gargamellae (Fauré–Fremiet, 1922)	x	x					
Bursellopsis truncata (Kahl, 1927)	x						
Caenomorpha medusula Perty, 1852			xx	*	xx		
Caenomorpha uniserialis (Levander, 1894)				x			

Table A1. *Cont.*

Sites	Pelagial	Bottom		Metaphytal	Periphytal of Macrophytes **
		Littoral	Profundal		
Campanella umbellaria (Linnaeus, 1758)		x		x	xx
Carchesium polypinum (Linnaeus, 1758)					xx
Caudiholosticha viridis (Kahl, 1932)	x	x			
Chaenea sp.					x
Chilodonella uncinata (Ehrenberg, 1838)		x			xx
Chilodontopsis depressa (Perty, 1852)					x
Cinetochilum margaritaceum (Ehrenberg, 1831)	xx	x		x	x	xx
Codonella cratera (Leidy, 1877)	x				
Coleps amphacanthus (Ehrenberg, 1833)	x	x	*			x
Coleps elongatus (Ehrenberg, 1830)		x	*	x	
Coleps hirtus (Müller, 1786)	xx	xx	*	xx	*	x	xx
Coleps nolandi (Kahl, 1930)	1			x	*	x	x
Colpidium colpoda (Losana, 1829)					x
Cristigera media (Kahl, 1928)		x			
Cristigera phoenix (Penard, 1922)		x		x	
Cristigera setosa (Kahl, 1928)		x	x	xx		x
Ctedoctema acanthocryptum (Stokes, 1884)	x	x	x			xx
Cyclidium glaucoma (Müller, 1773)	x	x	x			x
Cyclotrichium gigas (Fauré–Fremiet, 1924)		x		x	
Cyclotrichium sp.	x	x			
Cyclotrichium viride (Gajewskaja, 1933)	x				
Dexiostoma campylum (Stokes, 1886)					x
Dexiotricha granulosa (Kent, 1881)		xx	xx	x	xx		x
Disematostoma bütschlii (Lauterborn, 1894)	x	x		*	x	x
Disematostoma tetraedricum (Fauré–Fremiet, 1924)					x
Enchelyodon elegans (Kahl, 1926)	xx	x	x			x
Enchelys pupa (Müller, 1786)	x	x			x

Table A1. *Cont.*

Sites	Pelagial			Bottom		Metaphytal	Periphytal of Macrophytes **
				Littoral	Profundal		
Enchelys sp.							x
Epalxella bidens (Kahl, 1932)					x		
Epalxella mirabilis (Roux, 1899)			x	x	xx		
Epalxella sp.					x		
Epistylis plicatilis (Ehrenberg, 1831)							x
Epistylis procumbens (Zacharias, 1897)	x						
Euplotes patella (Müller, 1773)		x	xx	xx	x	x	xx
Euplotopsis affinis (Dujardin, 1841)						x	xx
Frontonia acuminata (Ehrenberg, 1834)		x		x			x
Frontonia leucas (Ehrenberg, 1834)		x		x			
Frontoniella complanata (Wetzel, 1927)				x			x
Gastronauta membranaceus (Bütschli, 1889)							x
Glaucoma scintillans (Ehrenberg, 1830)		x		x			x
Halteria bifurcata (Tamar, 1968)				x			
Halteria grandinella (Müller, 1773)	xx	xx		x	*	x	xx
Halteria minuta (Tamar, 1968)	xx	x	*		*		x
Histiobalantium natans (Claparède, Lachmann, 1858)	x	xx	x	x	x		x
Holophrya discolor (Ehrenberg, 1834)		x	x		x		x
Holophrya ovum (Ehrenberg, 1831)	x	xx	xx	xx	xx		xx
Holophrya simplex (Schewiakoff, 1893)							x
Holophrya sp.		x			*		x
Holophrya teres (Ehrenberg, 1834)		x		x	x		x
Holosticha pullaster (Müller, 1773)		x	x	x		x	xx
Holosticha sp.		*					x
Kerona pediculus (Müller, 1773)							x
Lacrymaria filiformis (Maskell, 1886)				x			x
Lacrymaria olor (Müller, 1786)							x

Table A1. *Cont.*

Sites	Pelagial	Bottom		Metaphytal	Periphytal of Macrophytes **
		Littoral	Profundal		
Lacrymaria sp.					x
Lagynus elegans (Engelmann, 1862)		x	xx		
Lembadion bullinum (Müller, 1786)	x				x
Lembadion lucens (Maskell, 1887)					x
Lembadion magnum (Stokes, 1887)	x	x			x
Leptopharynx eurystomus (Kahl, 1931)					x
Limnostrombidium viride (Stein, 1867)	xx	x		x	x
Linostomella vorticella (Ehrenberg, 1834)	xx	x	x	x	
Litonotus armillatus (Penard, 1922)	x			x	x
Litonotus fasciola (Ehrenberg, 1838)					x
Litonotus fusidens (Kahl, 1926)					x
Litonotus hirundo (Penard, 1922)	x		*	x	x
Litonotus sp.					x
Longifragma obliqua (Kahl, 1926)	xx			x	x
Loxocephalus lucidus (Smith, 1897)					x
Loxocephalus luridus (Eberhard, 1862)	xx	xx	x		
Loxodes magnus (Stokes, 1887)	xx	x	x		*
Loxodes striatus (Engelmann, 1862)	x	*	x		
Loxophyllum meleagris (Müller, 1773)					x
Loxophyllum sp.					x
Mesodinium pulex (Claparède, Lachmann, 1859)	x				
Metopus caudatus (Da Cunha, 1915)			xx		
Metopus es (Müller, 1776)		x	x		
Metopus hasei (Sondheim, 1929)			x		
Metopus setosus (Kahl, 1927)			x		
Metopus sp.	x	*			
Metopus striatus (McMurrich, 1884)			x		

Table A1. *Cont.*

Sites	Pelagial			Bottom		Metaphytal	Periphytal of Macrophytes **
				Littoral	Profundal		
Microthorax pusillus (Engelmann, 1862)							x
Microthorax tridentatus (Penard, 1922)							x
Monochilum elongatum (Mermod, 1914)				x	*		x
Monodinium balbianii (Fabre–Domergue, 1888)	xx	x	*	x		x	xx
Mylestoma sp.				x			
Obertrumia aurea (Ehrenberg, 1834)				x			x
Onychodromus grandis Stein, 1859							x
Opercularia nutans (Ehrenberg, 1831)							xx
Ophrydium sessile (Kent, 1882)							x
Ophryoglena macrostoma (Kahl, 1928)							x
Ophryoglena maligna (Penard, 1922)							x
Ophryoglena oblonga (Gajevskaja, 1927)							x
Ophryoglena utriculariae (Kahl, 1931)							x
Opisthodon niemeccensis (Stein, 1859)					x		
Oxytricha chlorelligera (Kahl, 1932)		x	x		x		
Oxytricha parvistyla (Stokes, 1887)							x
Oxytricha setigera (Stokes, 1891)							x
Oxytricha sp.						x	x
Paradileptus elephantinus (Svec, 1897)	x						
Paradileptus sp.							x
Paramecium bursaria (Ehrenberg, 1831)		x					xx
Paramecium caudatum (Ehrenberg, 1833)				x			x
Paramecium putrinum (Claparède, Lachmann, 1859)				x			x
Paraurostyla weissei (Stein, 1859)		x					x
Paraurotricha discolor (Kahl, 1930)	x	x		x	x	x	xx
Pelagohalteria cirrifera (Kahl, 1932)		x	x			x	x
Pelagostrombidium mirabile (Penard, 1916)	x						

Table A1. *Cont.*

Sites	Pelagial			Bottom		Metaphytal	Periphytal of Macrophytes **
				Littoral	Profundal		
Pelagovorticella mayeri (Fauré–Fremiet, 1923)	x	x				x	
Pelagovorticella natans (Fauré–Fremiet, 1924)	x	x					
Pelodinium reniforme (Lauterborn, 1908)				*			
Phascolodon vorticella (Stein, 1859)	x						
Phialina coronata (Claparède & Lachmann, 1859)							x
Phialina pupula (Müller, 1773)		x			x		
Phialina sp.							x
Phialina vertens (Stokes, 1885)				x			x
Plagiocampa rouxi (Kahl, 1926)				x			
Plagiopyla nasuta (Stein, 1860)		x	x		xx		
Pleuronema coronatum (Kent, 1881)		x		x	x	x	x
Podophrya fixa (Müller, 1786)							x
Prorodon niveus (Ehrenberg, 1834)							x
Protocyclidium citrullus (Cohn, 1866)				x	x		x
Pseudochilodonopsis piscatoris (Blochmann, 1895)				x		x	
Pseudocohnilembus pusillus (Quennerstedt, 1869)	xx	xx		x		x	xx
Pseudohaplocaulus anabaenae (Stiller, 1940)	x						
Pseudomicrothorax agilis (Mermod, 1914)							x
Pseudomonilicaryon anser (Müller, 1773)							x
Rhagadostoma completum (Kahl, 1926)		x	*	x			x
Rimostrombidium humile (Penard, 1922)	xx	x				x	
Rimostrombidium velox (Fauré-Fremiet, 1924)	xx	x		x		x	x
Saprodinium dentatum (Lauterborn, 1901)				*	x		
Spathidium spathula (Müller, 1773)							x
Spirostomum minus (Roux, 1901)				x	*		x
Spirostomum teres (Claparède, Lachmann, 1858)	x	xx	x	x	x		x
Stentor coeruleus (Pallas, 1766)		x		x			xx
Stentor igneus (Ehrenberg, 1838)							xx

Table A1. *Cont.*

Sites	Pelagial	Bottom		Metaphytal	Periphytal of Macrophytes **
		Littoral	Profundal		
Stentor muelleri (Ehrenberg, 1832)				x	x
Stentor niger (Müller, 1773)					x
Stentor polymorphus (Müller, 1773)		xx		x	xx
Stentor roeselii (Ehrenberg, 1835)					xx
Stichotricha secunda (Perty, 1849)					x
Stokesia vernalis (Wenrich, 1929)	x				x
Strobilidium caudatum (Fromentel, 1876)	x	x	*	x	x
Strobilidium sp.	x	x	*	x	
Strombidium sp.	xx	x	*	x	x
Strongylidium labiatum (Kahl, 1932)				x	
Stylonychia mytilus (Müller, 1773)		x			xx
Stylonychia pustulata (Müller, 1786)					x
Stylonychia sp.					x
Stylonychia vorax (Stokes, 1885)	x				
Tachysoma pellionellum (Müller, 1773)					xx
Thuricola obconica (Kahl, 1933)					x
Thylakidium truncatum (Schewiakoff, 1893)	x				
Tintinnidium fluviatile (Stein, 1863)	xx				
Trachelius ovum (Ehrenberg, 1831)					x
Trachelophyllum apiculatum (Perty, 1852)					x
Trachelophyllum sigmoides (Kahl, 1926)					x
Trimyema compressum (Lackey, 1925)		x	xx		
Trithigmostoma cucullulus (Müller, 1786)					x
Trithigmostoma sp.					x
Urocentrum turbo (Müller, 1786)	x	xx	*	x	x
Uroleptus musculus (Kahl, 1932)					x
Uroleptus piscis (Müller, 1773)			*		x

Table A1. *Cont.*

Sites	Pelagial			Bottom		Metaphytal	Periphytal of Macrophytes **
				Littoral	Profundal		
Uronema halophila (Kahl, 1931)	x	xx	xx	xx	xx		x
Uronema nigricans (Müller, 1786)		x	x	x			
Uronema sp.	x			x		x	
Urostyla grandis (Ehrenberg, 1830)							x
Urotricha armata (Kahl, 1927)	x	x		x		x	x
Urotricha farcta (Claparède, Lachmann, 1859)	xx						
Urotricha furcata (Schewiakoff, 1892)	xx	xx		xx	*	x	xx
Urotricha ovata (Kahl, 1926)	x						x
Urotricha pelagica (Kahl, 1935)	xx	x		x		x	x
Urotricha sp.							x
Urotrichopsis saprophila (Kahl, 1930)			x				
Vaginicola crystallina (Ehrenberg, 1830)							x
Vaginicola sp.							x
Vorticella campanula (Ehrenberg, 1831)						x	xx
Vorticella convallaria-complex	x	x				x	xx
Vorticella infusionum-complex							x
Vorticella lutea (Stiller, 1938)				x			x
Vorticella microscopica (Fromentel, 1876)							x
Vorticella microstoma-complex				*			x
Vorticella sp.							x
Vorticella vernalis (Stokes, 1887)	*	x					x
Zoothamnium arbuscula (Ehrenberg, 1831)							x
Zoothamnium simplex (Kent, 1881)							x

Note: xx—frequency of occurrence close to 50% of samples from the site; x—frequency of occurrence less than 50% of samples from the site; *—very rare and in very small numbers in the site; ** Macrophytes: *Ceratophyllum demersum* L., *Phragmites australis* (Cav.) Trin. ex Steudel, *Lemna minor* L., *Spirodela polyrrhiza* (L.) Schleid., *Nuphar lutea* (L.) Sibth. & Sm.

References

1. Finlay, B.J.; Esteban, G.F. Freshwater protozoa: Biodiversity and ecological function. *Biodivers. Conserv.* **1998**, *7*, 1163–1186. [CrossRef]
2. Corliss, J.O. Biodiversity and biocomplexity of the protists and an overview of their significant roles in maintenance of our biosphere. *Acta Protozool.* **2002**, *41*, 199–219.
3. Foissner, W.; Chao, A.; Katz, L.A. Diversity and geographic distribution of ciliates (Protista: Ciliophora). *Biodivers. Conserv.* **2008**, *7*, 345–363. [CrossRef]
4. Baldock, B.M.; Sleigh, M.A. The ecology of benthic protozoa in rivers: Seasonal variation in numerical abundance in fine sediments. *Arch. Hydrobiol.* **1988**, *111*, 409–421.
5. Madoni, P. Community structure and distribution of ciliated Protozoa in a freshwater pond covered by Lemna minor. *Boll. Zool.* **1991**, *58*, 273–279. [CrossRef]
6. Foissner, W.; Unterweger, A.; Henschel, T. Comparison of direct stream bed and artificial substrate sampling of Ciliates (Protozoa, Ciliophora) in a mesosaprobic river. *Limnologica* **1992**, *22*, 97–104.
7. Sleigh, M.A.; Baldock, B.M.; Baker, J.H. Protozoan communities in chalk streams. *Hydrobiology* **1992**, *248*, 53–64. [CrossRef]
8. Finlay, B.J.; Téllez, C.; Esteban, G. Diversity of free-living ciliates in the sandy sediment of a Spanish stream in winter. *J. Gen. Microbiol.* **1993**, *139*, 2855–2863. [CrossRef]
9. Babko, R.V.; Kuzmina, T.M. Spatial distribution of ciliates (Protista, Ciliophora) in the River Bityza (Dnieper basin). *Vestn. Zool.* **1999**, *33*, 83–89.
10. Babko, R.V.; Kuzmina, T.M. Ciliata (Protista, Ciliophora) of epiphyton of higher aquatic plants in a small River. *Hydrobiol. J.* **2004**, *40*, 22–38. [CrossRef]
11. Kovalchuk, A.A. Epyphitic ciliata of the Dnieper reservoirs. *Hydrobiol. J.* **2001**, *3*, 45–64.
12. Kreutza, M.; Foissner, W. The Sphagnum Ponds of Simmelried in Germany: A Biodiversity Hot-Spot for Microscopic Organisms. In *Protozoological Monographs*; Shaker-Publishers: Düren/Maastricht, Germany, 2006; Volume 3, p. 267. ISBN 3-8322-2544-7.
13. Finlay, B.J.; Esteban, G.F. Ubiquitous microbes and ecosystem function. *Limnetica* **2001**, *20*, 31–43.
14. Hul, M. Formation of the structure of Ciliata seston communities in the River Lyna (Northern Poland). *Acta Hydrobiol.* **1987**, *29*, 203–218.
15. El Serehy, H.A.N.; Sleigh, M.A. Ciliates in the Plankton of the River Itchen Estuary, England. *Acta Protozool.* **1993**, *32*, 183–190.
16. Modenutti, B.E.; Balseiro, E.G.; Queimalinos, C.P. Ciliate community structure in two South Andean lakes: The effect of lake water on Ophrydium naumanni distribution. *Aquat. Microb. Ecol.* **2000**, *21*, 299–307. [CrossRef]
17. Mazei, Y.A.; Burkovsky, I.V. Vertical structure of the interstitial ciliate community in the Chernaya River estuary (the White Sea). *Protistology* **2003**, *3*, 107–120.
18. Mazei, Y.A.; Burkovsky, I.V. Species composition of benthic ciliate community in the Chernaya River estuary (Kandalaksha Bay, White Sea) with a total checklist of the White Sea benthic ciliate fauna. *Protistology* **2005**, *4*, 107–120.
19. Mieczan, T. Size spectra and abundance of planktonic ciliates within various habitats in a macrophyte-dominated lake (Eastern Poland). *Biologia* **2007**, *62*, 189–194. [CrossRef]
20. Kiss, A.K.; Acs, E.; Kiss, K.T.; Török, J.K. Structure and seasonal dynamics of the protozoan community (heterotrophic flagellates, ciliates, amoeboid protozoa) in the plankton of a large river (River Danube, Hungary). *Eur. J. Protistol.* **2009**, *45*, 121–138. [CrossRef]
21. Coyne, K.J.; Countway, P.D.; Pilditch, C.A.; Lee, C.K.; Caron, D.A.; Cary, S.C. Diversity and Distributional Patterns of Ciliates in Guaymas Basin Hydrothermal Vent Sediments. *J. Eukaryot. Microbiol.* **2013**, *60*, 433–447. [CrossRef]
22. Yeates, A.M.; Esteban, G.F. Local ciliate communities associated with aquatic macrophytes. *Int. Microbiol.* **2014**, *17*, 31–40. [CrossRef] [PubMed]
23. Kammerlander, B.; Koinig, K.A.; Rott, E.; Sommaruga, R.; Tartarotti, B.; Trattner, F.; Sonntag, B. Ciliate community structure and interactions within the planktonic food web in two alpine lakes of contrasting transparency. *Freshw. Biol.* **2016**, *61*, 1950–1965. [CrossRef] [PubMed]

24. Fenchel, T.; Finlay, B.J. Oxygen and the spatial structure of microbial communities. *Biol. Rev.* **2008**, *83*, 553–569. [CrossRef] [PubMed]

25. Fenchel, T. Protozoa and Oxygen. *Acta Protozool.* **2012**, *52*, 11–20.

26. Babko, R.V. Microbenthometer for protistological studies. *Hydrobiol. J.* **1989**, *25*, 78–80. (In Russian)

27. Babko, R.V.; Fyda, J.; Kuzmina, T.; Hutorowicz, A. Ciliates on the macrophytes in industrially heated lakes (Kujawy Lakeland, Poland). *Vestn. Zool.* **2010**, *44*, 483–493. [CrossRef]

28. Foissner, W. Basic light and scanning electron microscopic methods for taxonomic studies of Ciliated Protozoa. *Europ. J. Protistol.* **1991**, *27*, 313–330. [CrossRef]

29. Kahl, A. Urtiere oder Protozoa I: Wimpertiere oder Ciliata (Infusoria) 1. Allgemeiner Teil und Prostomata. *Tierwelt Dtl.* **1930**, *18*, 1–180.

30. Kahl, A. Urtiere oder Protozoa I: Wimpertiere oder Ciliata (Infusoria) 2. Holotricha außer den im 1. Teil behandelten Prostomata. *Tierwelt Dtl.* **1931**, *21*, 181–398.

31. Kahl, A. Urtiere oder Protozoa I: Wimpertiere oder Ciliata (Infusoria) 3. Spirotricha. *Tierwelt Dtl.* **1932**, *25*, 399–650.

32. Kahl, A. Urtiere oder Protozoa I: Wimpertiere oder Ciliata (Infusoria) 4. Peritricha und Chonotricha. *Tierwelt Dtl.* **1935**, *30*, 651–886.

33. Foissner, W.; Berger, H. A user-friendly guide to the ciliates (Protozoa, Ciliophora) commonly used by hydrobiologists as bioindicators in rivers, lakes, and waste waters, with notes on their ecology. *Freshw. Biol.* **1996**, *35*, 375–482.

34. Foissner, W.; Blatterer, H.; Berger, H.; Kohmann, F. *Taxonomische und Ökologische Revision der Ciliaten des Saprobiensystems. Band I: Cyrtophorida, Oligotrichida, Hypotrichida, Colpodea. Informationsberichte des Bayer;* Landesamtes für Wasserwirtschaft: Deggendorf, Germany, 1991; pp. 1–478.

35. Foissner, W.; Berger, H.; Kohmann, F. *Taxonomische und Ökologische Revision der Ciliaten des Saprobiensystems. Band II: Peritrichida, Heterotrichida, Odontostomatida. Informationsberichte des Bayer;* Landesamtes für Wasserwirtschaft: Deggendorf, Germany, 1992; pp. 1–502.

36. Foissner, W.; Berger, H.; Kohmann, F. *Taxonomische und Ökologische Revision der Ciliaten des Saprobiensystems. Band III: Hymenostomatida, Prostomatida, Nassulida. Informationsberichte des Bayer;* Landesamtes für Wasserwirtschaft: Deggendorf, Germany, 1994; pp. 1–548.

37. Foissner, W.; Berger, H.; Blatterer, H.; Kohmann, F. *Taxonomische und Ökologische Revision der Ciliaten des Saprobiensystems. Band IV: Gymnostomatea, Loxodes, Suctoria. Informationsberichte des Bayer;* Landesamtes für Wasserwirtschaft: Deggendorf, Germany, 1995; pp. 1–540.

38. Jankowski, A.W. Morphology and evolution of Ciliophora. III. Diagnoses and phylogenesis of 53 sapropelebionts, mainly of the order Heterotrichida. *Arch. Protistenkd.* **1964**, *107*, 185–194.

39. Warren, A. A revision of the genus Vorticella (Ciliophora: Peritrichida). *Bull. Br. Mus. Nat. Hist. Zool.* **1986**, *50*, 1–57.

40. Warren, A. A revision of the genus Pseudovorticella Foissner & Schiffmann, 1974 (Ciliophora: Peritrichida). *Bull. Br. Mus. Nat. Hist. Zool.* **1987**, *52*, 1–12.

41. R Core Team. *R: A Language and Environment for Statistical Computing;* R Foundation for Statistical Computing: Vienna, Austria, 2013; Available online: https://www.R-project.org/ (accessed on 15 April 2020).

42. Maechler, M.; Rousseeuw, P.; Struyf, A.; Hubert, M.; Hornik, K. Cluster: Cluster analysis basics and extensions. *R Package Version* **2012**, *1*, 56.

43. Kassambara, A.; Mundt, F. Factoextra: Extract and visualize the results of multivariate data analyses. *R Package Version* **2017**, *1*, 337–354.

44. Le, S.; Josse, J.; Husson, F. FactoMineR: An R package for multivariate analysis. *J. Stat. Softw.* **2008**, *25*, 1–18. [CrossRef]

45. Wickham, H. *Ggplot2: Elegant Graphics for Data Analysis;* Springer: New York, NY, USA, 2009.

46. Fenchel, T.; Finlay, B.J.; Gianni, A. Microaerophily in ciliates: Responses of an Euplotes species (Hypotrichida) to oxygen tension. *Arch. Protistenkd.* **1989**, *137*, 317–330. [CrossRef]

47. Fenchel, T.; Bernard, C. Behavioural responses in oxygen gradients of ciliates from microbial mats. *Eur. J. Protistol.* **1996**, *32*, 55–63. [CrossRef]

48. Babko, R.V. Communities of Free-Living Ciliated Protozoa (Chromista, Ciliophora) of Continental Waters. Thesis, D.Sci. of biol. sci., Institute of Hydrobiology, National Academy of Sciences of Ukraine, Kyiv, Ukraine, 2019. (In Ukrainian).

49. Ricklefs, R.E. Disintegration of the Ecological Community. *Am. Nat.* **2008**, *172*, 741–750. [CrossRef]

50. Vellend, M. Conceptual synthesis in community ecology. *Q. Rev. Biol.* **2010**, *85*, 183–206. [CrossRef] [PubMed]

51. Burkovsky, I.V. *Ecology of Free-Living Ciliates*; Moscow State University: Moscow, Russia, 1984; p. 208. (In Russian)

52. Burkovsky, I.V. *Structural and Functional Organization and Stability of Marine Communities (on the Example of the White Sea Sand Littoral)*; Moscow State University: Moscow, Russia, 1992; p. 208. (In Russian)

53. Alekperov, I.K. Free-living ciliates of Azerbaijan (ecology, zoogeography, practical value). In *Monograph*; Elm: Baku, Azerbaijan, 2012; p. 520. (In Russian)

54. Dubrovsky, Y.V.; Kovalchuk, A.A.; Monchenko, V.I.; Mylnikov, A.P. On the composition of the animal population in the hollows of trees of the Girkan Reserve (*Azerbaijan*). Biodiversity and the role of animals in ecosystems. In Proceedings of the Materials of the VII International Scientific Conference, Dnipropetrovsk, Ukraine, 22–25 October 2013; Adverta: Dnipropetrovsk, Ukraine, 2013; pp. 50–52. (In Russian).

55. Kovalchuk, A.A. Infusoria of fish-breeding ponds of the Kiev region. III. Benthos. *Fish. Sci. Ukr.* **2012**, *3*, 27–38. (In Russian)

56. Kovalchuk, A.A. Infusoria of fish-breeding ponds of the Kiev region. I. Plankton. *Fish. Sci. Ukr.* **2012**, *2*, 31–44. (In Russian)

57. Blatterer, H.; Foissner, W. Morphology and infraciliature of some cyrtophorid ciliates (Protozoa, Ciliohora) from freshwater and soil. *Arch. Protistenkd.* **1992**, *14*, 101–118. [CrossRef]

58. Bonatti, T.R.; Siqueira-Castro, I.C.V.; Franco, R.M.B. Checklist of ciliated protozoa from surface water and sediment samples of Atibaia River, Campinas, São Paulo (Southeast Brazil). *Rev. Bras. Zoocienc.* **2016**, *17*, 63–76.

59. Bereczky, M.C. Die ökologische Charakterisierung einiger Ciliaten-Organismen des ungarischen Donauabschnittes. *Annls Univ. Sci. Bpest.* **1975**, *17*, 123–136.

60. Detcheva, R.B. Distribution des especes de cilies dans certains affluents Bulgares du Danube, aux eaux polluees. *Annls Stn. Limnol.* **1972**, *6*, 261–269.

61. Detcheva, R.B. Caracteristiques ecologiques des cilies de la riviere Maritza. *Annls Stn. Limnol.* **1983**, *16*, 200–219.

62. Patrick, R.; Cairns, J., Jr.; Roback, S.S. An ecosystematic study of the fauna and flora of the Savannah River. *Proc. Acad. Nat. Sci. Phila.* **1967**, *118*, 109–407.

63. Foissner, W.; Adam, H.; Foissner, I. Daten zur Autökologie der Ciliaten stagnierender Kleingewässer im Grossglocknergebiet (Hohe Tauern, Österreich). *Ber. Nat. Med. Ver. Salzburg* **1982**, *6*, 81–101.

64. Wilhelm, S. Die Lebensgemeinschaften der Swist im Verlauf der Selbstreinigung. *Arch. Hydrobiol.* **1964**, *60*, 89–106.

65. Bick, H.; Kunze, S. Eine Zusammenstellung von autökologischen und saprobiologischen Befunden an Süsswasserciliaten. *Int. Rev. Ges. Hydrobiol.* **1971**, *56*, 337–384. [CrossRef]

66. Detcheva, R. Paramètres saprobiologiques et hydrochimiques pour les ciliés de certains affluents Bulgares de la Mer Noire. *Hydrobiology* **1979**, *9*, 57–73.

67. Mihailowitsch, B. Taxonomische und Ökologische Untersuchungen an Ciliaten (Protozoa, Ciliophora) in Solebelasteten Fliessgewässern. Ph.D. Thesis, Universität Bonn, Bonn, Germany, 1989.

68. Neidl, F. Die Räumliche und Zeitliche Verteilung des Ciliaten *Spirostomum teres* im Benthal und Pelagial des Piburger Sees (Tirol, Österr.) Während und Nach der Sommerschichtung. Master's Thesis, Universität Innsbruck, Innsbruck, Austria, 1989.

69. Reck, E.M. Zur Ökologie der Pelagischen Ciliaten des Plußsees. Ph.D. Thesis, Christian-Albrechts-Universität, Kiel, Germany, 1987.

70. Foissner, W.; Wilbert, N. Morphologie, Infraciliatur und Ökologie der limnischen Tintinnina: Tintinnidium fluviatile Stein, Tintinnidium pusillum Entz, Tintinnopsis cylindrata Daday und Codonella cratera (Leidy) (Ciliophora, Polyhymenophora). *J. Protistol.* **1979**, *26*, 90–103.

71. Sládeček, V.; Sládecková, A. The plankton community of the Hamry and Sec reservoirs after the spring overturn. *Sb. Vys. Sk. Chem. Technol. Praze (Technol. Water)* **1962**, *6*, 389–405.

72. Schönborn, W. Die Ziliatenproduktion in der mittleren Saale. *Limnologica* **1982**, *14*, 329–345.

73. Bick, H. *Ciliated Protozoa. An Illustrated Guide to the Species Used as Biological Indicators in Freshwater Biology*; World Health Organization: Geneva, Switzerland, 1972; 198p.

74. Detcheva, R.B. Les cilies de la riviere Kamtchia a l'eau polluee. *Annls Stn. Limnol.* **1976**, *10*, 299–304.

75. Detcheva, R.B. Aspects ecologiques des cilies de la riviere Mesta. *Annls Stn. Limnol.* **1978**, *11*, 300–309.

76. Detcheva, R.B. L'influence de la pollution sur la distribution des ciliés (Protozoa-Ciliata) de la rivière de Vit-affluent Bulgare du Danube. *Hidrobiologia* **1983**, *17*, 348–361.

77. Detcheva, R.B. Distribution des ciliés (Protozoa-Ciliata) par rapport à la pollution hydrochimique de la rivière d'Ossâm-affluent Bulgare du Danube. *Hidrobiologia* **1983**, *17*, 362–380.

78. Madoni, P.; Ghetti, P.F. Ciliated protozoa and water quality in the Torrente Stirone (Northern Italy). *Acta hydrobiol.* **1981**, *23*, 143–154.

Article

Phytoneuston and Chemical Composition of Surface Microlayer of Urban Water Bodies

Józef Piotr Antonowicz [1] and Anna Kozak [2,*]

[1] Department of Environmental Chemistry, Institute of Biology and Earth Sciences, Pomeranian University in Słupsk, ul. Arciszewskiego 22b, 76-200 Słupsk, Poland; jozef.antonowicz@apsl.edu.pl
[2] Department of Water Protection, Faculty of Biology, Adam Mickiewicz University, Poznań, Uniwersytetu Poznańskiego 6, 61-614 Poznań, Poland
* Correspondence: akozak@amu.edu.pl

Received: 31 March 2020; Accepted: 28 June 2020; Published: 3 July 2020

Abstract: The concentration of chemical and biological parameters in the ecotone of the surface microlayer (SML) occurring between the hydrosphere and the atmosphere of urban water bodies was investigated. Parallel, sub-surface water (SUB) analyses were carried out to compare the SML properties with the water column. The concentrations of trace metals, macronutrients, nutrients, chlorophyll *a*, pheophytin, abundance and biomass of phytoplankton and the number of heterotrophic bacteria in both studied layers were analyzed. Each of the studied groups of chemical parameters was characterized by specific properties of accumulation. Trace metals occurring in concentrations below 1 ppm, such as Al, Ba, Cd, Cr, Cu, Fe, Mn, Ni, Zn and metalloid As, were accumulated to a higher degree in SML than in SUB. Macroelement concentrations, with the exception of Mg, were lower in the SML compared to the SUB. Nutrients, autotrophic and heterotrophic microorganisms occurred in the SML to a higher degree than in the SUB. Bacillariophyceae dominated the analyzed water bodies, which are typical for the spring period, as well as Chrysophyceae, Chlorophyceae, Dinophyceae and Euglenophyceae. Cyanobacteria dominated in one of the ponds. The abundance of individual phytoplankton groups was significantly correlated with Ca, K, Na, P-org, SO_4^{2-}, F^-, Al and Sr.

Keywords: surface microlayer; phytoneuston; phytoplankton; metals; nutrients; urban water body

---

1. Introduction

The surface microlayer (SML) is an ecotone, which constitutes an interface between the hydrosphere and the atmosphere. It is a thin layer of maximum 1000 µm in thickness [1]. The SML is unique due to its physical, chemical and biological properties, which differ from those of subsurface water (SUB) [1,2]. The SML is capable of accumulating microorganisms and chemical substances at a rate as high as 100-fold greater than that observed in the SUB [3]. The SML is the zone of matter and energy exchange between the hydrosphere and the atmosphere, and, as such, it is affected by global climate changes [4].

The accumulation of chemical substances in the SML may vary, as it is dependent on many factors such as the chemical composition of water, salinity, the nutrient status and the presence of neustonic organisms (organisms that live upon the upper surface of waters or beneath its surface film, the SML) [5,6]. The SML may also accumulate toxic substances at greater amounts than observed in the pelagic zone [7]. At the same time, it is the zone that accumulates nutrients [5,8]. In turn, such substances as calcium, magnesium, potassium or sodium, being macroelements found at high concentrations in the pelagic zone, are usually not accumulated in the SML [9]. The SML is inhabited by neustonic autotrophic and heterotrophic microorganisms, which exhibit both enzymatic and high respiratory activity [5,10–12]. The SML structure may be disturbed as a result of physical non-equilibrium processes such as temperature fluxes, irradiance, wind and wave action that influence its biogeochemical properties. However, it is well fit for a rapid self-reconstruction of its original structure [3]. Physical forces such as adhesion, cohesion, surface tension, vortex motion and the Langmuir circulations, associated with its unique chemical and biological structure, form the specific properties of SML, including its stability [2].

Chemical contaminants are accumulated in the SML because of physical, chemical and biological processes. These processes include, e.g., hydrosphere–atmosphere exchange, convection, upwelling of underlying waters, transport by bubbles, atmospheric deposition, simple diffusion, buoyant particles, turbulent mixing, scavenging and chelation of inorganic components by organic matter [3]. Important sources of contaminants in urban objects are connected with urban pollutants supplied with wet and dry atmospheric deposition [13,14].

The taxonomic composition of phytoplankton is dependent on the chemical composition of the aquatic environment, which it colonizes [5,15]. Consequently, communities of the phytoneuston and phytoplankton vary in the SML and the SUB in terms of their taxonomic composition and population size [16].

We hypothesized that, regardless of the type of analyzed urban water body, i.e., a lake vs. ponds, concentrations of nutrients and heavy metals would be higher in the SML than in the SUB. In turn, macroelement concentrations would be comparable in both layers. Therefore, the counts of microorganisms, i.e., heterotrophic bacteria and phytoneuston, were also expected to be higher in the SML than in the SUB.

The aim of this study were: to determine the taxonomic composition, abundance and biomass of phytoplankton in the SML and SUB layers of a freshwater urbanized lake and ponds; to characterize the chemical structure of two different habitats: the SML and the SUB; and to ascertain relationships between the studied chemical and biological components.

2. Materials and Methods

The studies were conducted in water bodies located in the Pomerania Region (northern Poland) in areas of moderate anthropogenic pressure under the influence of urban infrastructure. Pollutants generated by heavy industry did not affect the studied lake or urban ponds. Additionally, the towns in which the analyzed water bodies are situated are located near protected areas, i.e., for Słupsk it is the Dolina Słupii Landscape Park, while Człuchów lies near the Bory Tucholskie National Park. As a result, heavy metal contamination is relatively low in both areas. Water samples collected from the SML and SUB were analyzed for a wide range of chemical elements, composition and abundance of phytoneuston and phytoplankton as well as counts of heterotrophic bacteria.

The sampling stations were situated in areas representative for the studied water bodies, i.e., within five ponds (1P–5P) in the town of Słupsk and two stations in the lake in the town of Człuchów (Figure 1). The coordinates of these sampling stations are: 1P, 54°27′00.38″ N, 17°02′23.86″ E (at Arciszewskiego Street, trees at the bank); 2P, 54°27′27.03″ N, 17°02′25.85″ E (at Nad Śluzami St., station next to the trees on the island); 3P, 54°28′29.05″ N, 17°01′58.82″ E (at Orzeszkowej St., station in the north part of the pond, near the trees on the island); 4P, 54°28′32.66″ N, 17°02′16.10″ E (at Kaszubskiej St., next to the trees on the island); 5P, 54°28′47.26″ N, 17°01′21.50″ E (at Portowa St., Słupsk, Poland); and Lake Człuchowskie, 53°39′25″ N 17°21′33″ E with Stations 6 L in its northern part and 7 L in the southern part of the lake.

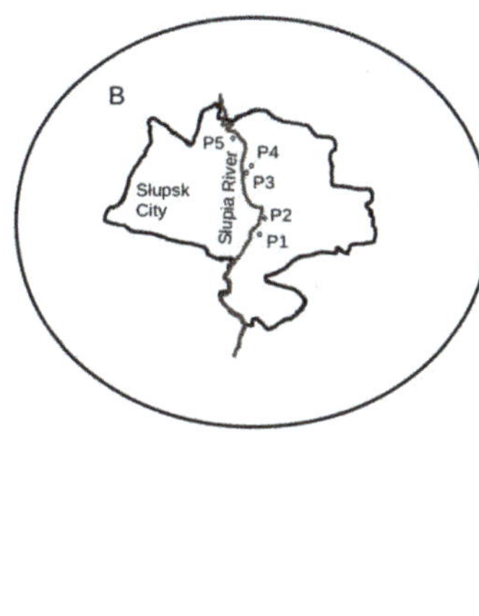

Figure 1. Locations of city ponds in Słupsk and Lake Człuchowskie on contour map of north Europe (**A**); and locations of city ponds in contour map of city Słupsk (**B**).

The water bodies are used for recreation and leisure activities. Ponds located in Słupsk are inhabited by such fish species as *Rutilus rutilus* (roach), *Carassius carassius* (crucian), *Abramis brama* (bream), *Tinca tinca* (tench) and *Esox lucius* (pike) [17]. Urban ponds in Słupsk are also habitats for waterfowl, e.g., swans and ducks [10]. The presence of these animals increases the value of leisure in urban ponds. For many years Lake Człuchowskie was the immediate receiver of municipal sewage, including that discharged by the hospital, containing high concentrations of nutrients; as a result, the lake waters did not meet water quality standards. A sewage treatment plant was commissioned [18]; sewage discharge to the lake was limited and the condition of the lake has now improved. The recorded concentrations of nutrients are lower. Nevertheless, the ecological status of the lake still needs to be improved.

Samples were collected in spring 2012. The SML samples (242 ± 40 µm in thickness) were collected using a Garrett screen [19]. Samples of the SUB layer were collected from a depth of 10 cm, using a sampler with a PET bottle. Two samples from each station (the SML and the SUB) were taken for analysis (n = 14). Collected samples were stored in clean PET containers [20]. Water samples for analyses of phytoplankton were fixed with formalin. A portion of the sample material was not fixed to conduct qualitative analyses. For microbiological analyses prior to sampling, the Garrett screen was rinsed with distilled sterile water (Hydrolab) and ethyl alcohol. The collected samples of water were transported to the laboratory in ice containers at a temperature of maximum 7 °C.

Water samples for analyses of metals were mineralized using a microwave mineralizer (Ertec Magnum II) using Suprapur® HNO_3 [21]. Metals were analyzed with a mass spectrometer (Thermo Scientific). Concentrations of dissolved Cl^-, F^- and SO_4^{2-} ions were assayed according to APHA [22]. Metals found in water at concentrations below 1 ppm according to Duffus [23] were classified as trace metals.

For nutrients, phosphate phosphorus (P-PO_4) was analyzed using the method with ascorbic acid, while ammonia nitrogen (N-NH_4) with the direct nesslerization method and the nitrate nitrogen (N-NO_3) using sulfanilic acid. Total phosphorus (TP) was assayed after mineralization. Absorption was measured using a UV–Vis spectrophotometer (Hitachi), while total Kjeldahl nitrogen (TN) was assayed with the distillation method. Organic nitrogen and phosphorus concentrations were calculated according to APHA [22].

The concentration of dissolved oxygen was determined measured (only in the SUB) using a Martini Instruments oxygen meter. Water temperature (temp., measured only in the SUB), electrolytic conductivity (EC) and pH were measured with Elmetron apparatuses. Abundance of heterotrophic bacteria (in CFU (colony forming units)) was established using the culture method. The analyses were described in detail by Mudryk et al. [24]. Concentrations of chlorophyll *a* (chl *a*) and pheophytin (pheo) were determined after extraction in ethanol [25].

The abundance and taxonomic composition of the phytoplankton and phytoneuston were examined under a biological microscope (Olympus with led lightning, the abundance of phytoplankton was determined in counting chambers and at 40–400× magnification). Phytoplankton biomass was calculated for each species according to Wetzel and Likens [26] and Hutorowicz [27]. Taxa were considered dominant when their percentage share in total phytoplankton abundance reached at least 5% in one sample. Statistical parameters such standard deviation (SD), means, medians, coefficient of variation (CV) and Spearman correlation coefficients between the SML and the SUB were conducted using Statistica 13 software [28]. The type of distribution was determined using the Shapiro–Wilk test. To show differences in the level of a given nutrient in the SML in relation to its level in the SUB, the enrichment factors (EF) were calculated according to the following formula presented by Estep et al. [29]: EF = C_{SML}/C_{SUB}, where C_{SML} is the level of a given nutrient in SML and C_{SUB} is the level of a given nutrient in SUB. EF were calculated for each pair of results and averaged [30].

The response of the phytoplankton taxonomical groups to environmental conditions was investigated using multivariate statistical analyses. A preliminary Detrended Correspondence Analysis (DCA) was used and its results allowed choosing Canonical Correspondence Analysis (CCA), as the most appropriate technique [31–35]. Analyses were conducted jointly for the SML and SUB using the CANOCO software package, version 4.5 [32].

3. Results

3.1. Physico-Chemical and Biological Parameters

Irrespective of the sampling station, nitrogen and phosphorus compounds were accumulated at a greater amount in the SML than in the SUB (EF from 1.40 to 4.38) (Figure 2 and Table 1). Trace elements, i.e., metals Al, Ba, Cd, Cr, Fe, Mn, Ni, Sr, V and Zn; metalloid As; and non-metal Se, were accumulated in the SML. The obtained enrichment factors (EF) ranged from 1.26 to 34.52. Trace elements were accumulated in the SML to a greater extent regardless of the water sample origin (lake vs. pond) (Figure 3). In turn, parameters recorded at concentrations exceeding 1 ppm, namely Ca, K, Mg, Na and Cl^- and SO_4^{2-}, generally showed comparable concentrations in the SML and SUB, or slight enrichment in the SML (EF = 1.10–1.35). An atypical pattern was observed for the accumulation of F^-, for which mean EF was high, whereas its median indicated a lack of accumulation (Table 1). Values of EC and pH were comparable in both layers (EF = 1.05 and 0.97, respectively), while pH values in the SML were lower than in the SUB.

Table 1. Mean, median and variation coefficients (CV) calculated for enrichment factors (EF) obtained from the examined urban water bodies (both city ponds and city lake, *n* = 7).

Parameter	Mean	Median	CV
biom	2.43	2.01	82.05
abund	1.45	1.54	40.07
chl *a*	2.00	1.77	47.91
pheo	2.22	2.24	32.95
CFU	6.49	5.30	82.37
Al	4.20	2.20	98.79
As	1.84	1.57	42.81
Ba	1.57	1.30	42.91
Ca	1.22	1.16	13.61
Cd	34.52	7.10	209.86
Cl^-	1.10	1.03	15.09
Cr	2.61	2.69	42.36
F	4.69	1.09	107.75

Table 1. *Cont.*

Parameter	Mean	Median	CV
Fe	3.27	2.85	65.69
K	1.23	1.24	18.88
Mg	1.35	1.17	37.10
Mn	2.16	2.57	65.90
Na	1.14	1.03	19.13
Ni	1.42	1.17	47.64
$N\text{-}NO_3$	1.40	1.28	34.77
Norg	1.91	1.79	25.14
TN	2.09	1.90	29.63
$N\text{-}NH_4$	4.38	4.69	53.26
Porg	2.09	2.11	25.95
$P\text{-}PO_4$	1.88	1.84	13.18
TP	2.00	1.96	19.35
SO_4^{2-}	1.03	1.02	2.53
Se	2.42	1.64	104.01
Sr	1.26	1.21	22.75
V	25.60	3.01	240.77
Zn	1.77	1.93	38.46
EC	1.05	1.04	4.12
pH	0.97	0.97	4.89

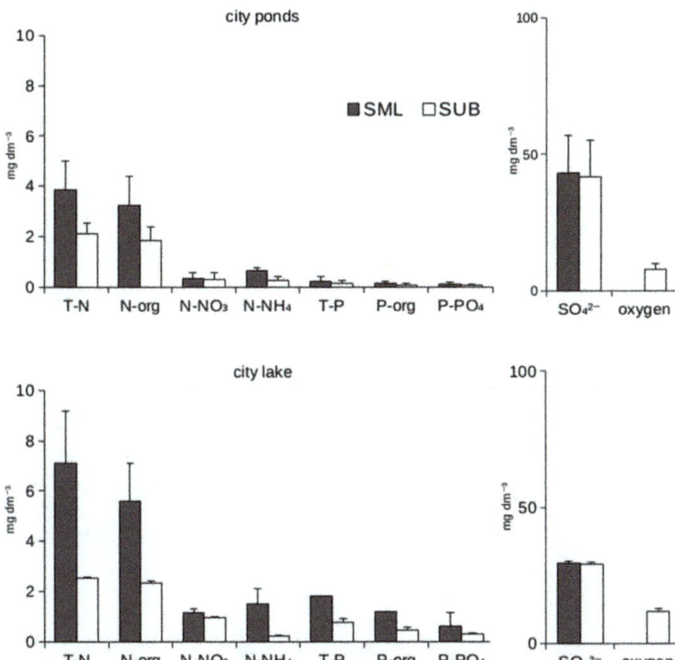

Figure 2. Comparison of the availability of nutrients in the SML and SUB of the studied urban water reservoirs (ponds in the city of Słupsk and Lake Człuchowskie). Standard deviation (SD) values are marked.

Values of the enrichment factors for biological parameters such as biomass and counts of phytoplankton, concentrations of chlorophyll a and pheophytin as well as counts of heterotrophic bacteria (in CFU) were higher in the SML than in the SUB (EF range of 1.45–6.49) (Table 1 and Figure 4).

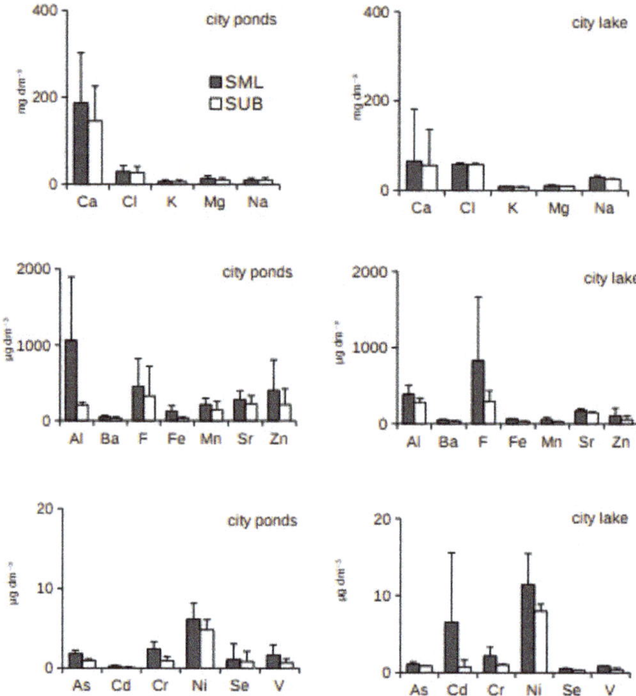

Figure 3. Comparison of the availability of macronutrients, trace metals and non-metals in the SML and SUB of the studied urban water bodies (ponds in the city of Słupsk and Lake Człuchowskie). For more information, see Figure 2.

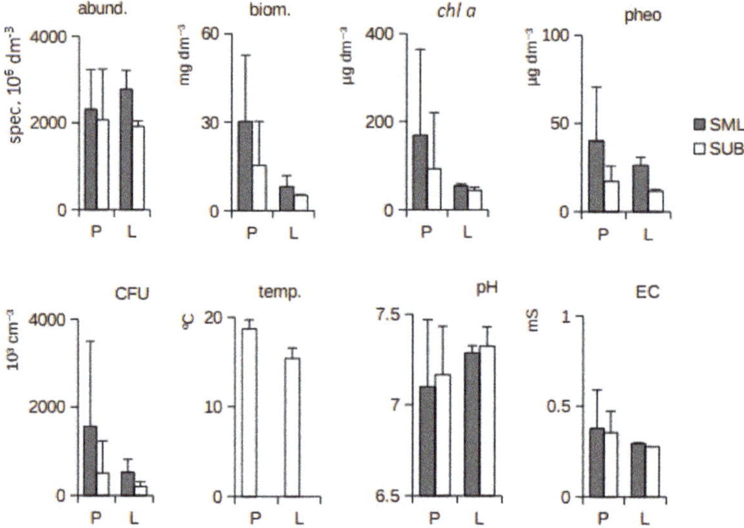

Figure 4. Comparison of biomass level, phytoplankton abundance, chlorophyll a concentration, pheophytin, heterotrophic bacteria, pH, electrolytic conductivity and temperature in the SML and SUB of the studied urban water bodies (P, ponds in Słupsk; L, Lake Człuchowskie). For more information, see Figure 2.

Analyzed nitrogen and phosphorus forms were found in most cases at higher concentrations at the sampling stations in the lake than in the urban ponds. Mean concentrations of TN, N-org and N-NH$_4$ in the SUB were comparable in the lake and in the ponds, while that of N-NO$_3$ was three-fold higher in the lake compared to the ponds. Phosphorus compounds were recorded at concentrations ranging from 6.0- to 6.3-fold higher in the SUB of the lake than in the ponds. Concentrations of all the analyzed nitrogen compounds in the SML were from 1.7- to 3.4-fold higher in the lake when compared to the ponds, and those of phosphorus compounds were from 5.3- to 9.7-fold higher. Nevertheless, the urban ponds were characterized by SO$_4{}^{2-}$ concentrations on average 1.4-fold higher than those in the lake waters. In addition, concentrations of Ca, Mn, Sr, V, Zn and Se in the SUB were higher in ponds than in the lake waters, but Cd, Cl$^-$, K, Na and Ni were higher in the lakes. Lake water was more oxygenated than ponds water (1.5 times), while water temperature in the ponds was by 3 °C higher than in the lake (Figure 4). Water pH in the analyzed water bodies was close to neutral (mean in SUB 7.1 in ponds and 7.3 in lake) and electrolytic conductivity was comparable at all studied stations (mean in SUB 0.38 in ponds and 0.30 in lake).

3.2. Phytoneuston and Phytoplankton Composition

Biomass of phytoplankton and phytoneuston and concentrations of chlorophyll a and pheophytin were higher in both analyzed layers in the urban ponds (Table 1 and Figure 4). Biomass of phytoneuston in the SML was larger by 3.7-fold in the pond waters, whereas biomass of phytoplankton in the SUB was three-fold higher. In addition, the count of heterotrophic bacteria was three-fold greater in the SML and 2.5-fold higher in the SUB in the ponds when compared to the respective levels in the lake. EF was higher for the phytoplankton biomass, chlorophyll a, pheophytin and bacterioneuston counts for the urban ponds than for the lake.

The identified phytoplankton species in all the water bodies represented seven taxonomic groups (Figure 5). Both in the case of phytoplankton abundance and biomass greater values were found in the SML. The highest abundance for phytoplankton species was recorded in Pond P3 in the SML and SUB. The greatest share in Ponds P1–P4 were reported for Bacillariophyceae, while in Pond P5 it was for Chrysophyceae both in the SML and SUB. In turn, in the urban lake Chlorophyceae accounted for the greatest share. When analyzing phytoplankton biomass in the ponds the greatest share was recorded for Dinophyceae in both layers of Pond P1. In Ponds P2 and in P4, it was for Chlorophyceae and Bacillariophyceae as well as Euglenophyceae (in Pond P2) and Chrysophyceae (in Pond P4). In, Pond P3, it was for Bacillariophyceae and Cyanobactera, whereas in Pond P5 it was Chrysophyceae. In contrast, Chlorophyceae were dominant in the lake waters.

Bacillariophyceae were abundant in the SML and SUB at all sampling stations (except for Pond P5) ranging from 1% to 67%. In the biomass, the percentage share ranged from close to 0% up to 50%. The species found in the ponds in greatest abundance included *Asterionella formosa*, *Fragilaria capucina*, *Melosira varians*, *Ulnaria acus* and *U. ulna* (Table 2). In contrast, a high abundance of *Nitzschia palea* and *Tabellaria flocculosa* was recorded in the lake.

The percentage share of Chlorophceae in the total abundance ranged from 0.6% to 70%. Its share in the biomass ranged from close to 0% to 71%. In the ponds, the greatest share was recorded for *Eudorina elegans* and *Chlorella vulgaris*, whereas in the lake waters it was for *Desmodesmus communis*, *D. magnus*, *Binuclearia lauterbornii* and *Tetrastrum glabrum*.

Representatives of Chrysophyceae ranged from 0% to 94% of the total abundance, while in the biomass from 0% to 98% at individual sampling stations. Among representatives of Chrysophyceae, the greatest share in the pond waters was recorded for *Synura uvella*, *Dinobryon divergens* and *D. sociale*.

Dinophyceae was reported at individual stations at 0–16% in terms of abundance and 0–77% in terms of biomass. *Peridinium* sp. was reported in the greatest numbers in Pond P1.

Cyanobacteria were represented in greatest abundance by *Limnothrix redekei*, while Euglenophyceae by the genus *Trachelomonas* sp.

Table 2. Dominant species (>5% in total abundance) in SML and SUB layers. Explanations: total abundance, ••• > 15%, •• 11–15%, • 5–10%, o < 5%.

Group	Taxon	P1 SML	P1 SUB	P2 SML	P2 SUB	P3 SML	P3 SUB	P4 SML	P4 SUB	P5 SML	P5 SUB	L1 SML	L1 SUB	L2 SML	L2 SUB
Bacillariophyceae	*Asterionella formosa* Hassall	•	•	••	•	•••	•••	•••	•••	o	o				o
	Aulacoseira granulata (Ehrenberg) Simonsen					•	•								
	Fragilaria capucina Desmazičres		••			•	o								
	Melosira varians C.Agardh	•	•••			•	o	•	•						
	Navicula tripunctata (O.F.Müller) Bory	o	o	o	o		o						o	•	
	Nitzschia acicularis (Kützing) W.Smith	o		o	o							o	•	•	•
	Nitzschia palea (Kützing) W.Smith								o					o	••
	Tabellaria flocculosa (Roth) Kützing	•				o						••			
	Ulnaria acus (Kützing) Aboal	o		••	•	•	•	o	••			o	•	•	
	Ulnaria ulna (Nitzsch) Compčre	o			••	o	o	o	o	o				•	•
Chlorophyceae	*Binuclearia lauterbornii* Proschkina-Lavrenko											••			
	Chlorella vulgaris Beyerinck	••		•								•	•	o	
	Crucigenia tetrapedia (Kirchner) Kuntze											o	•	o	•
	Desmodesmus armatus (Chodat) E.Hegewald											o	•		•
	Desmodesmus communis E.Hegewald	o		o	o	o	o		•			•	•	•••	••
	Desmodesmus magnus (Meyen) Tsarenko													••	•
	Eudorina elegans Ehrenberg		o					•••	•••		o	o		o	
	Scenedesmus arvernensis Chodat & F.Chodat														•
	Scenedesmus dimorphus (Turpin) Kützing												•	o	•
	Tetrastrum glabrum Ahlstrom & Tiffany		o									•	••	•	o
	Tetrastrum triangulare (Chodat) Komárek					o	o								•
Chrysophyceae	*Chrysococcus* sp.			o	•										
	Dinobryon divergens O.E.Imhof					•••	•••	•••	••						
	Dinobryon sociale (Ehrenberg) Ehrenberg	••	••	•	•••										
	Synura uvella Ehrenberg	o						o	o	•••	•••				
Cyanobacteria	*Merismopedia glauca* (Ehrenberg) Kützing											•			•
	Limnothrix redekei (Goor) Meffert			o				••							
	Microcystis aeruginosa (Kützing) Kützing					••	o								
Dinophyceae	*Peridinium* sp.	•	•••		o										
	Peridiniopsis sp.	•													
Euglenophyceae	*Euglena* sp.			•	•	o	o	•		o	o				
	Trachelomonas sp.		o	••	•										

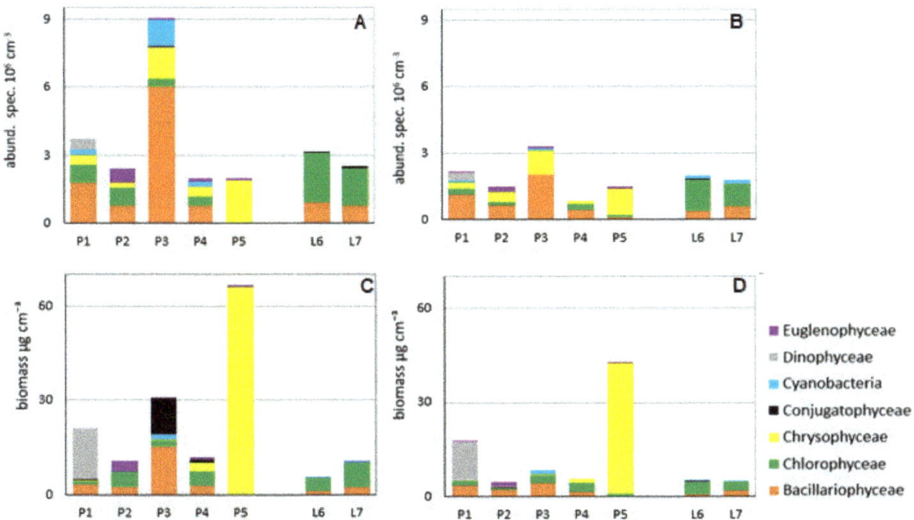

Figure 5. Phytoplankton abundance (in the SML (**A**) and in the SUB (**B**)) and biomass (in the SML (**C**) and the SUB (**D**)) and taxonomic groups at individual sampling stations.

3.3. Phytoplankton and Bacterioplankton Versus Chemical Parameters

Phytoplankton biomass, bacterioneuston and bacterioplankton number (CFU), chl *a*, EC, Sr, Zn, Ca, SO_4^{2-}, Cl^-, Na, K, $N\text{-}NO_3$, N_{org}, TN, P_{org}, $P\text{-}PO_4$ and TP showed a statistically significant positive correlation between the SML and SUB (Table 3). In turn, the concentration of V showed a negative correlation between the SML and SUB.

Table 3. Statistically significant Spearman correlation coefficients calculated between the concentrations of the tested parameters in the SML and SUB ($n = 7$, $p < 0.05$, significant for one-sided critical area).

Parameter	r
biom	0.86
chl *a*	0.75
CFU	0.93
EC	0.89
Sr	0.93
Zn	0.86
V	−0.75
Ca	0.96
SO_4^{2-}	0.96
Cl^-	0.86
Na	0.86
K	0.96
$N\text{-}NO_3$	0.96
N_{org}	0.96
TN	0.82
P_{org}	0.93
$P\text{-}PO_4$	0.93
TP	0.93

Using CCA, a significant effect was found in the case of both macronutrients such as Ca, Na, K, SO_4^{2-} and P_{org} on the abundance of the phytoplankton taxonomic groups and heterotrophic bacteria (Figure 6A). Statistically significant correlations were also recorded between abundance of phytoplankton, CFU and trace elements such as Ni, Sr and F^- (Figure 6B).

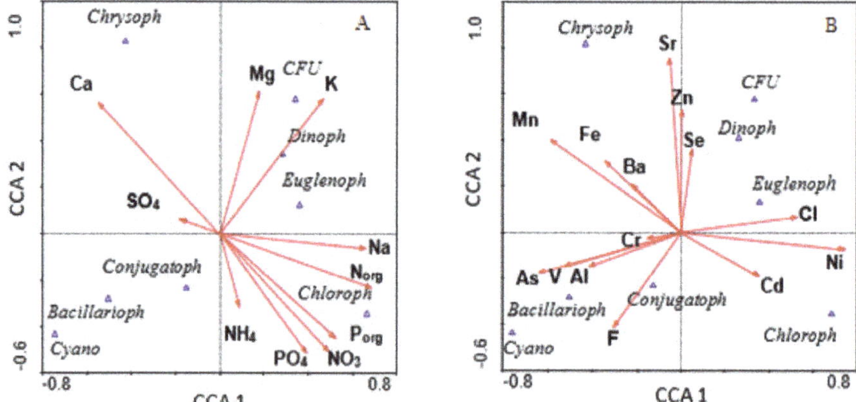

Figure 6. Ordination plot of canonical correspondence analysis (CCA) for phytoplankton abundance and chemical parameters: macroelements and nutrients (**A**); and trace metals (**B**) (in both layers).

4. Discussion

The investigated objects are located in an area of moderate human impact resulting from their immediate vicinity to urban infrastructure. Urban pollutants comprise a wide range of chemicals, including nutrients that cause eutrophication [18,36] of these water bodies, as well as trace metals [7]. In the SUB layer, the concentrations of N-org, TN and N-NH$_4^-$ were comparable both in the ponds and in the lake. Concentrations of N-NO$_3$ and phosphorus compounds were higher in the lake waters (Figure 1). When compared to the studies of this lake in the 1990s [18], the observed TN concentration was two-fold lower at a continuously high phosphorus concentration. In the 1990s, the lake water was classified as bad, which was the result of sewage inflow [18]. Moreover, high N-NO$_3$ concentrations were recorded in the lake, which may have been a consequence of N-NO$_3$ leaching from the catchment [37–40] and biological nitrification of ammonium nitrogen in the lake, particularly since good aerobic conditions were recorded there. Levels of nutrients in the studied water bodies indicate their eutrophic and hypereutrophic character, which is consistent with their levels reported in previous studies [10,18,41].

Concentrations of trace metals Mn, Sr, Zn, Se and V in the SUB (Figure 2) were higher in the ponds than in the lake; in contrast, higher concentrations of Cd, Ni and Al were recorded in the lake. The concentrations of trace metals found in the SUB layers of the investigated urban water bodies are comparable to those for other surface waters in Poland [42], while their values indicate moderate anthropogenic pollution associated with low household emissions [41].

The surface microlayers of the ponds and the urban lake accumulated the phytoneuston, trace metals and nutrients in comparison to the SUB layer. Chemical substances and organisms are supplied to the SML from subsurface water and the atmosphere. Greater accumulation of trace metals in the SML compared to the SUB has also been observed in other inland waters, both lotic and lentic as well as marine [6,7,10]. In the case of trace metals, the recorded EF values for the investigated urbanized water bodies were particularly high for Cd and V. They are the elements found at slight concentrations in the SUB layer. Low concentration of metals in the SUB may have contributed to their high accumulation in the SML.

In contrast, macroelements such as Ca, Na, K, Cl$^-$ and SO$_4^{2-}$ in waters of the analyzed urban water bodies showed comparable concentrations in the SML and SUB (EF values 1.03–1.23). Only Mg had a slightly higher enrichment factor (EF = 1.35). Macroelements are typically accumulated only periodically in the SML, e.g., in the period of spring phytoplankton and phytoneuston growth, Mg may have been taken up to form chlorophyll *a* [9].

In urban water bodies, the accumulation of trace metals in the SML may largely be the result of atmospheric deposition compared to in the case of water bodies located at a distance from urban areas.

The SML is the zone of matter exchange between the hydrosphere and the atmosphere [1,5]. Chemical substances from the atmosphere reach urban water bodies in the form of wet and dry deposition. Municipal transport and urban consumption result in the production of large amounts of gaseous pollutants as well as particulate matter subsequently deposited in urban areas [43,44], therefore they may be deposited in urban water bodies. Municipal and industrial pollutants include such metals as Fe, Mn, Cd and Ni [7]. Trace metals such as Pb, Cd and Ni, as well as the metalloid As may be bound by $PM_{2.5}$ and PM_{10} dusts and thus be deposited in water bodies; as such, they are common urban pollutants [13,45,46].

Statistically significant correlations between metal concentrations in the SML and SUB layers indicate the SUB as a source of the components in the SML, while in the case of a negative correlation a greater effect of the atmosphere is suggested. Substances can be delivered to the SML from both the atmosphere and the SUB. The SML supply from the water column occurs during the movement of matter from deeper layers to the surface together with gas bubbles raised from the bottom of the water body and the surface of plants and diffusion movements [3]. Moreover, neustonic organisms, usually derived from the SUB [2], have the ability to bioaccumulate metals [6] and thus can affect the enrichment and chemical composition of the SML. In view of the above, it may be assumed that Sr, Zn, Ca, Cl^-, Na and K mainly originated from the pool accumulated earlier in the SUB layer, particularly since the concentrations of these elements in the atmosphere are much lower when compared with those recorded in the SUB layer [13,17]. Similarly, parameters describing aquatic microorganisms, i.e., phytoplankton biomass, chlorophyll *a* and counts of heterotrophic bacteria, were correlated between the SML and the SUB.

A negative SML–SUB correlation was observed in the investigated urban water bodies for V at a lack of statistically significant correlations for the trace metals Al, As, Ba, Cd, Cr, Fe, Mn, Ni and Se as well as the nonmetal F^- (Table 3). It may be assumed in these cases that accumulation in the SML for V was determined by the load of atmospheric depositions, while for the other above-mentioned elements the share of the pollutants was more uniform between the sources from the hydrosphere and the atmosphere. The level of contaminants in urban water bodies such as ponds or lakes results from the sum of local factors as well as wet and dry deposition [17]. In this study, high accumulation of Ni and Cd was found in the SML layer in Człuchowskie Lake and Al in the ponds of the city of Słupsk. These accumulations did not correlate with concentration in the SUB. Hence it can be assumed that this accumulation could have resulted from atmospheric deposition, especially since Ni and Cd are typical urban pollution [7].

The SML is an unstable environment, on the one hand rich in nutrients and on the other hand dangerous for the neuston, as it may contain various substances, e.g., heavy metals, at toxic concentrations [7,17]. The presence of noxious substances, not exceeding toxic concentrations, may lead to the increased activity of neuston microorganisms [11,12], thus these microorganisms colonize the unstable SML environment even though it is harmful for them over a longer time perspective. Most research concerning the phytoneuston focuses on studies on primary production and analyses of photosynthetic pigments, primarily chlorophyll *a*, rather than the phytoneuston species composition or population size. The abundance of algae and cyanobacteria in the present study was higher in the SML than in the SUB, similar to other studies carried out in other places and conditions [10,29]; it was analogous to the contents of nutrients and trace metals [14,47].

Shares of taxonomic groups in individual water bodies varied in terms of their abundance and biomass. Both weather factors and catchment conditions affected this pattern [48,49]. Diatoms and chrysophytes were dominant in most water bodies [50,51]. *Asterionella formosa* was dominant among diatoms (particularly Ponds P4 and P5), which is characteristic of mixed, eutrophic small to medium water bodies [52]. Apart from this species, *Aulacoseira granulata*, *Fragilaria capucina* and *Melosira varians* were dominant in the ponds. These taxa prefer high-nutrient availability and water mixing [53,54]. The most abundant species in the lake were popular taxa reported both in small water bodies and in lakes, e.g., *Navicula tripunctata*, *Nitzschia acicularis* and *N. palea* [55].

In turn, among chrysophytes, particularly in Pond P5, the dominant species was *Synura uvella*, typical of small water bodies rich in organic compounds [40,53,56], including also astatic ones [57] and those with high trophic status [58]. Other chrysophytes such as *Dinobryon divergens* and *D. sociale* were dominant in Ponds P1–P3. Representatives of this group are typically dominant in the spring season, as they prefer lower water temperatures [59–62]. Because of chromatic adaptation, chrysophytes may develop in shaded water bodies (such as Pond P4); they can lead a heterotrophic life and can supplement nutrient uptake by the phagotrophic ingestion of bacteria [63–65]. As indicated by CCA, chrysophytes preferred water bodies characterized by lower concentrations of nitrogen and phosphorus.

Chlorophytes constituted another numerously represented group of phytoplankton. A particularly high share in the analyzed lake was recorded for the abundance of such taxa as *Desmodesmus communis*, *D. magnus*, *Binuclearia lauterbornii* and *Tetrastrum glabrum*, characteristic of eutrophic lakes [66–68]. The dominance of *Eudorina elegans* was reported in the ponds, a cosmopolitan species [69], characteristic of small eutrophic water bodies with stagnant water [55].

In Pond P2, where concentrations of nutrients were high, particularly TN, euglenids were dominant, i.e., *Euglena* sp. and *Trachelomonas* sp. (Table 2). Dominance of these genera in the water bodies indicates their strong eutrophication [50,70]. They are common in small water bodies, rich in organic matter [71,72], as well as dammed reservoirs [73].

Cyanobacteria being dominant in Pond P3 pose a particularly serious threat to the recreational use of small water bodies because of toxin production [74–76]. Some reports indicate that the cyanotoxins may inhibit algal growth, which increases the chance for Cyanobacteria development [77]. In contrast to the other groups, Cyanobacteria respond to the anthropogenic effect of the catchment, which is particularly significant in the case of small water bodies [78]. They are equipped with several specific adaptation mechanisms, making it possible for effective competition with other groups of autotrophic plankton [79]. They are also equipped with effective mechanisms, such as binding metals by proteins or polysaccharides, methylation, complexing by organic matter, development of energy-driven efflux pumps, precipitation, oxidation, vaporization and complexing with excreted metabolites, which protect them against toxic concentrations of heavy metals [80]. The dominant species were *Limnothrix redekei*, *Microcystis aeruginosa* and *Merismopedia glauca*, frequently reported in small eutrophic water bodies [81–83].

A high share of dinoflagellates from the genus *Peridinium*, recorded in Pond P1, indicated the presence of various trophic types, both those with higher trophic status such as in hypereutrophic water bodies [84], as well as those with lower trophic status, showing an improved water quality, e.g., during restoration processes [85,86]. Species from this genus may generate intensive blooms, such as those observed in urban lakes situated in northeastern Poland [87]. It needs to be mentioned here that species from the genus *Peridinium* are less numerously represented in the SML or they were reported only in the subsurface layer. They are characterized by large size, tend to sink to the bottom and are found in deeper layers of the photic zone [88,89].

Neuston bacterial counts expressed in CFU were more numerous compared to the bacterial populations in the SUB, which may be explained by the higher concentrations of organic substance recorded in the SML [90]. Analogous densities of heterotrophic bacteria have also been reported in other water bodies such as Jeziorak Mały [91] or Lake Łebsko [6]. The studied heterotrophic bacteria usually accumulated more in SML than in the SUB. It should be mentioned, that heterotrophic bacteria are only a part of total bacteria that can occur in the water environment [10,24]. Correlations (CCA) among phytoplankton, phytoneuston, heterotrophic bacteria and chemical parameters were observed (Figure 6A,B). Phosphorus is the basic component limiting growth of phytoplankton [92], thus abundance of chlorophytes was closely related with high values of organic phosphorus and phosphates [93,94]. Potassium is also an essential nutrient for phytoplankton [92].

Calcium is a macroelement, and its high concentration in the analyzed waters was correlated with the high abundance of Chrysophyceae, which in turn prefer a low phosphate content. Calcium may be found in the form of acid calcium carbonate ($Ca(HCO_3)_2$), which reacts with phosphates and precipitates to the bottom sediments [95], thus the high Ca content may result in a decreased concentration of phosphorus available for phytoplankton. Some chlorophytes and cyanobacteria utilize Ca as well as Sr and Ba to produce amorphous calcium carbonate inclusions [96]. Another macroelement, Na, showed a positive correlation with phytoplankton. Analyses were conducted in the spring period; the response of phytoplankton to Na may be connected with the leaching of this nutrient from the urban catchment in which together with the surface runoff Na, originating from salt used for the roads and urban street de-icing, may penetrate urban water bodies [9,97]. With surface runoff, sulfur compounds, while being nutrients, may penetrate water bodies, although they are harmful at excessively high concentrations. Sulfur compounds are emitted in cities during the heating season, particularly from coal-heated households.

Statistically significant correlations were also observed with elements found in trace concentrations. Al and F^- are elements constituting urban pollutants, supplied via the atmosphere together with PM_{10} and $P_{2.5}$ dusts [17]. This would explain the high content of Al and F^- in the SML ponds of the city of Słupsk. Fluorine is commonly used in cities, e.g., in cooling liquids, in electrical components and to produce halons, herbicides and pharmaceuticals [98]. Both fluorine and aluminum are considered toxic for phytoplankton organisms, causing a decrease in their populations [99]. It needs to be remembered that there are taxa, e.g., the diatom *Nitzschia palea*, dominant in the analyzed lake, which can tolerate high F^- concentrations [100]. In turn, a high Al content in water reduces the ability of algae to absorb phosphates, forming insoluble salts. At the pH ranges observed in the analyzed urban water bodies, Al was probably inactive since it forms sparsely soluble compounds [101,102]. Additionally, this element shows a high capacity to form organic complexes in solutions. It is worth noting that the molar ratio of Al to other ions is important with respect to Al toxicity and reduction of aluminum toxicity is observed when Ca ions increase [103].

5. Conclusions

In the investigated urban water bodies such chemical parameters as trace metals and nutrients were accumulated to a greater extent in the SML than in the SUB. Macroelements showed no such accumulation, with the exception of Mg, which was accumulated in the SML probably due to its incorporation into chlorophyll *a*. Heterotrophic bacteria and phytoplankton were found at greater abundance in the SML than in the SUB. The investigated water bodies were abundant in phosphorus compounds and macroelements. The analysis of the taxonomic structure of phytoneuston in the SML and phytoplankton in the SUB layer indicates that the SML is a separate habitat showing specific properties that determine its taxonomic composition.

Author Contributions: J.P.A. and A.K. designed the study, statistical analysis, designed and wrote the manuscript; J.P.A. field and laboratory analysis. All authors have read and agreed to the published version of the manuscript.

Funding: This research received no external funding.

Conflicts of Interest: The authors declare no conflict of interest.

References

1. Ebling, A.M.; Landing, W.M. Sampling and analysis of the sea surface microlayer for dissolved and particulate trace elements. *Mar. Chem.* **2015**, *177*, 134–142. [CrossRef]
2. Cunliffe, M.; Engel, A.; Frka, S.; Gašparović, B.; Guitart, C.; Murrell, J.C.; Salter, M.; Stolle, C.; Upstill-Goddardh, R.; Wurl, O. Sea surface microlayers: A unified physicochemical and biological perspective of the air-ocean interface. *Prog. Oceanogr.* **2013**, *109*, 104–116. [CrossRef]
3. Kostrzewska-Szlakowska, I. A surface microlayer of waters: Characteristics and research methods. *Wiad. Ekol.* **2003**, *49*, 183–203.
4. Gašparović, B.; Plavšić, M.; Ćosović, B.; Saliot, A. Organic matter characterization in the sea surface microlayers in the subarctic Norwegian fjords region. *Mar. Chem.* **2007**, *105*, 1–14. [CrossRef]
5. Hillbricht-Ilkowska, A.; Jasser, I.; Kostrzewska-Szlakowska, I. Airwater interface: Dynamic of nutrients and picoplankton in the surface microlayer of a humic lake. *Verh. Int. Ver. Limnol.* **1997**, *26*, 319–322.

6. Antonowicz, J.P.; Kozak, A. Chemical and phycological structure of surface microlayer and subsurface water in a southern Baltic Sea estuary. *J. Mar. Res.* **2018**, *76*, 93–118. [CrossRef]

7. Liu, Q.; Hu, X.; Jiang, J.; Zhang, J.; Wu, Z.; Yang, Y. Comparison of the water quality of the surface microlayer and subsurface water in the Guangzhou segment of the Pearl River, China. *J. Geogr. Sci.* **2014**, *24*, 475–491. [CrossRef]

8. Antonowicz, J. Daily cycle of variability contents of phosphorus forms in surface microlayer of a light salinity Baltic Sea Lagoon lake (North Poland)—Part II. *Open Chem.* **2013**, *11*, 817–826. [CrossRef]

9. Antonowicz, J.P.; Kubiak, J.; Machula, S. Macroelements in the surface microlayer of water of urban ponds. *Limnol. Rev.* **2016**, *16*, 115–120. [CrossRef]

10. Antonowicz, J.P.; Mudryk, Z.J.; Perliński, P.; Kubiak, J. Measurement of heavy metal concentrations and microbiological parameters in the surface microlayer of a downtown pond. *Ecohydrol. Hydrobiol.* **2017**, *17*, 283–296. [CrossRef]

11. Kostrzewska-Szlakowska, I.; Kiersztyn, B. Microbial biomass and enzymatic activity of the surface microlayer and subsurface water in two dystrophic lakes. *Pol. J. Microbiol.* **2017**, *66*, 75–84. [CrossRef] [PubMed]

12. Perliński, P.; Mudryk, Z.J.; Antonowicz, J. Enzymatic activity in the surface microlayer and subsurface water in the harbor channel. *Estuar. Coast. Shelf Sci.* **2017**, *196*, 150–158. [CrossRef]

13. Szefer, P. *Metals, Metalloids and Radionuclides in the Baltic Sea Ecosystems*; Elsevier: Amsterdam, The Netherlands, 2002.

14. Antonowicz, J. Air-water interface in an estuarine lake: Chlorophyll and nutrient enrichment. *Pol. J. Ecol.* **2018**, *66*, 205–217. [CrossRef]

15. Kawecka, B.; Eloranta, P.V. *Outline of Ecology of Freshwater and Terrestrial Algae*; Polish Scientific Publications: Warsaw, Poland, 1994; pp. 1–252.

16. Wang, Z.H.; Song, S.H.; Qi, Y.Z. A comparative study of phytoneuston and the phytoplankton community structure in Daya Bay, South China Sea. *J. Sea Res.* **2014**, *85*, 474–482. [CrossRef]

17. Antonowicz, J.P.; Panasiuk, D.; Machula, S.; Kubiak, J.; Opalińska, M. The anthropogenic pollutants in urban ponds based on the example of Słupsk. In Proceedings of the E3S Web of Conferences, Jakarta, Indonesia, 24–25 October 2018.

18. Trojanowski, J.; Trojanowska, C. Pollution state of Człuchów lakes. *Arch. Environ. Prot.* **1999**, *25*, 91–109.

19. Garrett, W. Collection of slick forming materials from the sea surface. *Limnol. Oceanogr.* **1965**, *10*, 602–605. [CrossRef]

20. American Public Health Association. *Standard Methods for the Examination of Water and Wastewater*, 21st ed.; American Public Health Association: Washington, DC, USA, 2005.

21. USA Environmental Protection Agency. Microwave assisted acid digestion of aqueous samples and extracts. In *SW-846 Manual: Methods for Evaluating Solid Waste, Physical/Chemical Methods*; USA Environmental Protection Agency: Washington, DC, USA, 1998.

22. American Public Health Association. *Standard Methods for the Examination of Water and Wastewater*, 18th ed.; American Public Health Association: Washington, DC, USA, 1992.

23. Duffus, J.H. "Heavy metals"—A meaningless term? *Pure Appl. Chem.* **2002**, *74*, 793–807. [CrossRef]

24. Mudryk, Z.; Trojanowski, J.; Antonowicz, J.; Skórczewski, P. Chemical and bacteriobiological studies of surface and subsurface water layers in estuarine Lake Gardno. *Pol. J. Environ. Stud.* **2003**, *12*, 199–206.

25. Arvola, L. Spectrophotometric determination of chlorophyll a and phaeopigments in ethanol extractions. *Ann. Bot. Fenn.* **1981**, *18*, 221–227.

26. Wetzel, R.G.; Likens, G.E. *Limnological Analyses*; Springer: Berlin/Heidelberg, Germany, 1991; pp. 1–391.

27. Hutorowicz, A. Standard Cell Volumes for Biomass Estimation of Selected Taxa of Planktonic Algae with the Method of Measurement and Estimation; 2009 (in Polish). Available online: http://www.gios.gov.pl/images/dokumenty/pms/monitoring_wod/objetosci_PMPL.pdf (accessed on 1 July 2020).

28. StatSoft Inc. *STATISTICA (Data Analysis Software System)*, 13rd ed.; StatSoft Inc.: Tulsa, OK, USA, 2013.

29. Estep, K.W.; Maki, J.S.; Danos, S.; Remsen, C. The retrieval of material form the surface microlayer with screen and plate samplers and its implications for partitioning of material within the microlayer. *Freshw. Biol.* **1985**, *15*, 15–19. [CrossRef]

30. Antonowicz, J.; Trojanowski, J.; Trojanowska, C. 24-hour cycle of variability in contents of nitrogen forms in the surface microlayer of the Baltic Sea lagoon lake (north Poland)—Part I. *Balt. Coast. Zone* **2010**, *14*, 37–47.

31. Ter Braak, C.J.F. Canonical correspondence analysis: A new eigenvector technique for multuvatiate direct gradient analysis. *Ecology* **1986**, *67*, 1167–1179. [CrossRef]

32. Ter Braak, C.J.F.; Smilauer, P. *CANOCO Reference Manual and CanoDraw for Windows User's Guide: Software for Canonical Community Ordination, Version 4.5*; TNO Institute of Applied computer Science: Wageningen, The Nederlands, 2002; pp. 1–29.

33. Ter Braak, C.J.F. Interpreting canonical correlation analysis through biplots of structure correlations and weights. *Psychometrika* **1990**, *55*, 519–531. [CrossRef]

34. Lepš, J.; Šmilauer, P. *Multivariate Analysis of Ecological Data Using CANOCO TM*; Cambridge University Press: Cambridge, UK, 2003; pp. 1–269.

35. Jongman, R.H.G.; Ter Braak, C.J.F.; Van Tongeren, O.F.R. *Data Analysis in Community and Landscape Ecology*; Cambridge University Press: Cambridge, UK, 1987; pp. 1–212.

36. Rzymski, P.; Sobczyński, T.; Klimaszyk, P.; Niedzielski, P. Sedimentary fractions of phosphorus before and after drainage of an urban water body (Maltański Reservoir, Poland). *Limnol. Rev.* **2015**, *15*, 31–37. [CrossRef]

37. Moczulska, A.; Antonowicz, J.; Krzyk, K. The influence of Słupsk agglomeration on the quality of the water of the River Słupia. *Słup. Prace Biol.* **2006**, *3*, 45–56.

38. Misztal, A.; Kuczera, M. Impact of Land Use Method in a Catchments Area on the Dynamics of the Nitrogen Compounds in the Outflowing Water. *Ecol. Chem. Eng.* **2010**, *17*, 643–649.

39. Ławniczak, A.E.; Zbierska, J.; Nowak, B.; Achtenberg, K.; Grześkowiak, A.; Kanas, K. Impact of agriculture and land use on nitrate contamination in groundwater and running waters in central-west Poland. *Environ. Monit. Assess.* **2016**, *188*, 172. [CrossRef]

40. Dondajewska, R.; Kozak, A.; Kowalczewska-Madura, K.; Budzyńska, A.; Gołdyn, R.; Podsiadłowski, S.; Tomkowiak, A. The response of a shallow hypertrophic lake to innovative restoration measures—Uzarzewskie Lake case study. *Ecol. Eng.* **2018**, *121*, 72–82. [CrossRef]

41. Antonowicz, J.P.; Wolska, M.; Libera, A. Seasonal changes in the accumulation of nutrients in the hydrosphere-atmosphere interface of urban ponds (Słupsk, northern Poland). *Balt. Coast. Zone* **2016**, *20*, 5–23.

42. Lis, J.; Pasieczna, A. *Geochemical Atlas of Poland*; Państwowy Instytut Geologiczny: Warszawa, Porland, 1995.

43. Niedźwiecki, E.; Protasowicki, M.; Wojcieszczuk, T.; Ciemniak, A.; Niedźwiecka, D. Content of Macroelements and Some Trace Elements in Dust Fallout within Szczecin Urban Area. In Proceedings of the International Workshop, Bratislava, Slovakia, 20–22 June 2001; pp. 119–125.

44. Nasali-Flores, L. Urban Lakes: Ecosystems at Risk, Worthy of the Best Care. In Proceedings of the 12th World Lake Conference, Rajasthan, India, 28 October–2 November 2007; pp. 1333–1337.

45. Siudek, P.; Frankowski, M. Atmospheric deposition of trace elements at urban and forest sites in central Poland—Insight into seasonal variability and sources. *Atmos. Res.* **2017**, *198*, 123–131. [CrossRef]

46. Barałkiewicz, D.; Chudzińska, M.; Szpakowska, B.; Świerk, D.; Gołdyn, R.; Dondajewska, R. Strom water contamination and its effect on the quality of urban surface waters. *Environ. Monit. Assess.* **2014**, *186*, 6789–6803. [CrossRef] [PubMed]

47. Antonowicz, J.; Mudryk, Z.; Trojanowski, J.; Dwulit, M. Concentrations of heavy metals and bacterial numbers and productivity in surface and subsurface water layers in coastal Lake Dolgie Wielkie. *Ecol. Chem. Eng.* **2008**, *15*, 873–879.

48. Kozak, A.; Gołdyn, R.; Dondajewska, R. Phytoplankton Composition and Abundance in Restored Maltanski Reservoir under the Influence of Physico-Chemical Variables and Zooplankton Grazing Pressure. *PLoS ONE* **2015**, *10*, e0124738. [CrossRef] [PubMed]

49. Toporowska, M.; Ferencz, B.; Dawidek, J. Impact of lake catchment processes on phytoplankton community structure in temperate shallow lakes. *Ecohydrology* **2018**, *11*, 1–12. [CrossRef]

50. Kozak, A. Changes in the structure of phytoplankton in the lowest part of the Cybina River and the Maltański Reservoir. *Oceanol. Hydrobiol. Stud.* **2010**, *39*, 85–94. [CrossRef]

51. Pełechata, A.; Pełechaty, M.; Pukacz, A. Winter temperature and shifts in phytoplankton assemblages in a small Chara-lake. *Aquat. Bot.* **2015**, *124*, 10–18. [CrossRef]

52. Reynolds, C.F. Resilience in aquatic ecosystems—Hysteresis, homeostasis and health. *Aquat. Ecosyst. Health Manag.* **2002**, *5*, 3–17. [CrossRef]

53. Celewicz-Gołdyn, S.; Boryca, I. The Phytoplankton Community in the Permanent Pond of the Dendrological Garden, Poznań University of Life Sciences in Springtime. *Bot. Steciana* **2012**, *16*, 61–74.

54. Padisak, J.; Crossetti, L.O.; Naselli-Flores, L. Use and misuse in the application of the phytoplankton functional classification: A critical review with updates. *Hydrobiologia* **2009**, *621*, 1–19. [CrossRef]

55. Kozak, A.; Kowalczewska-Madura, K. Phytoplankton Community Structure in Small Water Bodies in the Cybinka River. *Bot. Steciana* **2009**, *13*, 133–136.

56. Grabowska, M.; Glińska-Lewczuk, K.; Obolewski, K.; Burandt, P.; Kobus, S.; Dunalska, J.; Kujawa, R.; Goździejewska, A.; Skrzypczak, A. Effects of Hydrological and Physicochemical Factors on Phytoplankton Communities in Floodplain Lakes. *Pol. J. Environ. Stud.* **2014**, *23*, 713–725.

57. Poniewozik, M.; Płaska, W. Composition of Selected Phyto- and Zoocenozis in Small, Astatic Water Bodies. *Teka Kom. Ochr. Kszt. Środ. Przyr.* **2013**, *10*, 360–369.

58. Kalinowska, K.; Grabowska, M. Autotrophic and heterotrophic plankton under ice in a eutrophic temperate lake. *Hydrobiologia* **2016**, *777*, 111–118. [CrossRef]

59. Szeląg-Wasielewska, E. Seasonal changes in the structure of the community of phototrophs in a small mid-forest Lake. *Teka Kom. Ochr. Kszt. Środ. Przyr.* **2006**, *3*, 209–215.

60. Kozak, A.; Kowalczewska-Madura, K. Pelagic phytoplankton of shallow lakes in the Promno Landscape Park. *Pol. J. Environ. Stud.* **2010**, *19*, 587–592.

61. Kozak, A. Seasonal changes occurring over four years in a reservoir's phytoplankton composition. *Pol. J. Environ. Stud.* **2005**, *14*, 437–444.

62. Kozak, A.; Dondajewska, R.; Kowalczewska-Madura, K.; Gołdyn, R.; Holona, T. Water quality and phytoplankton in selected lakes and reservoirs under restoration measures in Western Poland. *Pol. J. Nat. Sci.* **2013**, *24*, 217–226.

63. Wiedner, C.; Nixdorf, B. Success of chrysophytes, cryptophytes and dinoflagellates over blue-greens (Cyanobacteria) during an extreme winter (1995/96) in eutrophic shallow lakes. *Hydrobiologia* **1998**, *229*, 369–370.

64. Reynolds, C.S.; Huszar, V.; Kruk, C.; Naselli-Flores, L.; Melo, S. Towards a functional classification of the freshwater phytoplankton. *J. Plankton Res.* **2002**, *24*, 417–428. [CrossRef]

65. Kamjunke, N.; Henrichs, T.; Gaedke, U. Phosphorus gain by bacterivory promotes the mixotrophic flagellate Dinobryon spp. during re-oligotrophication. *J. Plankton Res.* **2007**, *29*, 39–46. [CrossRef]

66. Kozak, A.; Kowalczewska-Madura, K.; Gołdyn, R.; Czart, A. Phytoplankton composition and physicochemical properties in Lake Swarzędzkie (midwestern Poland) during restoration: Preliminary results. *Arch. Pol. Fish.* **2014**, *22*, 17–28. [CrossRef]

67. Suchora, M.; Szczurowska, A.; Niedźwiecki, M. Trophic State Indexes and Phytoplankton in the Trophic Status Assessment of Waters of a Small Retention Reservoir at an Early Stage of Operation. *J. Ecol. Eng.* **2017**, *18*, 209–215. [CrossRef]

68. Kozak, A.; Gołdyn, R.; Dondajewska, R.; Kowalczewska-Madura, K.; Holona, T. Changes in Phytoplankton and Water Quality during Sustainable Restoration of an Urban Lake Used for Recreation and Water Supply. *Water* **2017**, *9*, 713. [CrossRef]

69. Dembowska, E.A. New and rare species of Volvocaceae (Chlorophyta) in the Polish phycoflora. *Acta Soc. Bot. Pol.* **2013**, *82*, 259–266. [CrossRef]

70. Brettum, P.; Andersen, T. *The Use of Phytoplankton as Indicators of Water Quality*; Norwegian Institute for Water Research: Oslo, Norway, 2005; pp. 1–33.

71. Poniewozik, M.; Juráň, J. Extremely high diversity of euglenophytes in a small pond in eastern Poland. *Plant Ecol. Evol.* **2018**, *151*, 18–34. [CrossRef]

72. Wołowski, K.; Poniewozik, M.; Walne, P.L. Pigmented euglenophytes of the genera Euglena, Euglenaria, Lepocinclis, Phacus and Monomorphina from the southeastern United States. *Pol. Bot. J.* **2013**, *58*, 659–685. [CrossRef]

73. Wołowski, K.; Grabowska, M. Trachelomonas species as the main component of the euglenophyte community in the Siemianówka Reservoir (Narew River, Poland). *Int. J. Limnol.* **2007**, *43*, 207–218. [CrossRef]

74. Dunalska, J.A.; Grochowska, J.; Winiewski, G.; Napiórkowska-Krzebietke, A. Can we restore badly degraded urban lakes? *Ecol. Eng.* **2015**, *82*, 432e441. [CrossRef]

75. Dziga, D.; Maksylewicz, A.; Maroszek, M.; Budzynska, A.; Napiorkowska-Krzebietke, A.; Toporowska, M.; Grabowska, M.; Kozak, A.; Rosińka, J.; Meriluoto, J. The biodegradation of microcystins in temperate freshwater bodies with previous cyanobacterial history. *Ecotoxicol. Environ. Saf.* **2017**, *145*, 420–430. [CrossRef]

76. Rosińska, J.; Kozak, A.; Dondajewska, R.; Gołdyn, R. Cyanobacteria blooms before and during the restoration process of a shallow urban lake. *J. Environ. Manag.* **2017**, *198*, 340–347. [CrossRef]

77. Babica, P.; Blaha, L.; Marsalek, B. Exploring the natural role of microcystins—A review of effects on photoautotrophic organisms. *J. Phycol.* **2006**, *42*, 9–20. [CrossRef]

78. Kozak, A.; Celewicz-Gołdyn, S.; Kuczyńska-Kippen, N. Cyanobacteria in small water bodies: The effect of habitat and catchment area conditions. *Sci. Total Environ.* **2019**, *646*, 1578–1587. [CrossRef] [PubMed]

79. Jaworska, B.; Dunalska, J.; Górniak, D.; Bowszys, M. Phytoplankton dominance structure and abundance as indicators of the trophic state and ecological status of Lake Kortowskie (northeast Poland) restored with selective hypolimnetc withdrawal. *Arch. Pol. Fish.* **2014**, *22*, 7–15. [CrossRef]

80. González–Dávila, M. The role of phytoplankton cells on the control of heavy metal concentration in seawater. *Mar. Chem.* **1995**, *8*, 215–236. [CrossRef]

81. Pełechata, A.; Pełechaty, M.; Pukacz, A. An attempt to the trophic status assessment of the lakes of Lubuskie Lakeland. *Limnol. Rev.* **2006**, *6*, 239–246.

82. Kozak, A.; Gołdyn, R. Variation in Phyto- and Zooplankton of Restored Lake Uzarzewskie, 2014. *Pol. J. Environ. Stud.* **2014**, *23*, 1201–1209.

83. Dembowska, E. Phytoplankton of shallow lakes in the Iławskie Lake District. *Limnol. Pap.* **2008**, *3*, 19–31. [CrossRef]

84. Toporowska, M.; Pawlik-Skowrońska, B. Fitoplankton hipertroficznego jeziora Syczyńskiego. *Fragm. Florist. Geobot. Pol.* **2011**, *18*, 409–426.

85. Kozak, A.; Rosińska, R.; Gołdyn, R. Changes in the phytoplankton structure due to prematurely limited restoration treatments. *Pol. J. Environ. Stud.* **2018**, *27*, 1097–1103. [CrossRef]

86. Kozak, A.; Budzyńska, A.; Dondajewska-Pielka, R.; Kowalczewska-Madura, K.; Gołdyn, R. Functional Groups of Phytoplankton and Their Relationship with Environmental Factors in the Restored Uzarzewskie Lake. *Water* **2020**, *12*, 313. [CrossRef]

87. Napiórkowska Krzebietke, A.; Dunalska, J.; Zębek, E. Taxa-specific eco-sensitivity in relation to phytoplankton bloom stability and ecologically relevant lake state. *Acta Oecol.* **2017**, *81*, 10–21. [CrossRef]

88. Kozak, A.; Graf, M. Phytoplankton Composition in Maltański Reservoir and the lowest part of Cybina River. *Bot. Steciana* **2013**, *17*, 85–94.

89. Wilk-Woźniak, E.; Żurek, R. Phytoplankton and its relationships with chemical parameters and zooplankton in the meromictic Piaseczno reservoir, Southern Poland. *Aquat. Ecol.* **2006**, *40*, 165–176. [CrossRef]

90. Engel, A.; Sperling, M.; Sun, C.; Grosse, J.; Friedrichs, G. Organic Matter in the Surface Microlayer: Insights From a Wind Wave Channel Experiment. *Front. Mar. Sci.* **2018**. [CrossRef]

91. Donderski, W.; Walczak, M.; Mudryk, Z. Neustonic bacteria of lake Jeziorak Mały (Poland). *Pol. J. Environ. Stud.* **1998**, *7*, 125–129.

92. Spijkerman, E.; Bissinger, V.; Meister, A.; Gaedke, U. Low potassium and inorganic carbon concentrations influence a possible phosphorus limitation in Chlamydomonas acidophila (Chlorophyceae). *Eur. J. Phycol.* **2007**, *42*, 327–339. [CrossRef]

93. Annadotter, H.; Cronberg, G.; Aagren, R.; Lundstedt, B.; Nilsson, P.-Å.; Ströbeck, S. Multiple techniques for lake restoration. *Hydrobiologia* **1999**, *395*, 77–85. [CrossRef]

94. Rosińska, J.; Kozak, A.; Dondajewska, R.; Kowalczewska-Madura, K.; Gołdyn, R. Water quality response to sustainable restoration measures—Case study of urban Swarzędzkie Lake. *Ecol. Indic.* **2018**, *84*, 437–449. [CrossRef]

95. Frau, D.; Spies, M.E.; Battauz, Y.; Medrano, J.; Sinistro, R. Approaches for phosphorus removal with calcium hydroxide and floating macrophytes in a mesocosm experiment: Impacts on plankton structure. *Hydrobiologia* **2019**, *828*, 287–299. [CrossRef]

96. Martignier, A.; Filella, M.; Pollok, K.; Melkonian, M.; Bensimon, M.; Barja, F.; Langenhorst, F.; Jaquet, J.-M.; Ariztegui, D. Marine and freshwater micropearls: Biomineralization producing strontium-rich amorphous calcium carbonate inclusions is widespread in the genus Tetraselmis (Chlorophyta). *Biogeosciences* **2018**, *15*, 6591–6605. [CrossRef]

97. Czarna, M. Overwiew of chemicals applied to winter road maintenance in Poland. *Zesz. Nauk. Inż. Środ.* **2013**, *151*, 18–25.

98. US Department of Health and Human Service. *Toxicological Profile for Fluorides, Hydrogen Fluoride, and Fluorine*; US Department of Health and Human Service: Washington, DC, USA, 2003.

99. Upreti, N.; Sharma, S.; Sharma, S.; Sharma, K.P. Toxic Effects of Aluminium and Fluoride on Planktonic Community of the Microcosms. *Nat. Environ. Pollut. Technol.* **2013**, *12*, 523–528.

100. Joy, C.M.; Balakrishnan, K.P. Effect of fluoride on axenic cultures of diatoms. *Water Air Soil Pollut.* **1990**, *49*, 241. [CrossRef]

101. Kubiak, J.; Machula, S.; Stepanowska, K.; Biernaczyk, M. The influence of the content of aluminium on the biocenosis of the waters of lakes with poorly urbanized reception basins. *Inżynieria Ekol.* **2013**, *35*, 95–105. [CrossRef]

102. Gworek, B. Glin w środowisku przyrodniczym a jego toksyczność. *Ochr. Śr. Zasobów Nat.* **2006**, *29*, 27–38.

103. Rosseland, B.O.; Eldhuset, T.D.; Staurnes, M. Environmental effects of aluminum. *Environ. Geochem. Health* **1990**, *12*, 17–27. [CrossRef]

Article

Riparian Ground Beetles (Coleoptera) on the Banks of Running and Standing Waters

Marina Kirichenko-Babko [1,*], Yaroslav Danko [2], Małgorzata Franus [3], Witold Stępniewski [4] and Roman Babko [1]

1   Department of Invertebrate Fauna and Systematics, Schmalhausen Institute of Zoology NAS of Ukraine, 01030 Kyiv, Ukraine; rbabko@ukr.net
2   Faculty of Natural Sciences and Geography, Sumy Makarenko State Pedagogical University, 40002 Sumy, Ukraine; yaroslavdanko@gmail.com
3   Faculty of Civil Engineering and Architecture, Lublin University of Technology, 20-618 Lublin, Poland; m.franus@pollub.pl
4   Faculty of Environmental Engineering, Lublin University of Technology, 20-618 Lublin, Poland; w.stepniewski@pollub.pl
*   Correspondence: kirichenko@izan.kiev.ua

Received: 29 April 2020; Accepted: 19 June 2020; Published: 23 June 2020

Abstract: Rivers and their floodplains offer a wide variety of habitats for invertebrates. River ecosystems are subject to high anthropic influence: as a result the channel morphology is changed, swamps are drained, floodplains are built up, and rivers are polluted. All this has radically changed the environment for the inhabitants of the floodplains, including riparian stenotopic species. Although riparian arthropods are oriented primarily to the production of hydro-ecosystems, the type of water body—lentic or lotic—has a determining effect in the structure of communities. Most riparian arthropods have evolutionarily adapted to riverbanks with significant areas of open alluvial banks. This paper considered the structure of assemblages of ground beetles associated with the riverbanks and the shores of floodplain lakes and their differences. The banks of rivers and the shores of floodplain lakes were considered separately due to the differences in the habitats associated with them. Our results showed that riverbanks, which experience significant pollution, were actively colonized by vegetation and were unsuitable for most riparian ground beetles. The shores of floodplain lakes, being an optional habitat for riparian arthropods, cannot serve as refugia. Thus, the transformation of floodplain landscapes and river pollution creates a problem for the biological diversity of floodplain ecosystems, since riparian stenotopic species of the riverbanks become rare and disappear.

Keywords: riverbanks; floodplain lakes; Carabidae; stenotopic species; assemblage; overgrown

1. Introduction

Rivers with their floodplains are some of the most diverse and biologically productive ecosystems on Earth [1,2]. At the same time, they are among the most vulnerable [3–5]. Unlike the seas and oceans, river ecosystems are highly dependent on the state of their floodplains. This dependence is magnified in small rivers.

The dependence of the water quality in rivers on the state of the floodplain ecosystems is obvious today. In the 20th century, rivers, having lost their function as suppliers of water and food (a civilizational function), were turned into objects for recreation and sport fishing, and became receptors of wastewater from enterprises and municipal treatment facilities. River valleys include residential areas, agricultural lands, and industrial enterprises.

Since the 1950s, as a result of drainage, a strong degradation of floodplain ecosystems, with loss of their characteristic mosaic of water and land habitats and a decrease in biodiversity, has been observed in Ukraine. Effluent has led to a drop in the water quality and an increase in the trophic status of most rivers, resulting in a significant decrease in their self-cleaning potential. River pollution has also promoted a change in the quality of the banks and overgrowth of vegetation.

A feature of intact lowland rivers is the presence of various coastal elements, with a significant part composed of open alluvial sandbars, constituting a unique biotope [6]. Over millions of years, wandering riverbeds, changing direction, left numerous meanders, which, over time, lost their connection with the channel, turning into floodplain lakes.

The banks of rivers and floodplain lakes are unique ecotones which form a complex gradient between aquatic and terrestrial ecosystems [7,8]. Ecotones in floodplains exist at the boundary between land and water, and between surface waters and groundwaters [9,10]. Unique communities are found on the banks of the rivers [11,12], the representatives of which are spatially limited by the shoreline and mainly trophically oriented to the production of the hydro-ecosystem [13–16]. Among riparian arthropods, ground beetles, staphylinids, and spiders have great species diversity [17–19]. Riparian arthropods are an important component of the diversity of floodplain ecosystems [20,21]. At the same time, most specialized riparian arthropods, unlike floodplain species, are unable to survive under the conditions of anthropically monotypic banks, which lack a habitat for riparian species, i.e., alluvial deposits. Since the end of the 1990s, regulation of river flows to increase the stability of bars has contributed to the succession of vegetation and has affected the communities of open sand sediments [22,23]. On anthropically altered rivers, many species, especially highly specialized ones, have become rare and disappeared in the absence of refugia. Since the end of the 1980s, there has been a significant reduction in the number of species typical of open riverbanks, some of which have become endangered in many European countries [24–27]. The changes caused by anthropic pressures on river ecosystems (water pollution, loss of riverine habitats) in many cases do not apply to floodplain lakes and oxbows. As a rule, wastewater is discharged mainly into flowing water bodies (rivers and streams), while the floodplain water bodies experience mainly recreational pressure and are rarely used for fish farming. This suggests that the shores of lakes could potentially act as refugia for riparian species.

Understanding how riparian species can adapt to different types of water bodies and their banks and shores, whether there are species common to the shorelines of riverbed and lakes (riparian and littoral zones), and whether riparian species can survive unfavorable periods on the littoral of lakes is important for conservation of biodiversity of the river valleys.

Most works to date are devoted to the study of the diversity of riparian arthropods of lowland and mountain rivers [16,28–33] and to the influence of changes in river hydrology on soil invertebrates [34–37], and little is known about the terrestrial fauna in the banks of stagnant water bodies [38–42].

In this study, using ground beetle assemblages, the following questions are addressed. The first relates to the extent to which banks of flowing and stagnant reservoirs of one flat river system differ in the composition and structure of the arthropod communities. The second is whether the shores (littoral) of floodplain lakes act as refugia for riparian species, i.e., whether they can preserve the diversity not only of floodplain species, but also of specialized riparian species.

2. Materials and Methods

2.1. Study Area

The study was conducted in a 80-km section of the Psel River (starting with Zapsillya), a first-order tributary of the Dnieper River, in the Sumy region (northeastern Ukraine, Figure 1). In the transverse profile of the studied section, alder forests grow on the wetlands situated near terraces, while oak forests, wet meadows, and riparian shrubs grow in the central part of the floodplain. There also are patches of artificial plantations: trees (birch, oak, pine) and shrubs (hazel, common ninebark). Part of the floodplain has been reclaimed and turned into meadows.

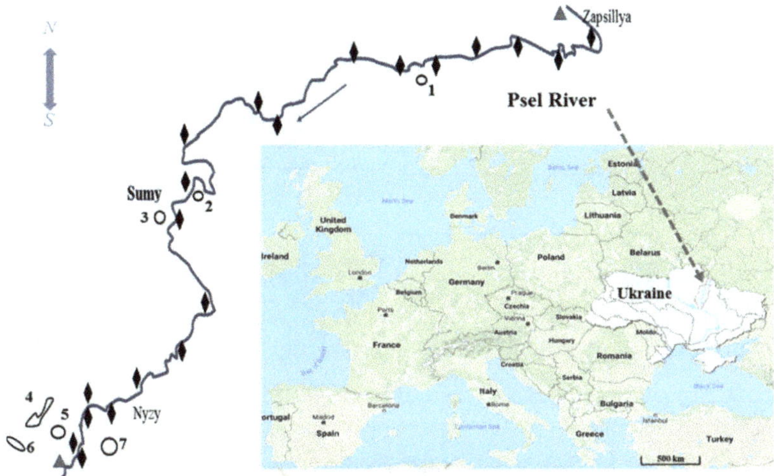

Figure 1. The study area, with the location of river stations (marked by black rhombs) and floodplain lakes (1–7) surveyed in the Psel River. The arrow indicates the river flow direction; the grey triangles are marked the study section; the dashed arrow shows the study area in the river basin of Psel.

The study included various habitats of the channel of the Psel River, covering the riverbed form units (open sand bars, referred to as exposed riverine sediments, ERS, by other researchers), which alternate with sections of the bank overgrown with different levels of vegetation (vegetated bars). A part of the open banks has been colonized by vegetation. At the same time, the open sandy riverbanks accounted about 40%.

Considering the vegetation cover on sandy and silted riverbanks (or bars) as a gradient, the following three stages of overgrowth by vegetation were distinguished [43]: stage 1—appearance of small spots of higher aquatic vegetation along the shoreline and solitary land plants; stage 2—expansion of terrestrial grassy vegetation on the surface of the bank and plant coverage of about 50% of the shoreline; and stage 3—complete coverage of the surface of the bank by grassy and woody–shrubby vegetation and almost complete separation of the dry land from the shoreline by the higher aquatic vegetation (Figure 2).

The choice of stations along the river was determined by the presence open bars and banks with varying degrees of vegetation coverage development. Accordingly, the distances between the 20 stations on the riverbanks (Figure 1) were not equal, ranging from 5 to 20 km.

The floodplain lakes under study are separate from the riverbed by different distances—from 100 to 300 m (Figure 1). These lakes are either mesotrophic or eutrophic and are 1.5 to 7 m deep. The lakes shores are mostly covered with vegetation and the few sections of the open shore are either silted alluvial sediments or alluvium covered with a layer of plant debris. The shores of oxbow lakes were analyzed in terms of their openness/closeness and categorized into shores overgrown with meadow hygrophytic vegetation (opened habitats) and those with trees and shrubs (closed).

Thus, the banks of the riverbed at different stages of vegetation cover development and the shores of floodplain lakes (opened and closed) were considered as separate types of habitats.

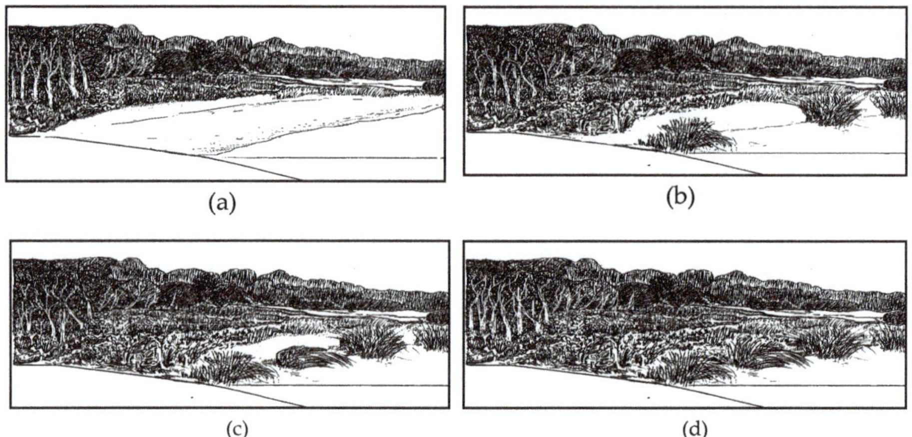

Figure 2. A graphical representation of the stages of overgrowth by vegetation of open alluvial bars along the riverbanks: (**a**) open sand bar (ob), (**b**) stage 1 (os1), (**c**) stage 2 (os2), and (**d**) stage 3 (os3) of overgrowth.

2.2. Sampling Methods

Sampling was conducted from April to October in 2008–2010 at 27 stations, with 20 along the banks of the river and 7 in the shores of the floodplain lakes. The beetles were collected using pitfall traps with 0.3-L plastic cups (diameter 7.2 cm) partly filled with a preserving solution. The traps were set parallel to the shoreline: 15 traps in three rows in accordance with the width of each station. Samples were collected 7 days after the installation of the trap, twice a month.

Carabid beetles were identified to the species level [44,45]. The species were classified as "riparian" based on the previous studies [43]. The concept of riparian, as used here, means that the species mainly or exclusively occurs in riverbanks with open riverine sediments [20,46]. The groups of species associated with the floodplain and the shores of floodplain lakes were also distinguished.

2.3. Data Analysis

A matrix containing the data on the abundance of 95 species of ground beetles in the 27 stations studied was used for analysis. The abundance for each species was averaged over habitats and years to obtain the average statistical density of each population.

Non-metric multidimensional scaling (NMDS) was used to uncover variation in beetle assemblages among stations and habitats and to analyze the patterns of the spatial distribution of ground beetles across the floodplain Psel River. This method is especially suitable for environmental data [47–50]. The Kulczynski distance was used, because it has shown good results in the biological and environmental applications [51–55]. The data on ground beetles abundances were Hellinger-transformed. The data analysis was performed in R through the core and vegan packages [56]. The results were plotted through the ggplot2 package [57].

3. Results

During the study period, a total of 95 species of ground beetles from 32 genera were recorded (Table 1). The largest numbers of species were found for the genera *Bembidion* (21 species), *Pterostichus* (9), *Agonum* (7), *Dyschirius* (5), and *Badister* (5). In overall, 81 species were recorded on the banks of the river and 51 species on the shores of floodplain lakes. Moreover, 36 species (38% of the total number of species) were found in both types of habitats.

Table 1. Ground beetles species caught on the banks of the river Psel and the shores of floodplain lakes, with their frequencies of occurrence (%) in the habitats and codes. Species are grouped according to their habitat requirements. rip: riparian, lit: littoral.

Species and Their Codes		River Banks		Shores of Lakes		Habitat Requirements
		Open Bars	Stages 1–3, Backwaters	Open	Closed	
Agonum impressum	Ag.impr	-	100	-	-	rip
Asaphidion flavipes	As.flav	22.1	77.4	-	0.5	rip
Bembidion argenteolum	Be.arge	100	-	-	-	rip
Bembidion azurescens	Be.azur	55.7	44.3	-	-	rip
Bembidion cruciatum	Be.andr	100	-	-	-	rip
Bembidion femoratum	Be.femo	86.8	13.2	-	-	rip
Bembidion laticolle	Be.lati	100	-	-	-	rip
Bembidion litorale	Be.lito	96.5	3.5	-	-	rip
Bembidion ruficolle	Be.rufi	100	-	-	-	rip
Bembidion semipunctatum	Be.semi	66.7	33.3	-	-	rip
Bembidion tenellum	Be.tene	25.0	75.0	-	-	rip
Bembidion tetracolum	Be.tetr	43.4	54.1	-	2.4	rip
Bembidion varium	Be.vari	73.1	27.7	-	0.9	rip
Chlaenius nitidulus	Ch.niti	43.	57	-	-	rip
Cicindela hybrida	Ci.hybr	-	100	-	-	rip
Dyschirius aeneus	Dy.aene	7.3	53.3	39.5	-	rip
Dyschirius arenosus	Dy.aren	82.5	16.7	-	0.8	rip
Dyschirius neresheimeri	Dy.nere	100	-	-	-	rip
Dyschirius nitidus	Dy.niti	72.7	27.3	-	-	rip
Elaphrus riparius	El.ripa	75.6	24.4	-	-	rip
Omophron limbatum	Om.limb	65.4	34.6	-	-	rip
Badister dilatatus	Ba.dila	-	-	-	100	lit
Badister peltatus	Ba.pelt	-	47.3	29.5	23.2	lit
Badister sodalis	Ba.soda	-	-	-	100	lit
Bembidion assimile	Be.assi	1.1	7.1	85.6	6.1	lit
Bembidion biguttatum	Be.bigu	0.7	6.7	-	92.6	lit
Bembidion dentellum	Be.dent	10.6	9.2	-	80.2	lit
Bembidion doris	Be.dori	11.4	13.6	49.2	25.8	lit
Elaphrus cupreus	El.cupr	9.2	6.9	20.6	63.3	lit
Philorhizus spilotus	Ph.spil	-	-	-	100	lit
Stenolophus skrimshiranus	St.skri	-	-	66.7	33.3	lit
Abax parallelopipedus	Ab.ater	-	59.0	-	41.0	
Abax parallelus	Ab.para	-	19.4	-	80.6	
Acupalpus flavicollis	Ac.flav	31.0	69.0	-	-	
Agonum duftschmidi	Ag.duft	-	100	-	-	
Agonum fuliginosum	Ag.fuli	1.8	2.3	19.2	76.7	
Agonum moestum	Ag.moes	-	23.1	-	76.9	
Agonum sexpunctatum	Ag.sexp	-	100	-	-	
Agonum versutum	Ag.vers	100	-	-	-	
Agonum viduum	Ag.vidu	-	100	-	-	
Amara communis	Am.comm	-	48.7	51.3	-	
Amara eurynota	Am.eury	-	100	-	-	
Amara ovata	Am.ovat	-	21.1	-	78.9	

Table 1. *Cont.*

Species and Their Codes		River Banks		Shores of Lakes		Habitat Requirements
		Open Bars	Stages 1–3, Backwaters	Open	Closed	
Anisodactylus binotatus	An.bino	-	-	100	-	
Anisodactylus nemorivagus	An.nemo	-	-	100	-	
Anisodactylus signatus	An.sign	-	100	-	-	
Anthracus consputus	An.cons	47.8	52.2	-	-	
Badister bullatus	Ba.bull	-	62.7	-	37.3	
Badister unipustulatus	Ba.unip	-	-	-	100	
Bembidion articulatum	Be.arti	45.1	30.3	23.8	0.7	
Bembidion lampros	Be.lamp	33.9	66.1	-	-	
Bembidion octomaculatum	Be.octo	26.4	73.6	-	-	
Bembidion obliquum	Be.obli	100	-	-	-	
Bembidion properans	Be.prop	-	100	-	-	
Bembidion quadrimaculatum	Be.quad	23.5	76.5	-	-	
Blemus discus	Bl.disc	37.2	62.8	-	-	
Broscus cephalotes	Br.ceph	-	-	-	100	
Carabus cancellatus	Ca.canc	-	100	-	-	
Carabus convexus	Ca.conv	-	100	-	-	
Carabus granulatus	Ca.gran	-	51.6	13.3	35.1	
Carabus menetriesi	Ca.mene	-	100	-	-	
Chlaenius nigricornis	Ch.nigr	89.6	10.4	-	-	
Chlaenius vestitus	Ch.vest	34.7	65.3	-	-	
Clivina collaris	Cl.coll	-	100	-	-	
Clivina fossor	Cl.foss	41.1	38.4	-	20.5	
Cymindis axillaris	Cy.axil	-	-	100	-	
Dyschiriodes globosus	Dy.glob	23.0	23.5	6.0	47.6	
Epaphius secalis	Ep.seca	2.0	10.9	-	87.1	
Harpalus affinis	Ha.affi	50	50	-	-	
Harpalus distinguendus	Ha.dist	-	100	-	-	
Leistus terminatus	Le.term	-	100	-	-	
Loricera pilicornis	Lo.pili	-	18.0	51.3	30.8	
Notiophilus palustris	No.palu	-	100	-	-	
Oodes gracilis	Oo.grac	-	-	-	100	
Oodes helopioides	Oo.helo	1.0	8.3	70.3	20.4	
Oxypselaphus obscurum	Ox.obsc	-	100	-	-	
Panagaeus bipustulatus	Pa.bipu	-	-	-	100	
Panagaeus cruxmajor	Pa.crux	-	30.9	27.6	41.4	
Patrobus atrorufus	Pa.atro	-	-	-	100	
Poecilus cupreus	Po.cupr	100	-	-	-	
Platynus assimile	Pl.assi	4.4	64.0	3.5	32.5	
Pterostichus anthracinus	Pt.anth	8.9	23.0	14.7	53.4	
Pterostichus gracilis	Pt.grac	-	-	-	100	
Pterostichus melanarius	Pt.mela	-	69.2	-	30.8	
Pterostichus minor	Pt.mino	5.0	17.0	52.6	25.3	
Pterostichus niger	Pt.nige	1.7	98.3	-	-	
Pterostichus nigrita	Pt.nigr	1.4	17.2	32.4	50.0	
Pterostichus oblongopunctatus	Pt.oblo	1.8	59.4	-	39.8	
Pterostichus strenuus	Pt.stre	5.8	72.6	-	21.6	

Table 1. *Cont.*

Species and Their Codes		River Banks		Shores of Lakes		Habitat Requirements
		Open Bars	Stages 1–3, Backwaters	Open	Closed	
Pterostichus vernalis	Pt.vern	-	15.6	-	84.4	
Stenolophus mixtus	St.mixt	11.2	13.8	75.0	-	
Stenolophus teutonus	St.teut	54.5	-	-	45.5	
Trechus quadristriatus	Tr.quad	-	-	-	100	
Trichocellus placidus	Di.plac	-	100	-	-	
Trichocellus rufithorax	Di.rufi	-	100	-	-	

The NMDS on the abundances of ground beetles among stations showed major differences between assemblages in the banks of the river and in the shores of the floodplain lakes (Figure 3). It also highlighted that the conditions in the backwaters of the river, where the flow rate is slowed down and the substrate is silted, are close to those on the shores of lakes.

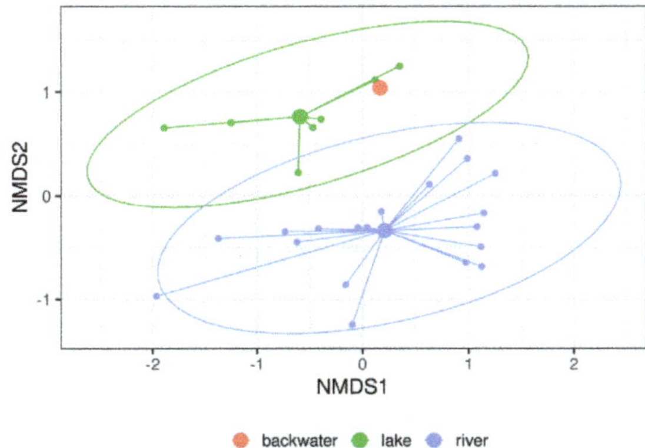

Figure 3. Non-metric multidimensional scaling (NMDS) of the Kulczynski distance matrix among ground beetle assemblages on the banks of the Psel River, backwater, and shores of floodplain lakes.

NMDS showed considerable variation between the shores of floodplain lakes (Figure 4, in the upper left quarter of the graph) and the habitats along the riverbanks (in the lower right part of the graph), with riverbanks completely covered by vegetation laying between them (Figure 4, os3). Indeed, the assemblage structure of ground beetles along the banks of the river and floodplain lakes is significantly different (Figure 5).

NMDS confirmed that the shores of floodplain lakes clearly differed in the composition of ground beetles from the banks of the channel (Figure 6). Open alluvial banks and sections of the banks with vegetation in the first and second stages of overgrowth were similar in composition. Overgrown riverbanks (stage 3) occupied an intermediate position, while both open and closed habitats on the shores of floodplain lakes are located in the left upper part of the plot (Figure 6).

The riparian and littoral species are represented in variable numbers on the banks of the Psel river and the shores of floodplain lakes (Figure 7). On the banks of the river, 21 riparian species were recorded (22% of the total number of species), of which only five species were found on the banks of floodplain lakes: *Asaphidion flavipes*, *Bembidion tetracolum*, *Bembidion varium*, *Dyschirius aeneus*, *Dyschirius arenosus* (Table 1, Figure 7). Of the 10 littoral species found on the littoral of lakes, six were found in the riverbanks, but in a smaller abundance: *Badister peltatus*, *Bembidion assimile*, *Bembidion biguttatum*, *Bembidion dentellum*, *Bembidion doris*, and *Elaphrus cupreus* (Figure 7).

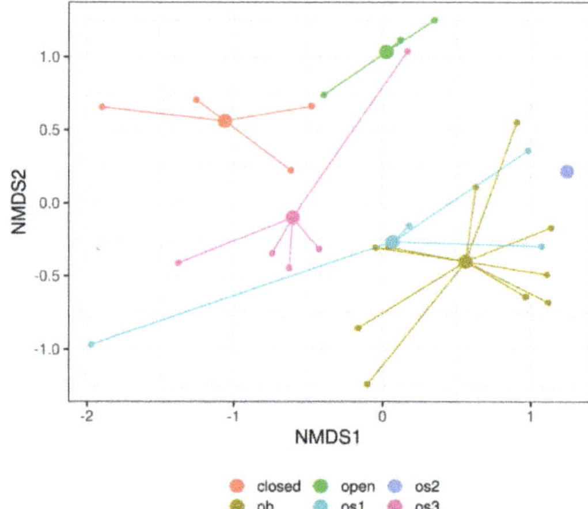

Figure 4. NMDS of the Kulczynski distance matrix among habitats located on the Psel riverbanks and oxbow lakes, classified according to plant overgrowth. ob, os1–3: as in Figure 2, open: shores of lakes with meadow vegetation, closed: shores of lakes covered with trees and shrubs.

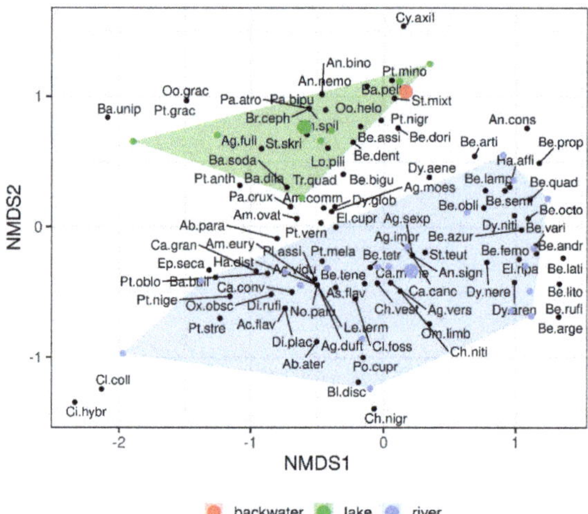

Figure 5. Species superimposed on the NMDS plain of the Kulczynski distance matrix among stations located on the riverbanks of the Psel, the shores of lakes, and in backwaters. Code abbreviations of the species are presented in Table 1.

The proportion of riparian and littoral species in the various habitats of riverbanks and shore lakes is significantly different. The proportion of specialized riparian species decreased in succession from the open banks of the river through the three stages of their overgrowth and to the shores of floodplain lakes (Figure 8). Littoral species prevailed in the shores of lakes, but showed tolerance to habitats on the banks of the river (Figure 8). The abundance of riparian species clearly decreased on the shores of floodplain lakes, while that of the littoral species decreased on the open riverbanks (Figure 9).

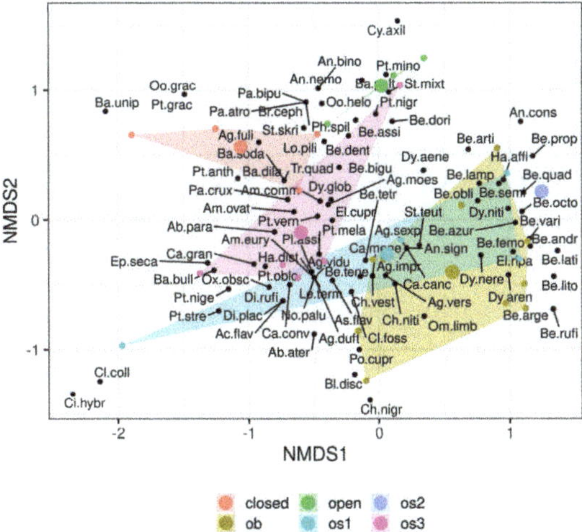

Figure 6. Species superimposed on the NMDS plain of the Kulczynski distance matrix among habitats located on the riverbanks of the Psel, oxbow lakes, and backwaters, classified according to vegetation overgrowth (as in Figures 2 and 4). Code abbreviations of the species are given in Table 1.

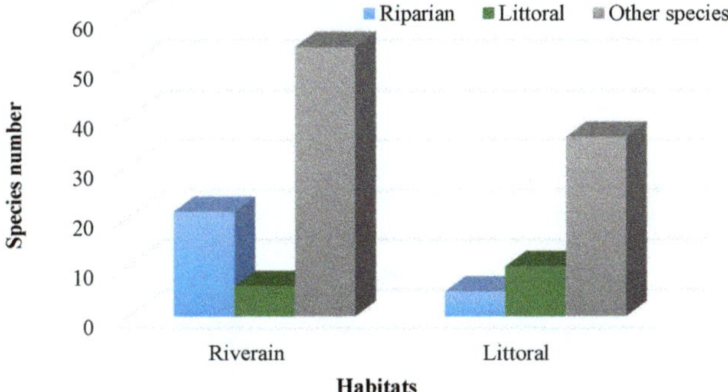

Figure 7. Number of riparian, littoral, and other species on the banks of the Psel River and the shores of floodplain lakes.

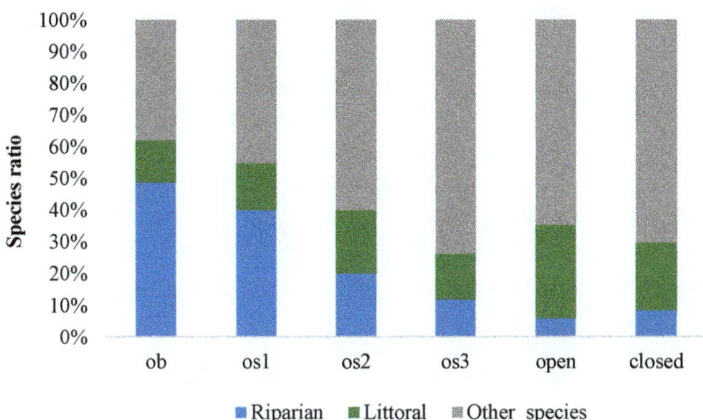

Figure 8. The proportion of riparian, littoral, and other species in the habitats of the Psel river and floodplain lakes. ob, os1–3, open, closed as in Figures 2 and 4.

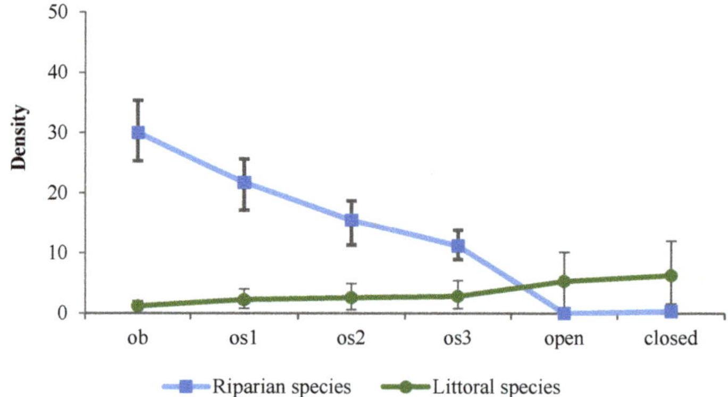

Figure 9. Density of populations of riparian and littoral species in open bars and the stages of vegetation overgrowth of the Psel riverbanks and the shores of floodplain lakes. ob, os1–3, open, closed as in Figures 2 and 4.

4. Discussion

Due to the constant drift of river channels within their floodplains, the landscape of river valleys is highly dynamic. The main factor determining the dynamics of channel processes and the transformation of shorelines is flood [58,59]. Flooding, regularly repeated on a geological time scale, constantly changes the configuration of the main elements of floodplains. The meanders provide a multitude of habitats that support multiple species with variable life strategies. The diversity of species along riverbanks is associated with a mosaic of habitats [60]. Fragmentation of the habitat also determines the structural organization of beetle assemblages, which usually show a spotted structure [35,61].

Despite the fact that the riparian species are focused on the production of hydroecosystems [13,15,62], and floodplain water bodies, especially shallow ones, are highly productive ecosystems, the structure of the ground beetle assemblages on the banks is determined not so much by the presence of food objects, but by the type (lotic or lentic) and the quality of habitats.

Among the riparian species, a highly specialized group (stenotopic) stands out as adapted to the conditions of river banks [17,20,43,46,63–65]. The riparian species are mainly winged forms that can spread along the river, small in size and with a flattened body [16,66–68].

According to the NMDS results, the banks of the river and the shores of the floodplain lakes clearly differ in the species composition of ground beetles (Figures 5 and 6). The assemblage of ground beetles on the banks of the Psel river contains 21 riparian species (Table 1). Among them, 11 species of the genus *Bembidion* predominate, as also noted by other researchers [36,68,69]. Specifically, in the open banks on the flat Hase River (Germany) 12 stenotopic species were noted [39], nine of which were also recorded on the open banks of the Psel River: *Asaphidion flavipes, Elaphrus riparius, Bembidion articulatum, Bembidion femoratum, Bembidion litorale, B. varium, Dyschirius arenosus, D. aeneus,* and *Omophron limbatum.*

Many researchers have noted that riparian species are good indicators of habitat quality and particularly in the structure and quality of alluvial sediments [70,71]. For example, *E. riparius* and *Bembidion semipunctatum* are indicators for sand bars less covered by vegetation, unvegetated, or with mud; *Dyschirius thoracicus* and *Bembidion litorale* are indicators for open sand bars; and *Bembidion cruciatum* and *B. femoratum* are characteristic of less vegetated or unvegetated cobble, gravel, and sand bars [71].

Hydrotechnical transformations of riverbeds and allochthonous organic and inorganic pollution significantly increased the trophic status of most rivers and triggered plant coverage development on the earlier open river sediments. These phenomena have become global. The above factors and the lack of floods lead to development of monotypic conditions along the river banks, which, in turn, lead to changes in the structure of ground beetle communities [27,72,73]. Most researchers confirm a significant reduction in the number of riparian species [22,23,39,74], many of which are classified as endangered [20,24–27].

As our results have shown, the expansion of vegetation on the open banks of the river significantly affects the species composition and spatial distribution of the riparian ground beetles (Figure 6). The number and density of riparian species decreases with increasing development of the vegetation cover (Figures 8 and 9).

Among 21 riparian species recorded on the banks of the Psel river, only *A. flavipes*, *D. arenosus*, *D. aeneus*, *B. tetracolum*, and *B. varium* tend to spread to the littoral of floodplain lakes. This ability has also been noted for *E. riparius*, *Bembidion articulatum*, and *O. limbatum*, which actively inhabited the open shores of the floodplain lake after its connection with the river was restored [39]. This is important and confirms the tolerance of these species to the conditions of floodplain water bodies.

Most studies distinguish a group of riparian species, the composition of which varies depending on the quality of the banks and the type of river (plain or mountain). There are few studies on species that prefer stagnant water bodies [38,40,41] and analyzing arthropods of river banks and their floodplain lakes in comparative way [39,42,61]. For example, Šustek [42], studying the banks of the Danube and Morava rivers and their lakes, identified 21 ripal species of ground beetles (pp. 136–138), but analyzed their preference for the banks of standing and flowing water. Indeed, the author exclusively analyzed the banks with thickets of *Phragmites australis*, which predetermined the low number of riparian species recorded [42].

During our research, 51 species of ground beetles were recorded on the banks of the floodplain lakes of the Psel river, among which 10 preferred the banks of floodplain lakes (Table 1). The littoral species identified by us were also noted on different types of shores of the floodplain lakes of the Dnieper [40]. Apparently, the possibilities of riparian and littoral species spreading beyond the floodplain are limited. According to Lott [38], no riparian species were recorded on the shore of the studied ponds in England, with the exception of the littoral species *Badister dilatatus*.

Thus, most riparian species are associated with the open banks of flowing water bodies, which makes them extremely vulnerable in the rivers with channels changed by man and lacking their natural characteristics. Riparian species of riverbanks are as a rule not found on the banks of floodplain lakes. They are also rare in the sections of river where riverbank sediments are covered with vegetation. A number of species tolerant to overgrowth, for example, *A. flavipes*, *E. riparius*, *B. articulatum*, *B. tetracolum*, *B. varium*, *D. arenosus*, and *D. aeneus*, are also found on the banks of lakes, especially if they are situated close to the river bed [39,61,75,76]. Obviously, these species use the shores of floodplain lakes as fodder areas. However, the shores of lakes for them are not fully adequate habitats.

5. Conclusions

The banks of rivers and floodplain lakes are different habitats, and different assemblages of ground beetles are formed within them. A total of 95 species of ground beetles were found, of which 81 species were found on the riverbanks of the Psel and 51 species were found on the shores of floodplain lakes. Based on the analysis of their spatial distribution, 21 species that prefer the riverbanks and 10 species that prefer the shores of lakes were identified.

Our findings confirm that typically riparian species are associated with open riverbanks, though some species also occurred on the shores of floodplain lakes. Changes in the quality of banks and overgrowth with vegetation due to river pollution leads to a critical reduction in the habitat of riparian species, making them extremely vulnerable. Moreover, since the shores of the floodplain water bodies are only an optional habitat for a sub-set of species, they cannot be considered as refugia for riparian ground beetles. This should be taken into account when developing measures for the renaturalization of rivers and their banks.

Author Contributions: Conceptualization, M.K.-B., R.B.; methodology, M.K.-B., R.B.; software and validation Y.D., R.B.; formal analysis, R.B., M.K.-B.; investigation, M.K.-B., R.B.; resources, M.K.-B., R.B.; data curation, M.K.-B., R.B.; writing—original draft preparation, M.K.-B., R.B., and Y.D.; writing—review and editing, M.K.-B., R.B., Y.D., M.F. and W.S.; visualization, M.K.-B., R.B., and Y.D.; supervision, M.K.-B., R.B. All authors have read and agreed to the published version of the manuscript.

Funding: This work was financially supported by National Academy of Science of Ukraine and Ministry of Science and Higher Education in Poland, within the statutory research of particular scientific units under subvention for science program.

Acknowledgments: The authors are much obliged to two anonymous reviewers for very valuable comments and linguistic revision of the manuscript.

Conflicts of Interest: The authors declare no conflict of interest.

References

1. Petts, G.E.; Amoros, C. (Eds.) *Fluvial Hydrosystems*; Chapman and Hall: London, UK, 1996.
2. Robinson, D.T.; Tockner, K.; Wad, J.V. The fauna of dynamic riverine landscapes. *Freshw. Biol.* **2002**, *47*, 661–677. [CrossRef]
3. Ward, J.V. Riverine landscapes: Biodiversity patterns, disturbance regimes, and aquatic conservation. *Biol. Conserv.* **1998**, *83*, 269–278. [CrossRef]
4. Naiman, R.J.; Décamps, H.; McClain, M.E. *Riparia: Ecology, Conservation, and Management of Streamside Communities*; Elsevier Academic Press Inc.: Cambridge, MA, USA, 2005.
5. Sabo, J.L.; Sponseller, R.; Dixon, M.; Gade, K.; Harms, T.; Heffernan, J.; Jani, A.; Katz, G.; Soykan, C.; Watts, J.; et al. Riparian zones increase regional species richness by harboring different, not more, species. *Ecology* **2005**, *86*, 56–62. [CrossRef]
6. UNECE. River Convention. In *Convention on the Protection and Use of Transboundary Watercourses and International Lakes*; Vlaams Parlement, Stuk 263 (No. 1); United Nations Economic Commission for Europe Geneva: Genève, Switzerland, 1992.
7. Ward, J.V.; Tockner, K.; Schiemer, F. Biodiversity of floodplain river ecosystems: Ecotones and connectivity. *Regul. Rivers Res. Mgmt.* **1999**, *15*, 125–139. [CrossRef]
8. Boscaini, A.; Franceschini, A.; Maiolini, B. River ecotones: Carabid beetles as a tool for quality assessment. *Hydrobiologia* **2000**, *422*, 173–181. [CrossRef]
9. Naiman, R.J.; Décamps, H. (Eds.) *The Ecology and Management of Aquatic Terrestrial Ecotones*; Pearl River: New York, NY, USA; Parthenon, Greece, 1990.
10. Ward, J.V.; Wiens, J.A. Ecotones of riverine ecosystems: Role and typology, spatio-temporal dynamics, and river regulation. In *Fish and Land/Inland Water Ecotones—The Need for Integration of Fisheries Science, Limnology and Landscape Ecology*; Zalewski, M., Thorpe, J.E., Schiemer, F., Eds.; Pearl River: New York, NY, USA; Parthenon, Greece, 1999.
11. Ballinger, A.; Mac Nally, R.; Lake, P.S. Immediate and longer-term effects of managed flooding on floodplain invertebrate assemblages is south-eastern Australia: Generation and maintenance of a mosaic landscape. *Freshw. Biol.* **2005**, *50*, 1190–12025. [CrossRef]
12. Stanford, J.A.; Lorang, M.S.; Hauer, F.R. The shifting habitat mosaic of river ecosystems. *Verh. Internat. Verein. Limnol.* **2005**, *29*, 123–136. [CrossRef]
13. Hering, D.; Platcher, H. Riparian ground beetles (Coleoptera, Carabidae) preying on aquatic invertebrates: A feeding strategy in alpine floodplains. *Oecologia* **1997**, *111*, 261–270. [CrossRef]
14. Sabo, I.L.; Power, M.E. River-Watershed Exchange: Effects of Riverine Subsidies on Riparian Lizards and Their Terrestrial Prey. *Ecology* **2002**, *83*, 1860–1869. [CrossRef]
15. Paetzold, A.; Schubert, C.J.; Tockner, K. Aquatic-terrestrial linkages along a braided –river: Riparian arthropods feeding on aquatic insects. *Ecosystems* **2005**, *8*, 748–759. [CrossRef]
16. Jähnig, S.C.; Brunzel, S.; Gacek, S.; Lorenz, A.W.; Hering, D. Effects of re-braiding measures on hydromorphology, floodplain vegetation, ground beetles and benthic invertebrates in mountain rivers. *J. Appl. Ecol.* **2009**, *46*, 406–416. [CrossRef]
17. Bonn, A.; Kleinwächter, M. Microhabitat distribution of spider and ground beetle assemblages (Araneae, Carabidae) on frequently inundated river banks of the River Elbe. *Z. Ökolgie Nat.* **1999**, *8*, 109–123.

18. Manderbach, R.; Hering, D. Typology of riparian ground beetle communities (Coleoptera, Carabidae, Bembidion spec.) in Central Europe and adjacent areas. *Arch. Hydrobiol.* **2001**, *152*, 583–608. [CrossRef]

19. Lambeets, K.; Lewylle, I.; Bonte, D.; Maelfait, J.-P. The spider fauna (Aranea) from gravel banks along the Common Meuse: Riparian assemblages and species conservation. *Nieuwsbr. Belg. Arachnol. Ver.* **2007**, *22*, 16–30.

20. Andersen, J.; Hanssen, O. Riparian beetles, a unique, but vulnerable element in the fauna of Fennoscandia. *Biodivers. Conserv.* **2005**, *14*, 3497–3524. [CrossRef]

21. Eyre, M.D.; Lott, D.A. *Invertebrates of Exposed Riverine Sediments*; R&D Technical Report W11; National Rivers Authority: Marlow, UK, 1996.

22. Manderbach, R.; Reich, M. Auswirkungen großer Querbauwerke auf die Laufkäferzönosen (Coleoptera, Carabidae) von Umlagerungsstrecken der Oberen Isar. *Arch. Hydrobiol. Suppl.* **1995**, *101*, 573–588. [CrossRef]

23. Niemeyer, S.; Reich, M.; Plachter, H. Ground beetle communities (Coleoptera: Carabidae) on the banks of two rivers in the eastern Carpathians, the Ukraine. *Verhandl. Gesellsch. Ökol.* **1997**, *27*, 365–372.

24. Marggi, W.A. Faunistik der Sandlaukäfer und Laukäfer der Schweiz (Cicindelidae and Carabidae) Coleoptera. Teil 1/Text. *Doc. Faun. Helv.* **1992**, *13*, 463.

25. Trautner, J.; Müller-Motzfeld, G.; Bräunicke, M. Rote Liste der Sandlaufkäfer und und Laukäfer Deutschlands (Coleoptera: Cicindelidae et Carabidae). 2. Fassung. Stand Dezember. *Nat. Landsch.* **1996**, *29*, 261–273.

26. Bräunicke, M.; Trautner, J. Die Ahlenläufer-Arten der Bembidion-Untergattungen Bracteon und Odontonium—Verbreitung, Bestandssituation, Habitate und Gefährdung charakteristischer Flussaue-Arten in Deutschland. *Angewande Carabidol. Suppl.* **1999**, *1*, 79–94.

27. Sadler, J.P.; Bell, D.; Fowles, A. The hydroecological controls and conservation value of beetles on exposed riverine sediments in England and Wales. *Biol. Cons.* **2004**, *118*, 41–56. [CrossRef]

28. Dieterich, M. Methods and preliminary results from a study on the habitat functions of the gravel bar interior in alluvial floodplains. *Verhandl. Gesellsch. Ökol.* **1996**, *26*, 363–367.

29. Antvogel, H.; Bonn, A. Environmental parameters and microspatial distribution of insects: A case study of carabids in an alluvial forest. *Ecography* **2001**, *24*, 470–482. [CrossRef]

30. Adis, J.; Junk, W.J. Terrestrial invertebrates inhabiting lowland river floodplains of Central Amazonia and Central Europe: A review. *Freshw. Biol.* **2002**, *47*, 711–731. [CrossRef]

31. Bonn, A.; Hagen, K.; Wohlgemuth-von Reiche, D. The significanse of flood regimes for carabid beetle and spider communities in riparian habitats–a comparison of three major rivers in Germany. *River Res. Appl.* **2002**, *18*, 43–64. [CrossRef]

32. Gerisch, M.; Schanowski, A.; Figura, W.; Gerken, B.; Dziock, F.; Henle, K. Carabid beetles (Coleoptera, Carabidae) as indicators of hydrological site conditions in floodplain grasslands. *Int. Rev. Hydrobiol.* **2006**, *91*, 326–340. [CrossRef]

33. Gobbi, M.; Fontaneto, D. Biodiversity of ground beetles (Coleoptera: Carabidae) in different habitats of the Italian Po lowland. *Agric. Ecosyst. Environ.* **2008**, *127*, 273–276. [CrossRef]

34. Lytle, D.; Poff, N. Adaptation to natural flow regimes. *Trends Ecol. Evol.* **2004**, *19*, 94–100. [CrossRef]

35. Bates, A.J.; Sadler, J.P.; Fowles, A.P. Condition-dependent dispersal of a patchily distributed riparian ground beetle in response to disturbance. *Oecologia* **2006**, *150*, 50–60. [CrossRef]

36. Lambeets, K.; Vandegehuchte, M.; Maelfait, J.-P.; Bonte, D. Understanding the impact of flooding on trait-displacements and shifts in assemblage structure of predatory arthropods on river banks. *J. Anim. Ecol.* **2008**, *77*, 1162–1174. [CrossRef]

37. Paetzold, A.; Yoshimura, C.; Tockner, K. Riparian arthropod responses to flow regulation and river channelization. *J. Appl. Ecol.* **2008**, *45*, 894–903. [CrossRef]

38. Lott, D. Ground beetles and rove beetles be associated with temporary ponds in England. *Freshw. Forum* **2001**, *17*, 40–53.

39. Gunter, J.; Assmann, T. Restoration ecology meets carabidology: Effects of floodplain restitution on ground beetles (Coleoptera, Carabidae). *Biodivers. Conserv.* **2005**, *14*, 1583–1606. [CrossRef]

40. Kirichenko, M.B.; Babko, R.V. Assemblages of ground beetles in the riverside reservoirs of Kyiv. In *Ecological State of the Reservoirs of Kyiv*; Phytocenter: Kyiv, Ukraine, 2005; pp. 68–74. (In Ukrainian)

41. Brauns, M.; Garcia, X.; Pusch, M.T. Potential effects of water-level fluctuations on littoral invertebrates in lowland lakes. *Hydrobiologia* **2008**, *613*, 5–12. [CrossRef]

42. Šustek, Z. Succession of carabid communities in different types of reed stands in Central Europe. *Oltenia. Stud. Comunicări Ştiinţele Nat.* **2010**, *26*, 127–138.
43. Kirichenko, M.B. The Carabid Fauna (Coleoptera, Carabidae) of the River Banks, Lakes Shores and Marshes of the Forest and Forest-Steppe of Eastern Part of the Ukraine. Ph.D. Thesis, Schmalhausen Institute of Zoology NAS of Ukraine, Kiev, Ukraine, 1999. (In Ukrainian).
44. Hůrka, K. *Carabidae of the Czech and Slovak Republics*; Kabourek: Zlín, Czech Republic, 1996; 565p.
45. Fedorenko, D. *Reclassification of World Dyschiriini with a Revision of the Palearctic Fauna (Coleoptea, Carabidae)*; Pensoft, Sofia, Moscow & ST: Sofia, Petersburg, 1996; p. 224.
46. Eyre, M.D.; Luff, M.L.; Phillips, D.A. The ground beetles (Coleoptera, Carabidae) of exposed riverine sediments in Scotland and northern England. *Biodivers. Conserv.* **2001**, *10*, 403–426. [CrossRef]
47. Jongman, R.H.G.; Ter Braak, C.J.F.; Van Tongeren, O.F.R. *Data Analysis in Community and Landscape Ecology*; Cambridge Univ. Press: Cambridge, UK, 1995.
48. Beals, M.L. Understanding community structure: A data-driven multivariate approach. *Oecologia* **2006**, *150*, 484–495. [CrossRef]
49. Shepard, R.N. The analysis of proximities: Multidimensional scaling with an unknown distance function. I. *Psychometrica* **1962**, *27*, 125–140. [CrossRef]
50. Kruskal, J.B. Multidimensional scaling by optimizing goodness of fit to a nonmetric hypothesis. *Psychometrica* **1964**, *29*, 1–27. [CrossRef]
51. Cao, Y.; Williams, W.P.; Bark, A.W. Similarity measure bias in river benthic Aufwuchs community analysis. *Water Environ. Res.* **1997**, *69*, 95–106. [CrossRef]
52. Mountford, M.D. An index of similarity and its application to classification problems. In *Progress in Soil Zoology*; Butterworths: London, UK, 1962; pp. 43–50.
53. Faith, D.P.; Minchin, P.R.; Belbin, L. Compositional dissimilarity as a robust measure of ecological distance. *Vegetatio* **1987**, *69*, 57–68. [CrossRef]
54. Hwang, C.M.; Yang, M.S.; Hung, W.L. New similarity measures of intuitionistic fuzzy sets based on the Jaccard index with its application to clustering. *Int. J. Intell. Syst.* **2018**, *33*, 1672–1688. [CrossRef]
55. Wo, J.; Mou, X.X.; Xu, B.D.; Xue, Y.; Zhang, C.L.; Ren, Y.P. Interannual changes in fish community structure in the northern part of the coastal waters of Jiangsu Province, China in spring. *Chin. J. Appl. Ecol.* **2018**, *29*, 285–292. [CrossRef]
56. Oksanen, J.; Blanchet, F.G.; Friendly, M.; Kindt, R.; Legendre, P.; McGlinn, D.; Minchin, P.R.; O'Hara, R.B.; Simpson, G.L.; Solymos, P.; et al. Vegan: Community Ecology Package. R package version 2.4-6. 2018. Available online: https://CRAN.R-project.org/package=vegan (accessed on 13 September 2019).
57. Wickham, H. *Ggplot2: Elegant Graphics for Data Analysis*; Springer: New York, NY, USA, 2016.
58. Middleton, B. *Flood Pulsing and Disturbance Dynamics*; John Wiley & Sons Ltd.: New York, NY, USA, 1999.
59. Junk, W.J. Flood pulsing and linkages between terrestrial, aquatic, and wetland systems. *Verh. Internat. Verein. Limnol.* **2005**, *29*, 11–38. [CrossRef]
60. Ward, J.; Tockner, K. Biodiversity: Towards a unifying theme for river ecology. *Freshw. Biol.* **2001**, *46*, 807–819. [CrossRef]
61. Gerken, B.; Dörfer, K.; Buschmann, M.; Kamps-Schwob, S.; Berthelmann, J.; Gertenbach, D. Composition and distribution of carabid communities along rivers and ponds in the region of Upper Were (NW/NDS/FRG) with respect to protection and management of a floodplain ecosystem. *Reg. Riv. Res Mngm.* **1991**, *6*, 313–320. [CrossRef]
62. Henshall, S.E.; Sadler, J.P.; Hannah, D.M.; Bates, A.J. The role of microhabitat and food availability in determining riparian invertebrate distributions on gravel bars: A habitat manipulation experiment. *Ecohydrology* **2011**, *4*, 512–519. [CrossRef]
63. Mazzei, A.; Bonacci, T.; Zetto, T.; Brandmayr, P. La carabidofauna dell'ecotopo fluviale del crati (Cosenza, Italia) (Coleoptera Carabidae). *Nat. Sicil.* **2010**, *34*, 187–199.
64. Naiman, R.J.; Décamps, H. The ecology of interfaces: Riparian zones. *Ann. Rev. Ecol. Syst.* **1997**, *28*, 621–658. [CrossRef]
65. Ward, J.V.; Tockner, K.; Arscott, D.B.; Claret, C. Riverine landscape diversity. *Freshw. Biol.* **2002**, *47*, 517–540. [CrossRef]
66. Andersen, J. Ecomorphological adaptations of Riparian Bembidiini species (Coleoptera: Carabidae). *Ecol. Generalis* **1985**, *11*, 41–46. [CrossRef]

67. Desender, K.; Turin, H. Loss of habitats and changes in the composition of the ground and tiger beetle fauna in four west European countries since 1950 (Coleoptera: Carabidae). *Biol. Conserv.* **1989**, *48*, 277–294. [CrossRef]

68. Desender, K. Ecomorphological adaptations of riparian carabid beetles. In *Verhandelingen Van Het Symposium 'Invertebraten Van België'*; Royal Institute of Natural Sciences: Brussels, Belgium, 1989; pp. 309–314.

69. Desender, K.; Maelfait, J.-P.; Stevens, J.; Allemeersch, L. Loopkevers langs de Grensmaas (carabid beetles along the Common Meuse). *Jaarb. LIKONA* **1993**, *3*, 41–49.

70. Van Looy, K.; Vanacker, S.; Jochems, H.; de Blust, G.; Dufrêne, M. Ground beetle habitat templets and riverbank integrity. *River Res. Appl.* **2005**, *21*, 1133–1146. [CrossRef]

71. Jachertz, H.; Januschke, K.; Hering, D. The role of large-scale descriptors and morphological status in shaping ground beetle (Carabidae) assemblages of floodplains in Germany. *Ecol. Indicat.* **2019**, *103*, 124–133. [CrossRef]

72. Nilsson, C. Conservation management of riparian communities. In *Ecological Principles of Nature Conservation*; Hansson, L., Ed.; Elsevier: Amsterdam, The Netherlands, 1991; pp. 352–372.

73. Ward, J.V.; Stanford, J.A. Ecological connectivity in alluvial river ecosystems and its disruption by flow regulation. *Reg. Rivers Res. Mgmt.* **1995**, *11*, 105–119. [CrossRef]

74. Kirichenko, M.; Babko, R. The spatial population distribution of Omophron limbatum Fabricius 1777 (Coleoptera, Carabidae) in the condition of regulated rivers. *Teka Kom. Ochr. Kszt. Srod. Przyr.* **2009**, *6*, 129–137.

75. Januschke, K.; Brunzel, S.; Haase, P.; Hering, D. Effects of stream restorations on riparian mesohabitats, vegetation and carabid beetles. *Biodivers. Conserv.* **2011**, *20*, 3147–3164. [CrossRef]

76. Kirichenko, M.; Łagód, G.; Majerek, D.; Franus, M.; Babko, R. The effect of landscape on the diversity in urban green areas. *Ecol Chem Eng S* **2017**, *24*, 613–625. [CrossRef]

Article

Spatial Changes in Invertebrate Structures as a Factor of Strong Human Activity in the Bed and Catchment Area of a Small Urban Stream

Robert Czerniawski [1,2,*], Łukasz Sługocki [1,2], Tomasz Krepski [1,2], Anna Wilczak [1] and Katarzyna Pietrzak [1]

[1] Department of Hydrobiology, University of Szczecin, PL-70-017 Szczecin, Poland; lukasz.slugocki@usz.edu.pl (Ł.S.); tomasz.krepski@usz.edu.pl (T.K.); ania.wilczak@wp.pl (A.W.); kasia930914@poczta.onet.pl (K.P.)

[2] Centre of Molecular Biology and Biotechnology, University of Szczecin, PL-70-017 Szczecin, Poland

* Correspondence: robert.czerniawski@usz.edu.pl

Received: 3 February 2020; Accepted: 19 March 2020; Published: 24 March 2020

Abstract: The threats to small urban streams lead to a decrease in their water quality and dysregulate their ecological balance, thereby affecting the biodiversity and causing degradation of indicators that determine the ecological potential. The aim of our study was to determine the impact of abiotic conditions induced by intensive human activity on the community structures of invertebrates (zooplankton and macroinvertebrates) in the small urban stream Bukówka in the Szczecin agglomeration (NW Poland). This stream exhibits the same characteristics as a large river, in which the mass of live organic matter increases with their length. The composition of invertebrates (zooplankton and macroinvertebrates) was strongly influenced by the changes caused by humans in the stream bed. The construction of small reservoirs and bed regulation in this small urban streams had a similar effect on the quality of the water and ecological potential as in large rivers, but at a lower scale.

Keywords: macroinvertebrates; zooplankton; water quality; anthropogenic impact; stream ecology; riverbed regulation

---

1. Introduction

Small urban watercourses are a common and desirable component of the landscape of agglomerations [1]. However, urban running waters are intensively affected by humans, most often due to the pollution caused by them and their constant interference in the shape and continuity of the riverbed [1]. These threats lead to a decrease in the water quality of streams and rivers and dysregulate the ecological balance, thereby affecting the biodiversity and causing a reduction in the value of indicators that determine the ecological potential [2]. In addition, these exert harmful effects on other basins. Urban streams flow into larger rivers or lakes, often affecting the abiotic and biotic conditions of the water bodies [3]. These recipients can constitute sites of permanent existence of resident fauna or act as a transitional shelter to native or migratory fauna [3,4]. They might also function as a refuge for fauna that colonize the whole bed from the headwaters to the mouth.

Micro(zoo)plankton and macroinvertebrates are good indicators of the physical and chemical influence on the running water bed [2,5–7]. Any spatial change in the invertebrate composition of running waters reflects the physical and chemical effects induced by dam impoundments and bed regulation. Regulated, dammed, and polluted streams or ditches are characterized by smaller numbers and lower diversity indexes of macroinvertebrate taxa compared to forest streams [2,8,9]. On the other hand, zooplankton almost always respond positively to dam impoundments and new conditions that are conducive to their development and reproduction [10]. Among the studies on the presence of micro- and macroinvertebrates in small streams, most have focused on a common topic—the impact of environmental factors (current velocity, water retention time, inorganic nutrients concentration, and other physico-chemical compounds) on their taxonomy and abundance [11–15]. Small urban streams undergo very rapid environmental changes, as these are under the strong influence of local factors that disrupt their functioning, such as human activity in the catchment and in the bed or rapid water runoff from the urban surface.

The urban impact on the spatial composition of invertebrates can be studied via small streams, such as Bukówka located in the Szczecin agglomeration (NW Poland). This watercourse is subject to intensive human activity over its entire length, such as regulation, bed damming, and pollution. The water of the stream is characterized by a high concentration of conductivity, total dissolved solids (TDS), and chlorides; these being very common parameters for indicating human activity. In addition, the lower section of the Bukówka stream is exposed to water inlet from the recipient, the Oder river, within the estuary.

The aim of our study was to determine the impact of abiotic conditions induced by intensive human activity on the structures of invertebrates in this small human-mediated urban stream.

2. Methods

2.1. Study Area

This study was performed in the Bukówka stream (and drainage of the lower stretch of the Oder river, NW Poland) (Figure 1). Bukówka is a regulated and non-natural stream flowing in the Szczecin agglomeration. It is the left tributary of the estuary of the Oder river. Bukówka is 9-km long, has a catchment area of 69 km^2, and is characterized by a mean slope of 0.075%. The stream provides water to the recreational, strongly eutrophicated reservoir "Słoneczne lake", also called a pond; therefore, a dam and its impoundments are located along the stream. The total area of this reservoir is 7 ha, and its maximum depth is 1.5 m.

Each section of the stream was regulated and channelized. It means that the bed of the stream was extremely changed by humans. The bed of the upper section (Site 1—U1; Site 2—U2) was covered by sand and dead organic matter. The sediments of the reservoir—"Słoneczne Lake" (POND)—were marshy and also contained dead organic matter. The bottom of the reservoir was densely covered by emerging macrophytes (dominants: *Glyceria maxima*, *Phragmites australis*, and *Sparganium erectum*). Due to the abundance of phytoplankton in the reservoir and low visibility, there was no occurrence of submerged macrophytes. The dominant components of the sediments in the middle section (Site 3—M3; Site 4—M4) were sand and gravel, together with a small amount of organic matter. No macrophytes were observed in this section. The lower downstream sections (Site 5—D5; Site 6—D6) were characterized by a high amount of dead organic matter at the bottom, without the presence of minerals. The riparian zone of the bed in the lower section was covered by emerging macrophytes (dominants: P. communis, G. maxima, and S. erectum). Moreover, this section contained many small floodplains and slack water areas.

The percentage of land use in the catchment was calculated using the Corine Land Cover inventory. The land use in the catchment was separately estimated in the total catchment area and the buffer zone, which was located within 100 m of the stream shoreline. For our calculations, we used the QGIS Wien software [2,7,8] (QGIS Development Team). The land use of the Bukówka total catchment is occupied by agricultural areas (52%), artificial surfaces (44%), seminatural areas (3%), and water bodies (<1) (GIS data). In the 100 m buffer zone the catchment is mostly covered by artificial surfaces (94%) and definitely less by seminatural areas (5%) and water bodies (<1). Along the entire stream there are a numerous rainwater runoffs.

No field sampling permits with regard to the site locations were needed.

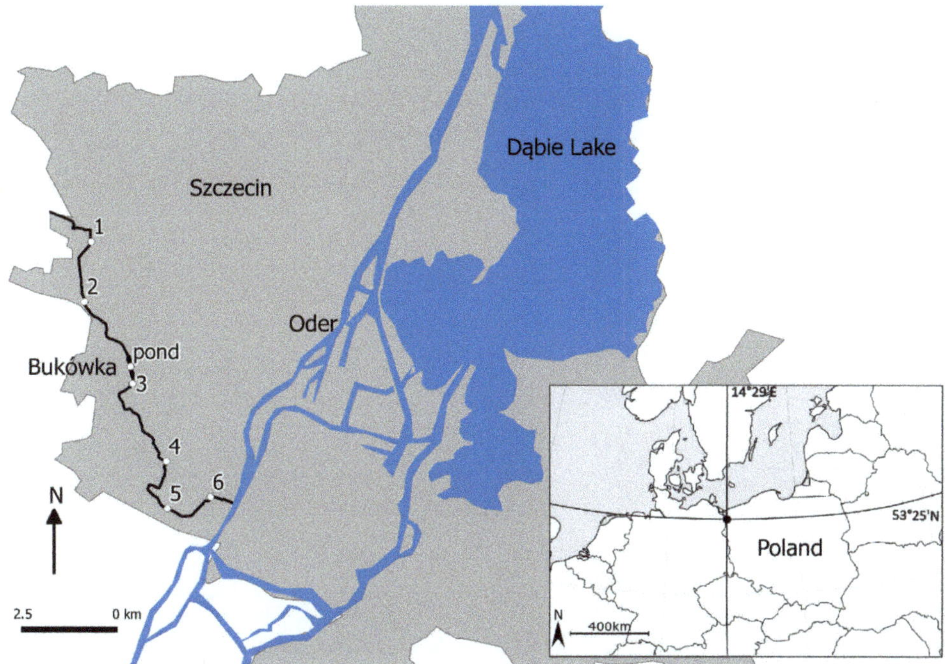

Figure 1. Study area. Numbers on the Bukówka stream indicate the sampling sites.

2.2. Sampling Methods

2.2.1. Environmental Factors

Environmental factors were measured monthly from March to October in 2016 ($n = 8$) in the same day when the invertebrate samples were collected. Temperature, dissolved oxygen, pH, and conductivity, as well as the content of chlorophyll *a*, total dissolved solids (TDS), and chloride were measured in situ using the Hydrolab DS5 multiparameter probe (USA). In addition, water velocity, width, and depth were measured at each site using an OTT electromagnetic water flow sensor (Germany). The values of the measured environmental parameters are reported in Table 1.

Table 1. Mean ± SD of environmental variables at the examined sites in the Bukówka stream. Temp—temperature; Cond—conductivity.

Site	Temp	Dissolved Oxygen	Ph	Cond	Chlorophyll A	Chloride	Total Dissolved Solids	Width	Depth	Current Velocity
	($^{\circ}$C)	(mg L^{-1})		(μS cm^{-1})	(mg L^{-1})	(mg L^{-1})	(mg L^{-1})	(m)	(m)	(m s^{-1})
U1	14.0 ± 7.3	7.1 ± 2.8	8.5 ± 0.5	866 ± 118	3.4 ± 1.7	2141 ± 3860	0.5216 ± 0.1174	0.88 ± 0.48	0.10 ± 0.05	0.030 ± 0.020
U2	13.8 ± 7.2	5.8 ± 3.0	8.5 ± 0.3	792 ± 161	5.3 ± 3.2	2150 ± 4128	0.4932 ± 0.1147	0.74 ± 0.41	0.12 ± 0.07	0.053 ± 0.037
POND	17.4 ± 7.6	8.9 ± 3.5	8.8 ± 0.4	536 ± 306	193.0 ± 108.9	1769 ± 2933	0.4217 ± 0.1285	-	-	-
M3	16.7 ± 6.1	7.1 ± 3.2	8.6 ± 0.4	608 ± 312	222.9 ± 122.6	1530 ± 2328	0.4277 ± 0.1339	0.94 ± 0.41	0.10 ± 0.04	0.285 ± 0.145
M4	15.0 ± 6.1	8.3 ± 2.0	8.6 ± 0.3	727 ± 160	81.1 ± 42.8	1980 ± 3359	0.4623 ± 0.1033	1.63 ± 0.47	0.11 ± 0.02	0.392 ± 0.166
D5	14.2 ± 6.0	7.1 ± 1.9	8.5 ± 0.3	1026 ± 231	39.1 ± 17.8	1669 ± 2224	0.6504 ± 0.1373	3.08 ± 0.30	0.18 ± 0.05	0.265 ± 0.058
D6	15.4 ± 7.4	7.9 ± 3.2	8.5 ± 0.4	903 ± 246	42.6 ± 25.5	1951 ± 3034	0.6392 ± 0.0636	4.19 ± 1.04	0.73 ± 0.19	0.038 ± 0.033

2.2.2. Zooplankton

Zooplankton were collected monthly from March to October in 2016 ($n = 8$). At each site a 50 L water sample was collected from the surface of the river current using a 10-L bucket. The water was filtered through a plankton net having a mesh of 30 μm. The samples were concentrated to 150 mL and fixed in a solution of 4%–5% formalin. Then, they were divided into ten subsamples measuring 3 mL each, and the contents of each were counted using a Sedgewick–Rafter counting chamber. A Nikon Eclipse 50i microscope was used for the identification of organisms. Species were identified using the keys described by Radwan [16] and Rybak and Błędzki [17]. Rotifers were divided into two categories according to their habitat preference—pelagic species (plankton) and benthic, epiphytic, and epilithic species attached to the substratum or otherwise known altogether as benthic species [16].

The sites for zooplankton sampling were selected in the lotic, free-flowing sections of Bukówka. Two sites were located in the upper section (U1, U2) before the reservoir, two in the middle section (M3, M4) below the reservoir, and two in the lower-downstream section (D5, D6) of the stream. In addition, a control site was chosen in the open water zone of the reservoir (POND). In the lower section, there was the inlet of waters from the recipient, the Oder river.

The Lagrangian scheme was followed in the Bukówka stream, according to which the cross sections were sampled in a downstream sequence with the zooplankton sampling interval approximating the time of travel between the sites. The sampling sites of Bukówka were chosen taking into account the following: (1) influence of the reservoir, (2) influence of the water inlet from the Oder river, (3) differences in environmental conditions, and (4) ease of access.

2.2.3. Macrozoobenthos—Macroinvertebrates

Macroinvertebrates were collected twice in 2016—in April and October. Sampling was conducted using a hydrobiological net, by scraping a fragment sized 0.5 m^2 from the bottom. The material was preserved in 96% ethyl alcohol. In the laboratory, the organisms were separated and identified to the family level before they were counted, excluding Oligochaeta, which was counted at the same level. A specialist key was used for the identification of the organisms, and their abundance was expressed in ind·m^2.

The following indexes and metrics were used to estimate the quality of water: BMWP-PL (Biomonitoring Working Party for Polish Standards), Margalef biodiversity index, ASPT (Average Score Per Taxon), Shannon–Weaver biodiversity index, TBI (Trent Biotic Index), GOLD (% of Gastropoda, Oligochaeta, and Diptera in the total abundance), and %DIPTERA (% of Diptera in the total abundance) [18]. The Polish BMWP-PL (Biological Monitoring Working Party) index is based on the British BMWP, where the list of indicator taxa (families) has been adapted to Polish standards. Each family from the BMWP-PL list is assigned. Each family on the list is assigned a score, indicating the sensitivity of the taxon. The higher the score, the more sensitive the taxon is to pollution. The BMWP-PL score is obtained by summing the number of taxa points found in the sample. Value >100 = high water quality, 70–99 = good water quality, 40–69 = moderate water quality, 10–39 = poor water quality, and <10 = bad water quality. The Margalef biodiversity index is an index calculated from the formula: $D = S/\log N$ where S—the number of taxa in the family rank, and N—total number of individuals from all species. Values >5.5 indicated a high water quality, 4.0–5.49 = good water quality, 2.50–3.99 = moderate water quality, 1.0–2.49 = poor water quality, and <1.0 = bad water quality.

2.2.4. Statistical Analyses

Taxonomical similarity of zooplankton and macroinvertebrates between the sites was determined by calculating the Jaccard's index. The Mann–Whitney U test ($p < 0.05$) was used for verifying the significance of differences in the number of species of total zooplankton and the abundance of each zooplankton group between the sites ($p < 0.05$). The same test was also used to identify significant differences in the taxa number and density of macroinvertebrates between the examined sections

($p < 0.05$). To illustrate the ordination of the zooplankton groups and macroinvertebrates groups in terms of their abundance with regard to environmental factors, a canonical correspondence analysis (CCA) was used.

It was expected that, as a basin of stagnant water, a reservoir has a large positive effect on the reproduction and development of zooplankton. Hence, when making the CCA between the zooplankton abundance and the environmental factors of the stream as running water, the environmental conditions of the reservoir were not taken into consideration, as they do not characterize the running but standing water and could affect the final results.

3. Results

3.1. Zooplankton

From the upper section of the stream to the lower downstream section, the species number of zooplankton was found to increase (Figure 2). A significantly lower species number was observed at the sites selected in the upper section ($p < 0.05$).

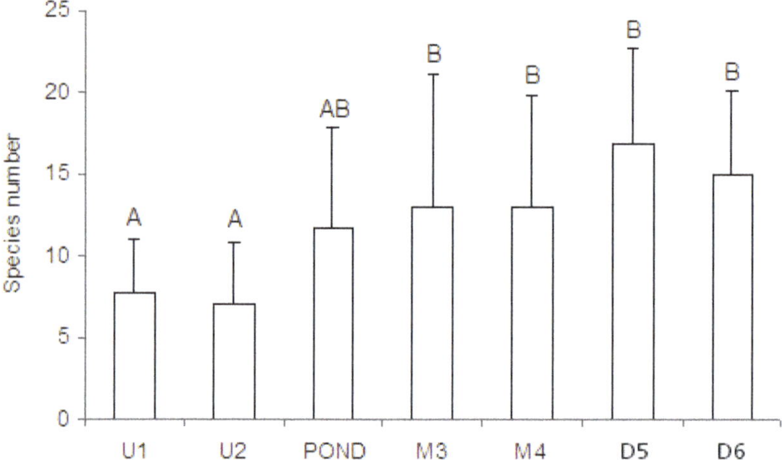

Figure 2. Spatial changes of mean + SD zooplankton species numbers at two sites in the upper section (U1, U2), in POND, at two sites in the middle section (M3, M4), and at two sites in the downstream section (D5, D6) of the Bukówka stream. Different letters on the columns show significant differences between sites ($p < 0.05$).

However, the Jaccard taxonomical similarity coefficient showed that the upper, middle, and lower downstream sections were clearly divided between themselves (Figure 3). This indicated that the taxonomical composition of each section was made up of different species, as described in Table 2.

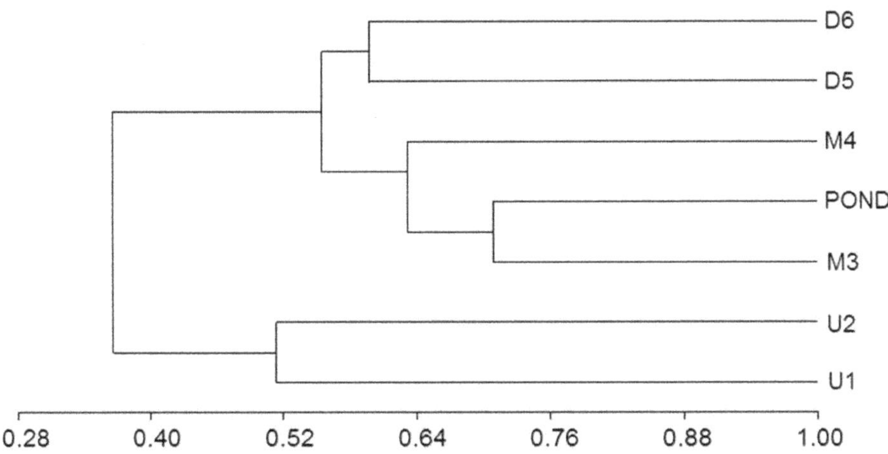

Figure 3. Jaccard taxonomical similarity between sites in the Bukówka stream.

Table 2. Taxonomic composition of zooplankton observed at sites of the Bukówka stream.

Species	U1	U2	POND	M3	M4	D5	D6
			Rotifera				
Anuraeopsis fissa		+	+	+	+	+	+
Asplanchna sieboldi			+	+		+	
Bdelloidea	+	+	+	+	+	+	+
Brachionus angularis			+	+	+	+	+
Brachionus calyciflorus		+	+	+	+	+	+
Brachionus quadridentatus			+	+	+	+	+
Brachionus sp.			+	+	+	+	+
Brachionus rubens			+	+	+	+	+
Brachionus urceolaris					+		
Cephalodella gibba	+	+		+		+	+
Cephalodella sp.	+	+	+	+	+	+	+
Cephalodella ventripes	+	+	+	+	+	+	+
Collotheca mutabilis					+		
Colurella adriatica	+	+	+	+	+	+	+
Colurella sp.	+			+		+	+
Colurella colurus	+	+		+	+	+	+
Colurella uncinata	+	+	+	+	+		
Conochilus unicornis			+	+	+	+	
Euchlanis dilatata	+	+	+		+	+	+
Filinia brachiata			+	+	+	+	+
Filinia longiseta			+	+	+	+	+
Filinia sp.				+			
Filinia terminalis			+	+	+	+	+

Table 2. *Cont.*

Species	U1	U2	POND	M3	M4	D5	D6
Keratella cochlearis cochlearis	+	+	+	+	+	+	+
Keratella cochlearis hispida						+	
Keratella cochlearis tecta	+	+	+	+	+	+	+
Keratella quadrata	+	+	+	+	+	+	+
Keratella testudo				-	+	+	
Keratella testudo gossei						+	
Keratella ticinensis				+	+		+
Keratella valga				+	+		
Lecane bulla	+						+
Lecane closterocerca	+	+	+	+	+	+	+
Lecane flexilis						+	+
Lecane hamata				+	+	+	
Lecane luna		+					
Lecane sp.	+		+		+	+	+
Lepadella acuminata		+			+	+	
Lepadella ovalis	+	+	+	+	+	+	+
Lepadella sp.	+					+	
Monommata longiseta						+	
Mytilina mucronata							+
Notholca acuminata					+	+	+
Notholca squamula	+					+	
Polyarthra dolichoptera			+	+	+	+	+
Polyarthra remata			+	+			
Polyarthra sp.			+	+	+	+	+
Polyarthra vulgaris		+	+	+	+	+	+
Proales sp.				-	+		
Rotifera not identified	+	+	+	+	+	+	+
Synchaeta pectinata				+	+		
Synchaeta sp.			+	+	+	+	+
Taphrocampa sp.	+						
Trichocerca capucina		+					
Trichocerca dixon-nutalli	+	+	+		+	+	+
Trichocerca elongata	+	+	+				
Trichocerca pusilla		+	+	+	+		+
Trichocerca rousseleti						+	
Trichocerca sp.	+		+	+	+	+	+
Trichocerca stylata				+			
Trichocerca weberi		+			+		

Table 2. *Cont.*

			Site				
Species	**U1**	**U2**	**POND**	**M3**	**M4**	**D5**	**D6**
Cladocera							
Acroperus harpae							+
Bosmina longirostris				+		+	+
Ceriodaphnia sp.							+
Chydorus sphaericus		+				+	
Diaphanosoma brachyurum			+				+
Moina micrura							+
Pleuroxus aduncus	+						
Polyphemus pediculus							+
Scaphaloberis mucronata			+	+			+
Cladocera juv	+						
Copepoda							
Eucyclops macruroides			+	+			+
Eucyclops macrurus							+
Eurytemora velox						+	+
Macrocyclops albidus		+					
Thermocyclops oithonoides							+
Nauplii Cyclopoida	+	+	+	+	+	+	+
Copepodid Calanoida						+	+
Copepodid Cyclopoida	+	+	+	+		+	+

All the zooplankton groups contributed to the increase in the abundance of zooplankton in the stream below the reservoir and the slight decrease of abundance in the last site in the middle section (Figure 4). This influence was especially noticeable in the case of pelagic rotifers, the abundance of which was significantly lower in the upper section than the other sections ($p < 0.05$). Further from the reservoir, the abundance of pelagic rotifers was lower. Although not very clear, a similar pattern was observed in the abundance of benthic rotifers.

A different pattern of abundance was observed in the case of crustaceans (at adult and larval stages). Their stream abundance was found to be increased below the reservoir, and then decreased, while at the last site in the downstream section, the abundance increased again (Figure 4). Cladocerans ($p < 0.05$) and adult copepods were characterized by a much higher abundance in the lowest site compared to the other sites. At both sites in the upper section, benthic rotifers were found to be dominant (Figure 5). On the other hand, from the pond to the last site in the lower downstream section, pelagic rotifers were found to be dominant. However, their abundance decreased in the lower downstream sections. In the last site of the stream, the abundance of adult crustaceans, especially cladocerans, was found to increase.

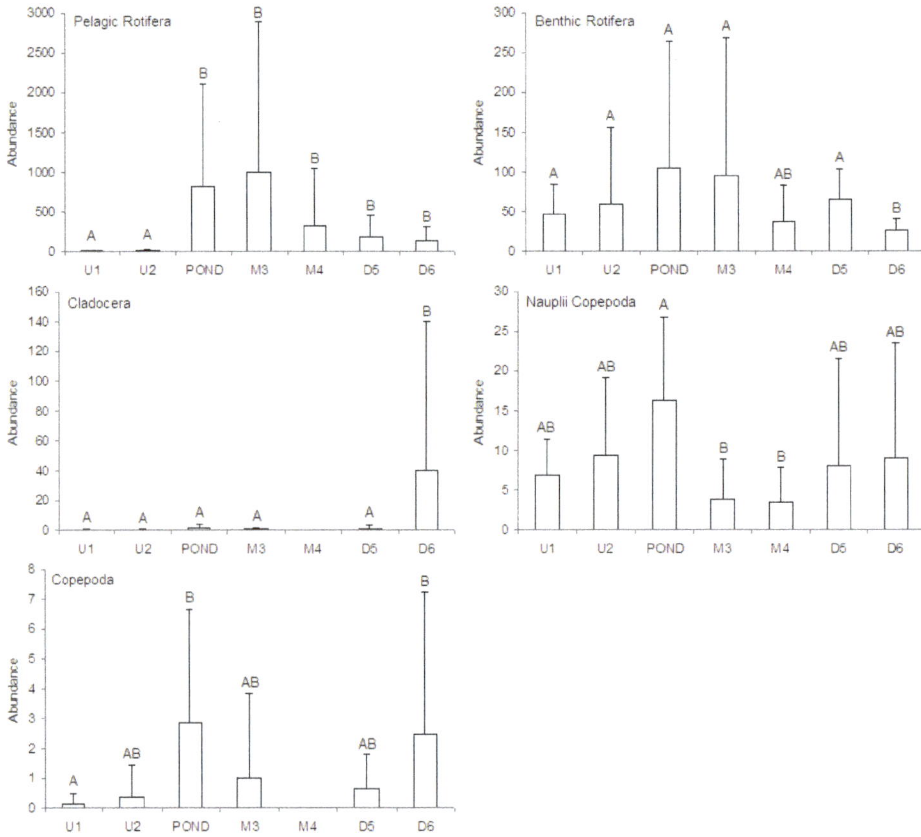

Figure 4. Spatial changes of the mean + SD zooplankton group abundance at two sites in upper section (U1, U2), in POND, at two sites in the middle section (M3, M4), and at two sites in the downstream section (D5, D6) of the Bukówka stream. The different letter on the column show significant differences between sites ($p < 0.05$).

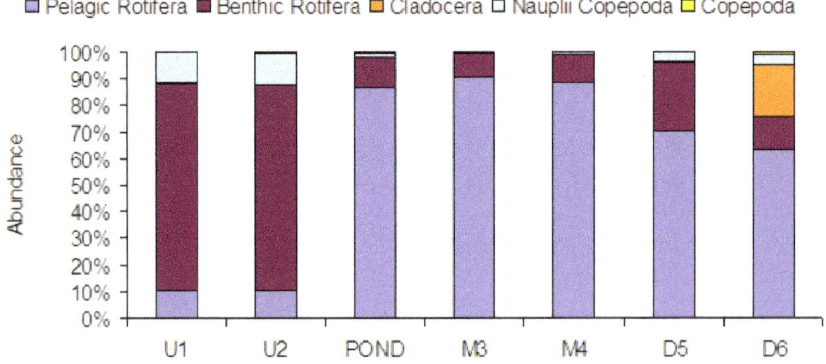

Figure 5. Spatial changes of percentage abundance of zooplankton groups at two sites in the upper section (U1, U2), in POND, at two sites in the middle section (M3, M4), and at two sites in the downstream section (D5, D6) of the Bukówka stream.

The first CCA axis explained 41% of the total variability in zooplankton group abundance with regard to the environmental factors (Figure 6). The second axis explained 9% of that variability. The morphological factors of stream (depth and width) correlated best with the first axis. The best correlation with the second axis was found for chlorophyll a and current velocity. Cladocerans and copepods (that the most correlated with the firs axis) correlated positively with depth and width and negatively with current velocity.

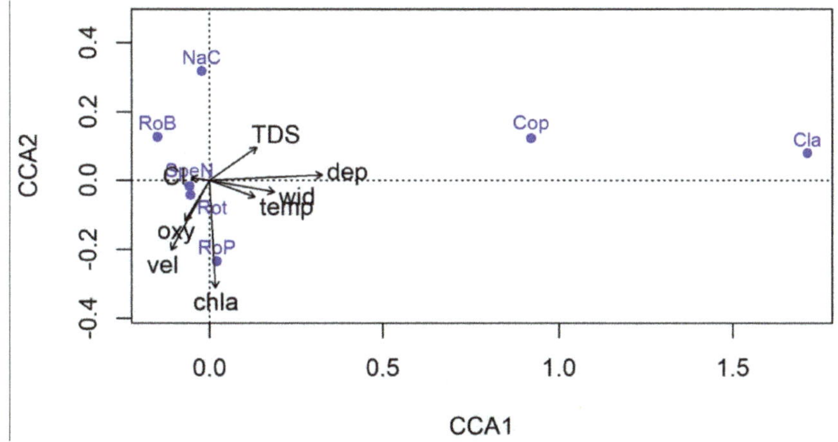

Figure 6. Zooplankton taxa abundance and species number of total zooplankton among environmental factors in the Bukówka stream. Canonical correspondence analysis (CCA) constrained ordination of taxa from sites with different environmental conditions. Environmental variables: chla—chlorophyll a, oxy—dissolved oxygen, temp—temperature, Cl—chloride, cond—conductivity, TDS—total dissolved solids, vel—current velocity, wid—width of bed, dep—depth of bed; Zooplankton taxa: RoP—pelagic rotifers, RoB—benthic rotifers, Rot—total rotifers, NaC—nauplii of copepods, Cla—cladocerans, Cop—copepodid stages and adult copepods, SpeN—species number of total zooplankton.

3.2. Macrozoobenthos—Macroinvertebrates

During the entire study period, 24 taxa of macroinvertebrates were identified at all sites. No significant differences were found in the number of taxa between the three sections of the Bukówka stream.

According to Jaccard's index, the most similar sites in terms of taxa were U1 and U2, which were in the upper section of the stream, and M3 and M4, which were in the middle section (Table 3). In addition, sites M4 and D5 were similar in taxa composition. The largest differences in the composition of taxa were observed between the site M3 and the POND.

Table 3. Taxonomic similarity of macroinvertebrates between sites (Jaccard's index).

Site	U1	U2	M3	M4	D5	D6
U2	76.5	X				
M3	58.8	52.6	X			
M4	50.0	44.4	64	X		
D5	64.3	56.3	47	73	X	
D6	44.4	47.4	32	47	50	X
POND	44.4	47.4	32	47	61.5	60

The Mann–Whitney *U* test used to test for differences in family abundances in different sections of the stream showed that the number of snails belonging to the Hydrobiidae family was significantly higher in the upper section of the stream than in the middle section ($p < 0.05$) (Figure 7). In the upper section, a higher number of Chironomidae larva and Oligochaeta were observed than the lower downstream section ($p < 0.05$). In the middle section, the Erpobdellidae family was more abundant than in the upper one ($p < 0.05$). No significant differences were found in the abundance of taxa between the middle and the lower downstream section ($p > 0.05$).

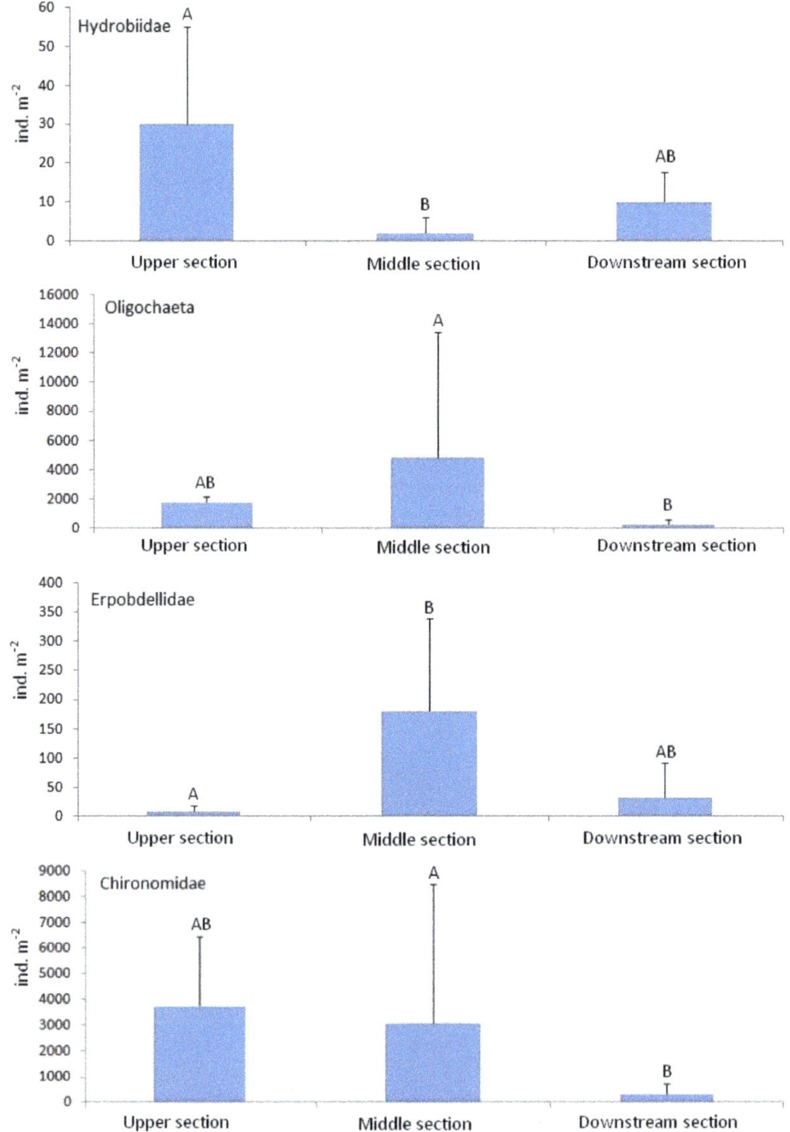

Figure 7. The significant differences between stream sections in abundance of Hydrobiidae, Oligochaeta, Erpobdellidae, and Chironomidae. The different letters on the columns show significant differences between sections ($p < 0.05$).

The first CCA axis explained 27% of the total variability in macroinvertebrates group abundance with regard to the environmental factors (Figure 8). The second axis explained 23% of that variability. The morphological factors of stream (depth and width) correlated best with the first axis. The best correlation with the second axis was found for temperature, chlorophyll a, and current velocity. Viviparidae, Platycnemidae, Notonectidae (that correlated the most with the first axis), and, to a lesser extent Coenagrionidae, correlated positively with depth and width and negatively with current velocity.

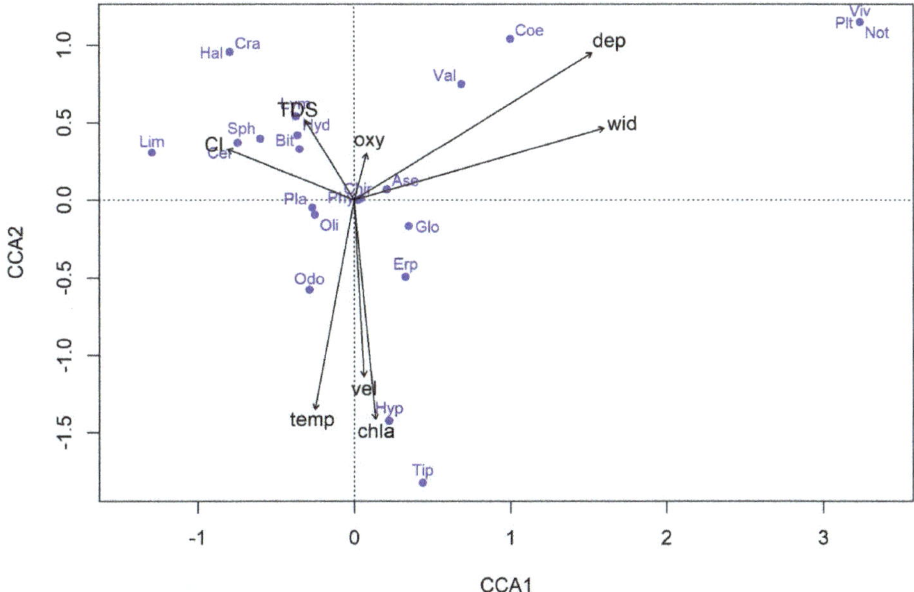

Figure 8. Macroinvertebrates taxa abundance among environmental factors in the Bukówka stream. Canonical correspondence analysis (CCA) constrained ordination of taxa from sites with different environmental conditions. Environmental variables: chla—chlorophyll *a*, oxy—dissolved oxygen, temp—temperature, Cl—chloride, cond—conductivity, TDS—total dissolved solids, vel—current velocity, wid—width of bed, dep—depth of bed; Macroinvertebrates taxa: Sph—Sphaeriidae, Bit—Bithyniidae, Lym—Lymnaeidae, Hyd—Hydrobiidae, Pla—Planorbidae, Phy—Physidae, Val—Valvatidae, Viv—Viviparidae, Oli—Oligochaeta, Erp—Erpobdellidae, Glo—Glossiphoniidae, Ase—Asellidae, Bae—Baetidae, Coe—Coenagrionidae, Plt—Platycnemidae, Not—Notonectidae, Hyp—Hydropsychidae, Odo—Odontoceridae, Hal—Haliplidae, Cer—Ceratopogonidae, Chi—Chironomidae, Lim—Limonidae, Tip—Tipulidae, Cra—Crambidae.

The values of the calculated indexes and metrics were characteristic of poor-quality water (Table 4). Both the BMWP-PL index and TBI pointed to class IV water quality for most of the sites, the biodiversity indexes (Margalef and Shannon–Weaver) were relatively low, and the GOLD and %DIPTERA metrics indicated a relatively high number of ubiquitous species in the studied stream.

Table 4. The values of the calculated indexes and metrics at the examined sites. Orange color stands for poor quality of water (IV class), yellow for moderate water quality (III class), green for good quality of water (II class), and blue for very good quality of water (I class).

Site/Month	BMWP-PL	Margalef Index	ASPT	Shannon-Weaver	TBI	GOLD (%)	% Diptera
U1/April	35	3.33	3.89	0.990	3	91	42
U1/October	45	4.01	3.75	0.990	4	90	49
U2/April	37	4.16	3.36	1.648	4	85	35
U2/October	47	5.86	2.94	0.796	6	94	74
M3/April	46	3.84	3.54	0.964	6	94	36
M3/October	31	3.40	3.44	1.684	5	36	10
M4/April	26	3.35	3.25	1.160	4	93	17
M4/October	25	3.09	3.57	1.476	5	53	40
D5/April	17	2.49	3.40	0.399	3	100	11
D5/October	31	3.45	3.44	1.236	4	43	23
D6/April	16	3.49	3.20	0.907	3	59	6
D6/October	43	5.78	4.30	0.989	7	79	73

4. Discussion

4.1. Zooplankton

The spatial pattern of the richness and abundance of zooplankton in the Bukówka stream showed that the value of these parameters increased with an increase in the distance from the river source. It was identified that these parameters were dependent on many environmental conditions that favored the growth of the zooplankton population. Thus, the abundance of zooplankton and their species number spatially changed in a similar manner as other running waters, including large rivers that were found to be affected by human activity [19,20]. The greatest impact on the development of typical planktonic organisms (pelagic rotifers, cladocerans, and some copepods) in Bukówka was posed by the reservoir as a basin due to its low current velocity and long water retention time. Hence, even a small dam that created a very shallow reservoir caused a significant difference in the richness and abundance of zooplankton between the upstream and the downstream sections. A similar pattern of the distribution of the zooplankton community was reported in small forest streams and rivers impounded by human-made dams or beaver dams in some studies [7,10,14,21]. We observed that in the impoundment (POND)—the "Słoneczne lake" of Bukówka—the richness and abundance of zooplankton were high, which was comparable to or even exceeded those in the typical temperate limnetic basins (e.g., eutrophic lakes or reservoirs) [22,23]. Our earlier studies on the increase of the zooplankton abundance in streams or rivers caused by damming did not show such high densities as observed in the Bukówka reservoir. It can be concluded that the reservoir was characterized by desirable trophic conditions for filter-feeding plankters because of the high concentration of chlorophyll *a*. Phytoplankton and other primary producers can reach high densities in Bukówka due to the exposure of the stream and reservoir to pollution in the form of excessive nutrient enrichment. This issue is often encountered in urban lakes in which zooplankton are found at very high densities [24]. Our present study shows that the abundance of zooplankton can also attain high values in urban reservoirs and streams.

A very common observation is that the abundance of zooplankton decreases along the section stretching from the dam (reservoir) or lake outlet to the lotic sections downstream to the mouth [20,23]. Probably this pattern of decreasing abundance is an effect of fish predation on zooplankton, grazing by suspension-feeding, filter-feeding, or net-spinning of macrozoobenthos, and settling of sediments [6,25,26]. However, in the lower downstream section of Bukówka, the abundance of some zooplankton groups (cladocerans and copepods) was higher than the sites in the middle section. This spatial pattern is much different from that observed in running waters and could result from the following: (1) low current velocity in the most downstream site, (2) occurrence of slack waters densely covered by macrophytes in the last section, and (3) water influx from the recipient.

It is generally known that zooplankton disperse passively from the outlet of stagnant basins to downstream and can colonize new habitats [27]. Drifting zooplankton can colonize the river if its bed offer zones with low current velocity or with stagnant water (e.g., slack waters, impoundments, and floodplains) [7,10]. The main variables that contribute to the increase in the richness and abundance of zooplankton in the Bukówka stream at the more downstream sites are as follows: a decrease in the current velocity, a longer water retention time or greater areas of open-water zones, higher number of slack water areas, covering of floodplains by macrophytes, and adjacent water bodies [14,28,29]. The majority of microfauna, including freshwater zooplankton, are unable to persist if the current velocity goes beyond 0.1 m s^{-1} [28,30]. The low values of current velocity noted at the last site of Bukówka (D6) were favorable for the occurrence of zooplankton and the increase in their populations. Moreover, crustaceans and other plankters could have migrated or been washed out from the riparian zones or slack waters connected with the stream water [12]. The highest width and depth of the stream bed at the last site might have also favored this phenomenon.

However, we hypothesize that the greatest impact on the higher densities of crustaceans at the most downstream site was posed by the influx of crustacean assemblages in the inlet waters from the recipient Oder river into the stream. This might have occurred during the high-water period. The reason for arriving at this conclusion is the occurrence of *Ceriodaphnia* sp., *Moina micrura*, *Polyphemus pediculus*, *Eucyclops macrurus*, *Thermocyclops oithonoides*, and *Eurytemora velox*, which were found only in the lower downstream sections of Bukówka. Most of these crustaceans are not typical of small-stream fauna. Crustaceans that could came from the Oder river could have migrated up the stream mainly in the last site (D6) because of low current velocities, which were too high at the D5 site for most of the crustaceans. Thus, it can be concluded that the highest impact on the zooplankton-shaping structures in the last section of the stream was posed by the inlet water from the Oder river; however, it is a hypothesis.

4.2. Macroinvertebrates

A relatively low number of taxa were found in the studied stream, regardless of the section considered. One of the reasons for this observation was the regulation of the stream. Sudduth and Meyer [31] reported a high number of pollution-tolerant taxa related to urban stream banks in their study. Such a pattern is caused by the regulation of stream banks, leading to the homogeneity of the environment [32,33]. Only an appropriate bioengineering mechanism could slightly increase the abundance of sensitive species [31,34]; however, a lot of other factors also decrease the diversity of urban streams. The stream examined in the present study flows through a part of the city that has no natural or forest area and does not have many trees adjacent to the stream. Couceiro et al. [35] showed that deforestation in the urban regions decreases the taxa richness of macroinvertebrates. In addition, the authors explained that water pollution (by nitrogen and phosphorus compounds) affects the composition of macroinvertebrates similar to deforestation. Similar pattern was observed by Gołdyn et al. [36] in small urban river in Poznań agglomeration (Poland), where stormwater runoff into this river was a source of significant contamination with heavy metals, organic matter and other pollutants, causing a strong transformation in the communities of benthic macroinvertebrates. Bukówka is a stream in the close neighborhood of allotments, from where a large amount of fertilizers is released. Moreover, the presence of a pond in the middle of the stream can be associated with the high values of TDS and conductivity and release of a large amount of total nitrogen and phosphorus, mostly from anthropogenic activities, such as the use of baits in fishing ground or feeding of aquatic birds. Many authors claim that biogenic compounds found in excess in urban streams negatively affect the macroinvertebrate composition [37–39]. Moreover, a channelization and regulation of Bukówka bed caused the low values of current velocity that affected the higher densities of taxa typical for ponds, lakes and large river, what was observed in lower section of stream. Therefore, low taxa richness and relatively high abundance in Bukówka are typical of a small, degraded stream, which is under the influence of high anthropogenic pressure.

All the biological indexes of macrozoobenthos are dependent on taxa compositions, and most of them take abundance into account. The values of TBI and BMWP-PL and its modification—ASPT index—were rather low for the examined stream, pointing at its poor water condition, and it is similar for many other Polish streams that are under anthropogenic pressure [40]. However, these indexes are basically calculated based on the taxa composition (the examined site has a lot of sensitive stenobiotic taxa). In addition, both Margalef and Shannon–Weaver biodiversity indexes were lower in Bukówka compared to other streams in Poland, and also correlated with abundance and taxa richness [41]. The last two metrics—GOLD and %DIPTERA—can shed some light on the whole pattern: these are based on ubiquitous taxa of macroinvertebrates, and thus, the higher their value, the poorer the conditions of the stream. In the case of Bukówka, the values of both metrics were high, which shows a high proportion of pollution-tolerant taxa in the abundance of macrozoobenthos. Taking all the indexes into account—and the fact that the statistical analysis of the basic environmental parameters did not reveal any significant biological correlations—we can state that streams like Bukówka represent a different type of running water, which are strongly affected by human activity. Therefore, urban streams should be given more attention, constantly monitored, and treated with special measures.

The macroinvertebrates of Bukówka were mainly represented by ubiquitous species, which are typical of streams flowing through urban agglomerations. Bukówka is a regulated stream flowing along the entire section through highly urbanized areas, carrying a large impact of the Oder River in its final section. Furthermore, the examined watercourse is not characterized by too many microhabitats for macroinvertebrates, and the riverbed is homogeneous what is non-natural for running waters. According to many authors, such a pattern leads to the degradation of the richness of bottom fauna, and consequently, to a decrease in the value of the biological indexes. So, it would be very important to pay attention to the need to rehabilitate the Bukówka stream to reach better values of ecological potential and biodiversity.

As Karr [42] claims, ecological integrity is the sum of the elements of biological diversity and processes occurring in an environment. Zooplankton and benthic macroinvertebrate taxonomic composition and abundances, macroinvertebrates metrics, and physicochemical parameters used to evaluate the state of the Bukówka stream can be considered together as biological and abiotic elements. It seems our results constitutes good ecosystem integrity in terms of the observed water quality and aquatic biology. Substantially, all these elements show that the worst conditions in the stream occurred in the last section in which pollution accumulated from the entire basin and where the riverbed was most transformed. However, all parameters react similarly on the changes that occurred in the stream.

5. Conclusions

In small urban streams the amount of live organic matter increases with their length. An additional and important factor that affects the increase of this matter is the intensive human activity in the catchment area and in the bed. The construction of reservoirs and bed regulation also have a similar effect on the ecological potential in small urban streams as in large rivers, but of course at a lower scale. The composition of invertebrates is also influenced by the influx of water from the recipient river (especially water with a different chemical composition). It seems that the structures of invertebrate assemblages in Bukówka are significantly governed by stream-scale variables, similar to the findings of Knott et al. [43].

Author Contributions: Conceptualization, R.C., Ł.S., and T.K.; methodology, R.C., Ł.S., and T.K.; software, R.C., Ł.S., and T.K.; validation, X.X., Y.Y. and Z.Z.; formal analysis, R.C., Ł.S., and T.K.; investigation, R.C., Ł.S., and T.K.; resources, R.C., Ł.S., T.K., A.W. and K.P.; data curation, R.C., Ł.S., and T.K.; writing—original draft preparation, R.C., Ł.S., and T.K.; writing—review and editing, R.C., Ł.S., and T.K.; visualization, R.C., Ł.S., and T.K.; supervision, R.C., Ł.S., and T.K.; All authors have read and agreed to the published version of the manuscript.

Funding: This research received no external funding.

Conflicts of Interest: The authors declare no conflict of interest.

References

1. Zalewski, M.; Wagner, I. Ecohydrology—The use of water and ecosystem processes for healthy urban environments. *Ecohydrol. Hydrobiol.* **2005**, *5*, 263.
2. Krepski, T.; Czerniawski, R. Shaping of macroinvertebrate structures in a small fishless lowland stream exposed to anthropopressure, including the environmental conditions. *Knowl. Manag. Aquat. Ecosyst.* **2018**, *419*, 19. [CrossRef]
3. Wohl, E. The significance of small streams. *Front. Earth Sci.* **2017**, *11*, 447–456. [CrossRef]
4. Williams, L.R.; Taylor, C.M. Influence of fish predation on assemblage structure of macroinvertebrates in an intermittent stream. *T. Am. Fish. Soc.* **2003**, *132*, 120–130. [CrossRef]
5. Böhmer, J.; Zenker, A.; Ackermann, B.; Kappus, B. Macrozoobenthos communities and biocoenotic assessment of ecological status in relation to degree of human impact in small streams in southwest Germany. *J. Aquat. Ecosyst. Stress Recov.* **2001**, *8*, 407–419. [CrossRef]
6. Chang, K.H.; Doi, H.; Imai, H.; Gunji, F.; Nakano, S.I. Longitudinal changes in zooplankton distribution below a reservoir outfall with reference to river planktivory. *Limnology* **2008**, *9*, 125–133. [CrossRef]
7. Czerniawski, R.; Kowalska-Góralska, M. Spatial changes in zooplankton communities in a strong human-mediated river ecosystem. *PeerJ* **2018**, *6*, e5087. [CrossRef]
8. Domagała, J.; Krepski, T.; Czerniawski, R.; Pilecka-Rapacz, M. Prey availability and selective feeding of sea trout (*Salmo trutta* L.; 1758) fry stocked in small forest streams. *J. Appl. Ichthyol.* **2015**, *31*, 375–380.
9. Lundquist, M.J.; Zhu, W. Aquatic insect diversity in streams across a rural–urban land-use discontinuum. *Hydrobiologia* **2019**, *837*, 15–30. [CrossRef]
10. Czerniawski, R.; Sługocki, Ł. A comparison of the effect of beaver and human-made impoundments on stream zooplankton. *Ecohydrology* **2018**, *11*, e1963. [CrossRef]
11. Souto, R.D.M.G.; Facure, K.G.; Pavanin, L.A.; Jacobucci, G.B. Influence of environmental factors on benthic macroinvertebrate communities of urban streams in Vereda habitats, Central Brazil. *Acta Limnol. Bras.* **2011**, *23*, 293–306. [CrossRef]
12. Czerniawski, R. Zooplankton community changes between forest and meadow sections in small headwater streams, NW Poland. *Biologia* **2013**, *68*, 448–458. [CrossRef]
13. Zhang, Y.; Zhang, J.; Wang, L.; Lu, D.; Cai, D.; Wang, B. Influences of dispersal and local environmental factors on stream macroinvertebrate communities in Qinjiang River, Guangxi, China. *Aquat. Biol.* **2014**, *20*, 185–194. [CrossRef]
14. Czerniawski, R.; Domagała, J. Small dams profoundly alter the spatial and temporal composition of zooplankton communities in running waters. *Int. Rev. Hydrobiol.* **2014**, *99*, 300–311. [CrossRef]
15. Kakouei, K.; Kiesel, J.; Kail, J.; Pusch, M.; Jähnig, S.C. Quantitative hydrological preferences of benthic stream invertebrates in Germany. *Ecol. Indic.* **2017**, *79*, 163–172. [CrossRef]
16. Radwan, S. *Freshwater Fauna of Poland*; Wydawnictwo Uniwersytetu Łódzkiego: Łódź, Polska, 2004; pp. 1–447.
17. Rybak, J.I.; Błędzki, L.A. *Planktonic Crustaceans of Freshwaters*; Wydawnictwo Uniwersytetu Warszawskiego: Warszawa, Poland, 2010; pp. 1–366.
18. Kownacki, A.; Soszka, H. *Guidelines for Assessing the State of the River on the Basis of Macroinvertebrates and for Collecting a Sample of Macroinvertebrates in Lakes*; Instytut Ochrony Środowiska, Zakład Ochrony Przyrody PAN: Warszawa–Kraków, Poland, 2004; pp. 1–54.
19. Akopian, M.; Garnier, J.; Pourriot, R. A large reservoir as a source of zooplankton for the river: Structure of the populations and influence of fish predation. *J. Plankton Res.* **1999**, *21*, 285–297. [CrossRef]
20. Pourriot, R.; Rougier, C.; Miquelis, A. Origin and development of river zooplankton: Example of the Marne. *Hydrobiologia* **1997**, *345*, 143–148. [CrossRef]
21. Karpowicz, M. Influence of eutrophic lowland reservoir on crustacean zooplankton assemblages in river valley oxbow lakes. *Pol. J. Environ. Stud.* **2014**, *23*, 2055–2061.
22. Gołdyn, R.; Kowalczewska-Madura, K. Interactions between phytoplankton and zooplankton in the hypertrophic Swarzędzkie lake in western Poland. *J. Plankton Res.* **2008**, *30*, 33–42.

23. Lair, N. A review of regulation mechanisms of metazoan plankton in riverine ecosystems: Aquatic habitat versus biota. *Riv. Res. Appl.* **2006**, *22*, 567–593. [CrossRef]

24. Ejsmont-Karabin, J.; Kuczyńska-Kippen, N. Urban Rotifers: Structure and Densities of Rotifer Communities in Water Bodies of the Poznań Agglomeration Area (Western Poland). In *Rotifer IX*; Springer: Dordrecht, The Netherlands, 2001; pp. 165–171.

25. Czerniawski, R.; Krepski, T. Zooplankton size as a factor determining the food selectivity of roach (*Rutilus rutilus*) in water basins outlets. *Water* **2019**, *11*, 1281. [CrossRef]

26. Jack, J.D.; Thorp, J.H. Impacts of fish predation on an Ohio River zooplankton community. *J. Plankton Res.* **2002**, *24*, 119–127. [CrossRef]

27. Havel, J.E.; Shurin, J.B. Mechanisms, effects, and scales of dispersal in freshwater zooplankton. *Limnol. Oceanogr.* **2004**, *49*, 1229–1238. [CrossRef]

28. Richardson, W.B. Microcrustacea in flowing water: Experimental analysis of washout times and a field test. *Freshw. Biol.* **1992**, *28*, 217–230. [CrossRef]

29. Kuczyńska-Kippen, N.M.; Nagengast, B. The influence of the spatial structure of hydromacrophytes and differentiating habitat on the structure of rotifer and cladoceran communities. *Hydrobiologia* **2006**, *559*, 203–212. [CrossRef]

30. Czerniawski, R.; Sługocki, Ł. Analysis of zooplankton assemblages in man-made ditches in relation to current velocity. *Oceanol. Hydrobiol. Stud.* **2017**, *46*, 199–211. [CrossRef]

31. Sudduth, E.B.; Meyer, J.L. Effects of bioengineered streambank stabilization on bank habitat and macroinvertebrates in urban streams. *Environ. Manag.* **2006**, *38*, 218–226. [CrossRef]

32. Grygoruk, M.; Frąk, M.; Chmielewski, A. Agricultural rivers at risk: Dredging results in a loss of macroinvertebrates. Preliminary observations from the Narew Catchment, Poland. *Water* **2015**, *7*, 4511–4522. [CrossRef]

33. Crook, D.A.; Robertson, A.I. Relationships between riverine fish and woody debris: Implications for lowland rivers. *Mar. Freshw. Res.* **1999**, *50*, 941–953. [CrossRef]

34. Acreman, M.; Arthington, A.H.; Colloff, M.J.; Couch, C.; Crossman, N.D.; Dyer, F.; Overton, I.; Pollino, C.A.; Stewardson, M.J.; Young, W. Environmental flows for natural, hybrid, and novel riverine ecosystems in a changing world. *Front. Ecol. Environ.* **2014**, *12*, 466–473. [CrossRef]

35. Couceiro, S.R.M.; Hamada, N.; Luz, S.L.B.; Forsberg, B.R.; Pimentel, T.P. Deforestation and sewage effects on aquatic macroinvertebrates in urban streams in Manaus, Amazonas, Brazil. *Hydrobiologia* **2007**, *575*, 271–284. [CrossRef]

36. Gołdyn, R.; Szpakowska, B.; Świerk, D.; Domek, P.; Buxakowski, J.; Dondajewska, R.; Barałkiewicz, D.; Sajnóg, A. Influence of stormwater runoff on macroinvertebrates in a small urban river and a reservoir. *Sci. Total Environ.* **2018**, *625*, 743–751. [CrossRef] [PubMed]

37. Luo, K.; Hu, X.; He, Q.; Wu, Z.; Cheng, H.; Hu, Z.; Mazumder, A. Impacts of rapid urbanization on the water quality and macroinvertebrate communities of streams: A case study in Liangjiang New Area, China. *Sci. Total Environ.* **2018**, *621*, 1601–1614. [CrossRef] [PubMed]

38. Elfritzson, M.P.; Batucan, L.S.; De Jesus, I.B.B.; Triño, E.M.C.; Ishida, Y.U.T.; Kobayashi, Y.; Ko, C.-Y.; Iwata, T.; Borja, A.S.; Briones, J.C.A.; et al. Nutrient loadings and deforestation decrease benthic macroinvertebrate diversity in an urbanised tropical stream system. *Limnologica* **2020**, *80*, 125744. [CrossRef]

39. Medupin, C. Spatial and temporal variation of benthic macroinvertebrate communities along an urban river in Greater Manchester, UK. *Environ. Monit. Assess.* **2020**, *192*, 84. [CrossRef] [PubMed]

40. Rybak, J.; Niedzielska, K. Analiza biologiczna jakości wody rzeki Rudna graniczącej z zbiornikiem osadów poflotacyjnych "Żelazny Most". The biological water quality assessment of the Rudna River situated near the post-flotation tailing pond "Żelazny Most" on the basis of communities of benthic invertebrates. *Inżynieria Ekologiczna* **2012**, *29*, 119–129.

41. Korycińska, M.; Królak, E. The use of various biotic indices for evaluation of water quality in the lowland rivers of Poland (exemplified by the Liwiec river). *Polish J. Environ. Stud.* **2006**, *15*, 419–428.

42. Karr, J.R. Defining and assessing ecological integrity: Beyond water quality. *Environ. Toxicol. Chem.* **1993**, *12*, 1521–1531. [CrossRef]
43. Knott, J.; Mueller, M.; Pander, J.; Geist, J. Effectiveness of catchment erosion protection measures and scale-dependent response of stream biota. *Hydrobiologia* **2019**, *830*, 77–92. [CrossRef]

Article

Response of Zooplankton Indices to Anthropogenic Pressure in the Catchment of Field Ponds

Natalia Kuczyńska-Kippen

Department of Water Protection, Faculty of Biology, Adam Mickiewicz University, Uniwersytetu Poznańskiego 6, 61-614 Poznań, Poland; kippen@hot.pl

Received: 31 January 2020; Accepted: 6 March 2020; Published: 10 March 2020

Abstract: As methods for assessing the environmental conditions in ponds are still not well developed, I studied zooplankton to identify a response of community indices to abiotic, biotic, and habitat type in two types of ponds differing in the level of human stress. Ponds of low human alterations (LowHI) harbored generally richer communities and a higher share littoral zooplankton, whose occurrence was associated with higher water transparency and complex macrophyte habitat, particularly the presence of hornworts and charoids. In high human-impact ponds (HighHI) planktonic communities prevailed. Their distribution was mainly related to the open water area and fish presence. Anthropogenic disturbance was also reflected in the frequency of rare species, which were associated with LowHI ponds. Higher diversity of zooplankton increased the chance for rare species to occur. Despite the fact that the majority of rare species are littoral-associated, they had no prevalence towards a certain ecological type of plants, which suggests that any kind of plant cover, even macrophytes typical for eutrophic waters (e.g., *Ceratophyllum demersum*) will create a valuable habitat for conservation purposes. Thus, it is postulated that a complex and dense cover of submerged macrophytes ought to be maintained in order to improve the ecological value of small water bodies.

Keywords: rotifers; microcrustaceans; aquatic vegetation; small water bodies; human-induced impact; ecological assessment

---

1. Introduction

Shallow and small aquatic environments, whose functioning may differ from better recognized ecosystems of lakes, consist of many different types, which include not only typical ponds, but also small ditches, relatively permanent standing or slow-flowing drainage canals, temporary pools, or puddles. Despite being prone to anthropogenic stress in the direct catchment area [1,2], ponds often create a refuge for diverse organisms [2]. The structure of the community and diversity of aquatic organisms can reflect environmental conditions [3]. Due to the high susceptibility of zooplankton to changes in the environment, especially to human impact in the catchment area of aquatic systems, these animals are often used to determine a variety of environmental features [4,5]. Knowledge of zooplankton response to changes is therefore very practical and necessary for both the assessment of habitat quality as well as the establishment of criteria for conservation.

Observing a constant global change of natural environment there is a pressing need to understand, and also assess, the speed and direction of changes. Therefore, some monitoring tools must be worked out and the application of biological communities can serve this purpose. Zooplankton are known to be sensitive to changes in the environment; thus, they have been proposed to be a very valuable indicator of water quality [6], including the trophic state of water [7]. I presume that it can also identify anthropogenic activity, contributing to varying levels of human-induced stress in the direct catchment of a pond, in water bodies located within the agricultural landscape [8,9].

Not only single species, but also communities of organisms can be a valid tool for monitoring changes in the ecosystem. Thus, for examining the response of zooplankton diversity to changes in environmental parameters, I employed biocoenotic features based on diversity indices (number of species and Shannon–Weaver index) in order to find out which method of diversity analysis will be more suitable for assessing the level of human impact. A biodiversity-based approach is very effective, especially for analyzing the total number of species in a community and/or measurement based on biodiversity indexes. The Shannon–Weaver index is one of the most commonly used in the examination of both the terrestrial environment and freshwater systems [10,11]. This is why these two approaches were taken into account in this study carried out on small water bodies.

Another approach that can be used for environmental assessment is the application of groups of organisms, which play a similar role in the ecosystem [12]. Studying functional categories can refer to niche occupation or habitat requirements, such as ecological groups of zooplankton. Pelagic and littoral species are those that are evolutionary adapted to living in the open water zone, or among macrophytes, respectively. The first group possesses specific adaptations to living in the homogenous conditions of the open water, where they can also be particularly prone to high numbers of potential predators. The second group is well adapted, both morphologically and metabolically, to living in a heterogenic area of macrophytes. Littoral species evolved strategies that enable them to move in often very complex habitats, but also have to forage for specific food resources in a highly qualified way [13]. They must also cope with unfavorable abiotic conditions that often prevail among plant beds [14], and this group of species is expected to greatly contribute to the overall diversity.

Using the data set from a survey of 296 small water bodies, I explore the responses of zooplankton community metrics (rotifers, cladocerans, and copepods), referring to the diversity and occurrence of two ecological groups (pelagic and littoral), to limnological variables, including abiotic and biotic features, and also to the site effect (macrophyte site occurrence and biometric features of a plant habitat). This is done to compare the effectiveness of these two approaches for distinguishing the effect of anthropogenic pressure in the catchment area of ponds located within an agricultural landscape.

Along with mid-field woodlots, ditches, oxbows, and wetlands—but also small rivers and streams—ponds are ecosystems of great importance for the preservation and enrichment of biodiversity. They provide an optimal habitat for various plants and animals [2], often being a source of rare and unique species, e.g., oligochaets, molluscs, beetles, gastropods, large branchiopods, annelids, as well as algae species and zooplankton [11,15]. A great range of habitats, especially created by hydromacrophytes, make ponds a valuable aquatic ecosystem, and despite their small area and depth, they will particularly contribute to the enrichment of biodiversity since favorable life conditions for organisms representing various life strategies [16] are created. Macrophyte beds offer pelagic zooplankters concealment from predators, both planktivorous fish and invertebrates, and provide littoral species with a number of various niches, which altogether increase total biodiversity. Plant build reflects the degree of spatial complexity and at the same time the availability of ecological niches for inhabiting organisms. Therefore, usually the most complex elodeids host the most diverse communities [17–19] compared to architecturally simpler stands of helophytes or nymphaeids. Thus, I hypothesized that species diversity will be highest in the habitats of the greatest complexity understood as plant density. There is also a high potential that along with the increase of ecological niches, due to variation in macrophyte cover, more species of high conservation value, such as those that are rare or infrequent on a local, regional, or a worldwide scale [20–22], will occur.

Rare species are often attributed to the littoral area [10]. Macrophyte-dominated zones hosting complex zooplankton assemblages may be found in any type of pond, as well as in anthropogenically-modified water bodies. Some researchers [23], examining only eutrophic ponds, have stated that extensive, often monospecific macrophyte stands, will provide an environment suitable not only for species associated with macrophytes, but also for migrating pelagic species; thus, maintaining high biodiversity of heleoplankton. A good example of a vegetation zone that is ascertained even in degraded areas, and can be an indicator of high trophic conditions, is *Ceratophyllum demersum* stands [24,25]. It can therefore be expected that rare species should occur not only in more natural ponds, but also in those with human-induced perturbation, although high anthropogenic impact may be a modifying force responsible for the occurrence of certain species or their prevalence. Therefore, I predicted that in two types of ponds, distinguished on the basis of anthropogenic impact, different frequencies of such species will be found where traces of human impact are stronger. It can also be presumed that more of such species will occur in ponds of low human impact.

The main purpose of the study was to define the response of zooplankton to human-induced transformation, where two groups of ponds were distinguished: (1) Ponds of low human alteration (LowHI) and (2) Ponds of high human alteration (HighHI). I aimed to define whether zooplankton community indices will be a responsive tool responding to environmental parameters in these two groups of ponds, differing in the level of human stress. This may prove that zooplankton, which is often used for trophic state assessment, can also be implemented to assess the level of anthropogenic impacts in the direct basin of a water body. As the type of catchment area, in terms of the level of agricultural coverage and forestation, can determine the specificity of zooplankton [8,9], only field ponds were taken into account in order to limit the effect of type of surroundings on the structure of the occurring organisms.

2. Materials and Methods

Research on the occurrence of zooplankton community indices was carried out in the optimum period of the summer season (July–August), between years 2011 and 2018 on 296 small water bodies situated within four large districts of central-western Poland (Provinces: Dolnośląskie, Kujawsko-Pomorskie, Lubuskie, Wielkopolskie). All of the ponds were located within rural catchment areas with the domination of agricultural land and also meadows, pastures, or within household and rural settlements. The water bodies that were chosen for the examination were divided into two main groups, depending on the level of observed anthropogenic impact in the direct catchment area. The first group consisted of ponds of a high level of anthropogenic transformation—HighHI (e.g., ponds situated within reach of rural settlements, such as farm buildings, cattle and pig farms, or manure storage areas). The second group of ponds was of a low level of anthropogenic transformation—LowHI (e.g., ponds situated within an agricultural landscape of very low human alterations, such as meadows, pastures, or wasteland meadows of extensive types of use). Ponds that were difficult to quantify in one of the groups were not adopted for this examination. Divisions of water bodies, based on a quantitative categorization of water bodies in reference, e.g., to trophic state exist. These methods assess the consequences of negative human impact, which can be measured by, e.g., specific levels of nutrients, primary production, or water transparency. However, in this study the division of two groups of ponds was done by assessing the observed level of anthropogenic impact in the surrounding area and was aimed to find out whether this will be reflected in the abiotic and biotic parameters of water. The studied water bodies also varied in respect to the surface area (between 0.001 and 3.4 ha) and to origin (post-glacial ponds, oxbows, and artificial water bodies, such as clay-pits, peat-pits). All of the studied ponds were shallow: LowHI—mean depth 1.4 m and HighHI—1.2 m. Fish were present in 75% of the investigated water bodies.

In order to provide a full spectrum of species inhabiting the ponds, samples were taken from various habitats. The open water area (total number of samples = 274) and vegetated sites (171) were studied. Among helophytes (59 sites) dominating stands of *Phragmites australis* (Cav.) Trin. ex Steud, *Schoenoplectus lacustris* (L.) Palla, *Typha angustifolia* L., and *Typha latifolia* L. were recorded; among elodeids (106 sites) *Ceratophyllum demersum* L., *Ceratophyllum submersum* L., stoneworts (among them e.g., *Chara hispida* L. and *Chara tomentosa* L.), milfoil (*Myriophyllum spicatum* L. and *Myriophyllum verticillatum* L.); nymphaeids (6 sites) with *Potamogeton natans* L.

The unequal number of stations, representing particular types, reflect the natural conditions prevailing in the water bodies, where the open water area was present in almost each water body, while particular vegetated sites occurred less frequently. Four types of stations (the open water area, elodeids, helophytes, nymphaeids) that were undergoing the examination had a similar proportion in reference to two groups of ponds – LowHI vs. HighHI ponds (elodeids—24% and %22% of all stations, helophytes—13% and 11%, nymphaeids—2 and 1%, open water—61% and 66%, respectively).

The method for collecting samples differed, depending on the type of habitat. Zooplankton material, of a volume of 10 L, was collected from among macrophytes with a plexiglass core sampler (Ø 50 mm), going vertically through a plant stand. The open water samples, of a volume of 20 L, were taken with a calibrated vessel. All samples were collected from the surface layer of water (0–30 cm). Collected samples were concentrated using a 45-μm plankton net and finally fixed with 96% ethanol. Specified methods for sample collection were described in previous papers [26,27].

Zooplankton was classified into two groups in reference to their evolutionary habitat requirements: pelagic and littoral groups. Within the pelagic group species indicating eutrophic conditions of water (e.g., *Anuraeopsis fissa* (Gosse), *Brachionus angularis* Gosse, *Filinia longiseta* (Ehrenberg), *Keratella cochlearis* f. *tecta* (Gosse), *Keratella quadrata* (Müller), *Pompholyx sulcata* Hudson, *Trichocerca pusilla* (Lauterborn), *Bosmina longirostris* (O.F. Müller), *Chydorus sphaericus* (O.F. Müller), and *Diaphanosoma brachyurum* (Liévin)) were separated, according to [7,28,29]. Moreover, due to the fact that cladocerans size varies greatly, and also because different responses of particular body-size groups can be expected [30], the plankton-associated fraction of this group of animals was divided into small-bodied cladocerans and large-bodied cladocerans. Moreover, rare species for the Polish fauna were established and their frequency was calculated. Only species that reached a high frequency (over 5% of the examined samples) underwent analyses.

Physical features of water (temperature, oxygen saturation, conductivity, and pH) were measured in-situ each time and at each sampling station. Water was also sampled for measurements of nutrient concentration (total phosphorus and dissolved inorganic nitrogen, which was the sum of nitrate, nitrite, and ammonium nitrogen) and chlorophyll a concentration. Detailed methods for nutrient determination and chlorophyll a concentration (corrected for pheopigments) were described in the previous paper [31].

The biometric character of each macrophyte bed was determined by calculating the stem density (length) and dry plant biomass (mass) occupied by plants in a water column (volume) in 1 L of water. Plant material, collected in triplicate, was obtained by cutting all stems from the surface of an area of 0.0625 m^2 and a height of 0.25 m. For a better description of the procedures connected with plant stand measurements, see [25].

The identification of rare species occurring in the studied ponds has been made on the basis of literature [32], where rare and infrequently occurring species are listed.

Species diversity, given as the Shannon–Weaver index [33], was used to measure the species diversity attributed to each pond and station.

On the basis of detrended correspondence analysis (DCA), which revealed that the gradients of length of species and the environmental variables were shorter than three, but longer than two standard deviations, Redundancy Analysis (RDA) was performed [34] using CANOCO 4.5 for Windows (Wageningen UR, Netherlands). RDA was applied to analyze zooplankton indices (number of species, Shannon–Weaver index, % participation of pelagic and littoral species in the total rotifer abundance, abundance) distribution patterns in relation to physical-chemical factors (conductivity, chlorophyll *a*, dissolved inorganic nitrogen, pH, total phosphorus, Secchi disc transparency), level of anthropogenic transformation, the presence of fish, the level of overshading (percent of pond's surface coverage created by nymphaeids and pleustophytes) and type of habitat (open water zone and dominating macrophyte species representing particular ecological groups of macrophytes). In order to obtain normality of variance, data used in the RDA analyses were log (x+1) transformed, except for the pH, as it is already a negative logarithm of the hydrogen ion. The statistical significance in RDA was assessed by Monte-Carlo permutation tests (9999 permutations) [34].

The relationships between environmental factors and the frequency of rare species in the studied water bodies were calculated using the Spearman rank correlation coefficient. The Mann–Whitney test was used in order to determine the effect of anthropogenic transformation in the vicinity of the ponds and the Kruskal–Wallis test to determine the effect of a specific type of habitat on the percent participation of rare species in the zooplankton total abundance in a particular water body. For these two analyses, only rare species, whose general frequency in the examined material was higher than 5%, were taken into consideration for statistical analyses.

3. Results

The study ponds revealed a wide range of measured features in respect to the two groups of small water bodies, distinguished on the basis of human impact in the direct catchment area. Neither their morphology (area), nutrient level, nor the spatial characteristics of their vegetated stands, reflected in the density of macrophyte bed, differed between the two types of ponds. Significantly higher values for HighHI ponds were obtained for pH, oxygen saturation (OS), conductivity, and water surface overshading caused by pleustophytes and nymphaeids. At the same time water transparency measured as Secchi disc visibility was higher in LowHI ponds (Table 1). Fish were present in three-quarters of the studied ponds and there was a similar share in two types of water bodies (73% in LowHI and 83% in HighHI).

Table 1. Descriptive statistics of zooplankton community indices and results of Mann–Whitney (M–W) test in two types of water bodies (Rot—rotifers, Cla—cladocerans, Cop—copepods; pel—% participation of pelagic species; lit—% participation of littoral species; N—number of species; C-large—abundance (ind L^{-1}) of large cladocerans; C-small—abundance (ind L^{-1}) of small cladocerans; Sh-W—Shannon Weaver index value) and environmental parameters of the studied small water bodies (length—macrophyte length; pH—water reactivity; TP—total phosphorus; DIN—dissolved inorganic nitrogen; Chl a—chlorophyll a concentration; OS—oxygen saturation; EC—conductivity; SDV—Secchi disc visibility; %surf—% overshading of the water surface caused by coverage of nymphaeids and pleustophytes; LowHI—low human alteration; HighHI—high human alteration; Min—Minimum; Max—Maximum.

Parameter	LowHI Ponds				HighHI Ponds				M–W Test	
	Mean	Min	Max	SD	Mean	Min	Max	SD	Z	p
pH	7.80	6.29	10.45	0.69	8.09	6.16	10.83	0.82	2.98	*
OS	77.65	0.15	216	48.53	94.03	3	227.1	57.99	2.11	*
EC	672	109	3620	386	944	105	3155	490	5.30	**
SDV	0.87	0.05	4	0.66	0.59	0.08	1.5	0.39	−4.00	**
area	0.60	0	17.54	1.38	0.37	0.01	3.2	0.50	-	ns
%surf	18	0	100	26	30	0	100	36	2.03	*
TP	0.58	0.01	10.31	1.06	0.53	0.01	2.85	0.61	-	ns
DIN	2.23	0.09	14.11	2.33	2.54	0.25	15.29	2.81	-	ns
Chl a	43.5	0.8	549.5	69.7	66.1	1.3	2030.6	226.1	-	ns
length	69	1	901	118	46	3	233	46	-	ns
N Rot	26	0	65	13	21	4	44	8.98	−3.48	**
Sh-W Rot	2.08	0	3.69	5	1.53	0.24	2.78	0.55	−3.81	**
Rot pel	59	0	100	37	67	1	100	37	-	ns
Rot lit	41	0	100	37	33	0	99	37	-	ns
N Cla	7	0	21	5	4	0	15	4	−3.88	**
Sh-W Cla	1.01	0	2.47	0.64	0.72	0	1.85	0.56	−3.75	**
C-large	9	0	100	25	2	0	67	9	−2.23	*
C-small	47	0	225	37	51	0	200	44	-	ns
Cla lit	44	0	100	34	38	0	100	39	-	ns
N Cop	3	0	9	2	2	0	7	2	−3.44	**
Sh-W Cop	0.62	0	2.01	0.57	0.37	0	1.66	0.44	−3.48	**
Cop pel	49	0	100	40	60	0	100	44	−	ns
Cop lit	51	0	100	40	40	0	100	40	−1.97	*

*: $p < 0.05$; **: $p < 0.001$; ns: not significant.

The number of species and Shannon–Weaver index values were significantly higher in LowHI ponds for each zooplankton group. Moreover, LowHI ponds had a significantly higher participation of large pelagic cladocerans and of the littoral fraction of copepods. In all other cases, the percent participation of pelagic and littoral zooplankton was evenly distributed between the two groups of ponds (Table 1). The pelagic group of zooplankton contained 40% of eutrophic species on average (34% of rotifers and 45% of crustaceans) in the open water area.

Redundancy analysis (RDA) used to define the relationships between zooplankton indices and abiotic parameters generated two axes, accounting for 51.0% and 23.4%. A total of twelve environmental variables (Table 2) as well as 13 zooplankton indices were chosen to perform the RDA.

Table 2. Results of Monte Carlo test (F and p values given) of relationships between zooplankton indices and analyzed environmental/habitat factors in two types of ponds: HighHI ponds—anthropogenically modified ponds and LowHI ponds—ponds of low level of human impact (for legend see Table 1, Figure 2). Bold font – significant ($p \leq 0.05$).

ENVIRONMENTAL	F	p	HABITATS	F	p
Fish	**26.92**	**0.002**	**Water**	**62.14**	**0.002**
% Surf	**12.64**	**0.002**	**length**	**49.64**	**0.003**
SDV	**8.40**	**0.002**	**Cdem**	**5.81**	**0.002**
OS	**5.92**	**0.002**	**Csub**	**4.76**	**0.004**
area	**3.42**	**0.010**	**Pnat**	**3.25**	**0.014**
DIN	**2.78**	**0.032**	**HighHI/LowHI**	**2.75**	**0.016**
TP	**2.39**	**0.038**	**Ch**	**2.08**	**0.050**
Chl a	**2.56**	**0.028**	Tlat	1.87	>0.05
HighHI/LowHI	2.01	>0.05	Tang	1.25	>0.05
pH	1.46	>0.05	Myr	1.19	>0.05
EC	0.92	>0.05	Phr	0.49	>0.05

The results of RDA analysis indicated a division of zooplankton indices in reference to environmental features (Figure 1). The first group gathered around LowHI ponds and consisted of metrics responsible for the high taxonomic structure of crustaceans (increasing number of species of cladocerans and copepods) and of the littoral community indices of all zooplankton groups. Among them, large-bodied cladocerans of pelagic origin (daphnids) also occurred. This group of indices was associated with total phosphorus concentration. Moreover, the rotifer littoral community was also attributed to ponds of a high level of shade caused by the layer of pleustophytes and nymphaeids. At the same time, these community metrics negatively correlated with chlorophyll a concentration and the presence of fish.

The second group of zooplankton features, which was in opposition to LowHI ponds, was clustered close to HighHI ponds. Both rotifer number of species and diversity, as well as densities of the rotifer pelagic community, and that of small pelagic cladocerans and pelagic copepods, revealed the strongest affinity with the presence of fish and high concentrations of chlorophyll a.

There was also an additional group of factors, which had a strong impact on the pelagic community of crustaceans. Both pelagic copepods and a small-sized fraction of pelagic cladocerans were found to be strongly structured by the increasing area of small water bodies as well as by high water transparency, measured by Secchi disc visibility. These communities were associated with waters of low overshading. Environmental parameters, such as water conductivity, pH, and dissolved oxygen content were of minor effect in reference to the zooplankton pelagic community (Figure 1).

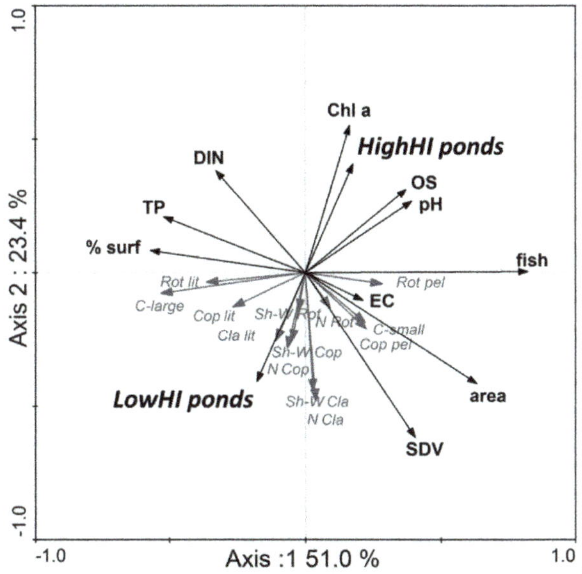

Figure 1. RDA diagram illustrating zooplankton community indices with physical-chemical variables in two types of water bodies: HighHI ponds—anthropogenically modified ponds and LowHI ponds—ponds of low level of human impact (legend: see Table 1).

The second RDA analysis with habitat type and biometric features, reflecting the heterogeneity level of a plant stand, (76.7 and 11.6% of explained variance) took into consideration twelve variables (with eight of high significance: open water area, length of macrophyte stems, stands of *Ceratophyllum demersum*, *Ceratophyllum submersum*, *Potamogeton natans*, charophytes, level of anthropogenic transformation) as well as 13 zooplankton indices. This analysis also showed a clear division of zooplankton indices in reference to human-induced changes in the direct catchment area (Figure 2). The open water area was attributed to HighHI ponds and pelagic rotifers and cladocerans (both the small-sized fraction as well as the large-bodied community) prevailed here. The second group of habitats, consisting of stands located within elodeids, was in close association with LowHI ponds. Community indices of all zooplankton groups, such as number of species, Shannon–Weaver diversity index, share of littoral community, reached their highest values, being particularly associated with the presence of *Ceratophyllum demersum*, *Ceratophyllum submersum*, and charoids stands. Increasing values of biometric features of macrophyte beds were responsible for the rise of the above-mentioned zooplankton metrics. The presence of helophytes had no effect on zooplankton diversity (Figure 2; Table 2).

In the samples 34, rotifer species and 11 of cladocerans were rare or of infrequent occurrence in the Polish fauna. No rare species among copepods were recorded. There were 9 rotifer species (*Cephalodella carina* Wulfert, *Cephalodella gibboides* Wulfert, *Euchlanis triquetra* Ehrenberg, *Lecane elsa* Hauer, Lecane furcata Murray, *Lecane inermis* Bryce, *Lecane nana* Murray, *Lecane pyriformis* Daday, *Trichocerca vernalis* Hauer) and 4 cladoceran species (*Alona rustica* Scott, *Chydorus gibbus* G.O. Sars, *Chydorus ovalis* Kurz, *Tretocephala ambigua* Lilljeborg) that reached a considerably high frequency (they were present in >5% of the samples) in the examined material. The majority of such species were littoral-associated taxa, e.g., representatives of the genus *Cephalodella*, *Euchlanis*, *Lecane*, *Lepadella*, *Alona*. However, some of them were also pelagic-associated organisms, e.g., *Brachionus falcatus* Zacharias, *Brachionus polyacanthus* (Ehrenberg), *Brachionus leydigi* Cohn, or *Keratella paludosa* (Lucks).

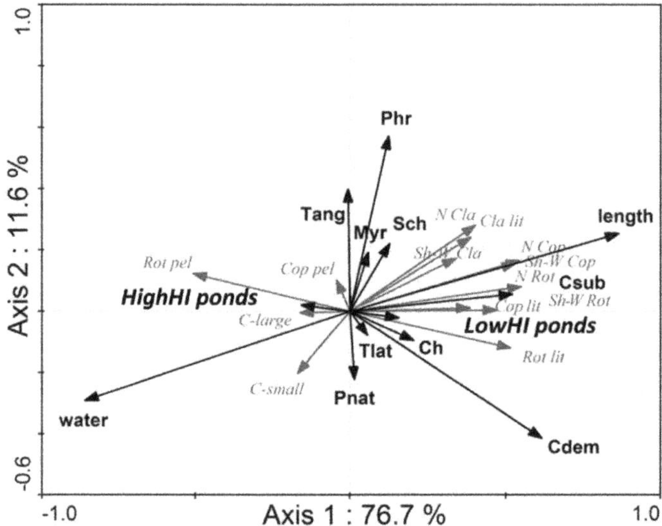

Figure 2. RDA diagram illustrating zooplankton community indices with habitats (water—the open water zone, Phr—Phragmites australis, Sch—*Schoenoplectus lacustris*, Tang—*Typha angustifolia*, Tlat—*Typha latifolia*, Cdem—*Ceratophyllum demersum*, Csub—*Ceratophyllum submersum*, Ch—*Chara hispida* and *Chara tomentosa*, Myr—*Myriophyllum spicatum* and *Myriophyllum verticilatum*, Pnat—*Potamogeton natans*) and presence of fish in two types of water bodies: HighHI ponds—anthropogenically modified ponds and LowHI ponds—ponds of low level of human impact (legend: see Table 1).

Rare rotifers contributed on average 1.46% of the total rotifer abundance, while crustaceans 2.26%. A higher frequency of both rotifers and cladocerans was recorded for LowHI ponds (MW-Z = −3.12; p < 0.01 and MW-Z = −2.49; p < 0.05, respectively).

Looking at the relationships between frequency of rare species and environmental parameters, it was noticed, especially in case of rotifers, that overshading and chlorophyll a concentration (r = 0.22; p < 0.05 and r = 0.22; p < 0.05, respectively) had a positive effect on the prevalence of such species, while level of water oxygenation and conductivity (r = −0.21; p < 0.05 and r = −0.30; p < 0.001, respectively) had a negative effect. The frequency of rare species also positively correlated with the total number of species and values of the Shannon–Weaver index for both rotifers and cladocerans (r = 0.67; p < 0.001, r = 0.36; p < 0.001 and r = 0.33; p < 0.001, r = 0.23; p < 0.01, respectively).

The frequency of both rotifer and cladoceran rare species revealed a significant variation between particular habitats in small water bodies (KW-H = 95.24; p < 0.01 and KW-H = 38.63; p < 0.01, respectively). It was noticed that rotifer rare species were, on average, most frequent among charoids, *Ceratophyllum submersum*, *Potamogeton natans*, and *Schoenoplectus lacustris*, while rare cladocerans most often chose macrophyte sites among *Typha angustifolia*, *Schoenoplectus lacustris*, and stoneworts.

4. Discussion

Zooplankton community indices were proved to be a sensitive tool for the identification of environmental features in two different types of ponds, distinguished on the basis of the level of human stress (LowHI vs. HighHI ponds) in the direct catchment area. It was found that both diversity measures, as well as the occurrence of the littoral fraction of zooplankton, revealed a similar response to environmental variables and to habitat type. Taxonomic richness and the Shannon–Weaver index values of each group of zooplankton were significantly higher in LowHI ponds, which were, on average, characterized by higher water transparency and a lower level of conductivity, pH, and OS. Moreover, large-bodied cladocerans and the littoral community of copepods reached higher values in less human-impacted ponds (LowHI). This shows that zooplankton, which is often used for trophic state assessment [7,28,35], can also be implemented to assess the level of anthropogenic impacts in the direct basin of a water body. The less the surroundings of the pond are transformed, the more diverse the zooplankton community that inhabit a pond will be, and there is also a greater chance that large-bodied cladocerans will occur. The occurrence of large cladocerans is often attributed to small and productive ecosystems and the majority of ponds that occurred in the central-western part of Poland were generally eutrophic. This has been proved by the lack of significant variation in the content of nitrogen (DIN) and phytoplankton biomass level measured as chlorophyll a concentration (Chl a), between the two types of ponds selected in this study (LowHI vs. HighHI). Neither did the content of phosphorus (TP), very important for the level of eutrophication, differ between these two groups of ponds (0.58 and 0.53 mg L^{-1}, respectively). A higher proportion of large cladocerans, such as representatives of the *Daphnia* genus (e.g., *Daphnia pulex*) in less human-impacted ponds, contributed to generally better water transparency. Moreover, [36], who examined a set of 33 interconnected ponds in Belgium, obtained clear differences in the structure of zooplankton communities that were attributed to local biotic and abiotic factors. They observed a variation in the occurrence of large daphnids between the clear-water and turbid state ponds, with ponds in the clear-water state having low fish densities being chosen by large *Daphnia* species. Moreover, the clear-water ponds had more species designated as littoral ones. At the same time, their group of ponds in the turbid-water state contained a high abundance of rotifers and cyclopoid copepods. The results of the RDA analysis that I conducted on 296 small water bodies has also confirmed the affinity of zooplankton diversity indices, and large-bodied cladocerans for LowHI, and a high level of shade caused by macrophyte cover built by pleustophytes and nymphaeids. These zooplankton indices were in opposition to the occurrence of fish in ponds. Even though fish had a similar share in the two types of water bodies, it was clear that their presence negatively affected zooplankton diversity measures and the occurrence of large-bodied cladocerans.

Moreover, the littoral fraction of all zooplankton groups was also found to prevail here. Limnetic communities inhabiting the open water areas of small water bodies have always received more scientific attention than communities inhabiting littoral areas. One of the main causes of the neglect of macrophyte-dominated areas in hydrobiological studies is associated with the difficult research methodology and labor-consuming identification of single littoral taxa, such as periphytic and benthic species. This can also be associated with littoral species being overlooked, mainly due to the improper methods applied for sampling [37], as well as the extremely difficult taxonomic features of animals subjected to determination. However, in this study, a whole littoral community was treated as a group. Such a method, without a need for distinctive species determination, is much easier to implement, and I have proved that littoral zooplankton can serve as a sensitive indicator for identifying a low level of human impact on the aquatic environment.

Unlike the zooplankton diversity indices and the occurrence of littoral community and large-bodied cladocerans, the pelagic fraction of rotifers, copepods, and small-bodied cladocerans occurred in association with different environmental conditions. They were found to prevail in larger ponds with fish presence. Pelagic rotifers and small-bodied cladocerans are equipped with morphological or behavioral anti-predator mechanisms [38], such as, e.g., strong lorica, long bristles, or they undertake migrations, which help them cope with predation and enable them to develop in water bodies with fish presence. At the same time, these organisms avoided ponds overgrown by macrophytes (pleustophytes and nymphaeids). The pelagic community of zooplankton contains typically eutrophic species, which confirms their prevalence in HighHI ponds. Therefore, it is evident that not only taxon-related indices (e.g., taxonomic composition), which are argued to be a valid instrument in environmental assessment [39], or functional-based metrics (e.g., feeding-type groups) will indicate changes in the environment [40]. Thus, it can be recommended that for examining small water bodies' basic biocoenotic features, such as diversity indices, as well as the share of pelagic and littoral species can be a sensitive tool that responds to varying levels of human stress in a pond's surroundings.

Because small water bodies usually create a mosaic of microhabitats, where the open water area is often divided by various patches of aquatic vegetation, I determined which habitats favored high zooplankton diversity and the occurrence of certain ecological groups of animals. The second RDA analysis revealed a distinct division of zooplankton metrics depending on anthropogenic transformation in the direct catchment area and the level of habitat heterogeneity, expressed in the length of macrophyte stems. The open water area, associated with HighHI ponds, favored the occurrence of pelagic zooplankton, which contain species of eutrophic origin. The association of areas of open water zones with HighHI ponds is probably connected with significantly lower water transparency in this type of pond compared with LowHI ponds. This is why macrophytes have worse light conditions and do not develop as well as in ponds of lower human impact, where water transparency is higher. At the same time, it had a negative effect on all diversity indices and the littoral community of zooplankton, which found the most advantageous conditions among aquatic vegetation, especially in elodeids. In particular, the presence of stands built by *Ceratophyllum demersum, Ceratophyllum submersum,* and charoids was of significant importance. Stoneworts as a group, despite the varying tolerance between certain species, are very sensitive to water quality and often maintain the stabilization of a clear-water state [41,42]. Therefore, the occurrence of charoids was associated with LowHI ponds and their presence positively influenced the diversity increase of zooplankton. Contrary to this, hornworts usually appear in nutrient-rich waters [43], such as the examined small water bodies located in the agricultural landscape of central-western Poland. Despite the clear niche partitioning of both *Ceratophyllum* species, where *C. demersum* occurs in clearer waters, and *C. submersum* in lower light conditions [44], both macrophyte species provided a very suitable habitat for diverse zooplankton communities and substratum for the development of littoral organisms. This shows that in case of small water bodies, which are often eutrophic systems, aquatic plants of a high level of complexity expressed by the length of macrophyte stems will contribute to the rise of overall zooplankton diversity, even when these plants are indicators of poor quality of water, such as hornworts. Other plant stands, such as helophytes (e.g., *Typha latifolia, Typha angustifolia, Phragmites australis*), which are of much lower structural complexity compared to elodeids [45,46], had a much lower impact on zooplankton diversity or the occurrence of the littoral community. Therefore, in order to maintain a high diversity of zooplankton and thereby improve the ecological status of small water bodies, it is recommended that a complex and dense cover of submerged macrophytes be encouraged and sustained.

It is not only the biocoenotic features of zooplankton that can be valuable for detecting the level of anthropogenic impact on a pond ecosystem; the occurrence of rare species can also serve this purpose. A higher frequency of rare species, which favored waters of low conductivity and a high level of overshading caused by the occurrence of pleustophytes and nymphaeids, was found in LowHI ponds. I also found a positive relationship between the total number of rotifer and crustacean species and values of the Shannon–Weaver index with the increasing frequency of rare species. This indicates that increased diversity in the zooplankton community will improve the chances that species of very high conservation value will occur. The occurrence of rare species in areas of a very low level of human transformation has also been ascertained for other small water bodies. The results of a study carried out on 126 ponds distributed over the area of Belgium show that a high share of forest in the immediate surroundings of ponds was associated with the clear water state. These ponds had a higher number of abundant water plant taxa, a higher number of water plant growth forms, and also were characterized by a higher diversity of biotope types [47]. However, some studies have also presented data that rare species occur in water bodies situated within areas of higher anthropogenic impact, such as farmland or urban areas. Although it has been shown that effective conservation of species diversity in such areas should include typical human-impacted as well as semi-natural habitats, which could both cover the full range of occurring species [48].

A minority of rare species were of pelagic origin, especially rotifers of the family Brachionidae, e.g., *Brachionus falcatus, B. polyacanthus, B. leydigi,* or *Keratella paludosa,* while littoral-associated taxa prevailed. This also confirms the major role of vegetated zones in even very small water bodies in increasing the overall biodiversity, and particularly supporting the occurrence of species of high ecological value.

The results of this study provide further evidence that diversity may play an extremely important role in maintaining the occurrence of rare species. In spite of the fact that macrophyte stations constituted 39% of the whole material in the case of LowHI ponds and 34% in HighHI ponds, rare species were mainly littoral organisms. Therefore, a complexity of habitat is of primary importance for the enrichment of the local fauna and finally for species which do not occur frequently. Analyzing particular habitats, where rare zooplankton was most frequent, it was noticed that they displayed significant preferences towards certain habitats. They chose stations located in elodeids, such as stoneworts (both rotifers and cladocerans) and *Ceratophyllum submersum* (rotifers), also among helophytes, such as *Schoenoplectus lacustris* (both rotifers and cladocerans) and *Typha angustifolia* (cladocerans), and finally nymphaeids such as *Potamogeton natans* (rotifers). The lack of correlations between biometric features of macrophyte stands, referring to their level of complexity, indicates the fact that rare species will frequently occur in various macrophyte-dominated habitats, irrespective of their build and density. These findings also support the importance of maintaining diverse macrophyte cover even within very small and shallow aquatic systems, such as ponds.

5. Conclusions

The species richness of zooplankton may differ in respect to various factors, including geographical region, trophic type, morphometric features, or even the origin of a water body [49]. However, it has been demonstrated that it also depends on the level of anthropogenic transformation in the vicinity of a pond, with high zooplankton diversity, and high frequency of rare species in ponds of low human impact, where littoral communities also prevailed, and large-bodied cladocerans predominated. Furthermore, a major role of a mosaic of habitats prevailing in ponds of low anthropogenic transformation was also found to be responsible for zooplankton occurrence. Particularly dense patches of elodeids, even those typical of highly eutrophic waters, such as *Ceratophyllum demersum,* were responsible for the overall rise in zooplankton diversity and supporting the development of a littoral community of both rotifers and microcrustaceans. The key role of macrophytes in ponds was also demonstrated in the case of rare species whose frequency was much higher in ponds less impacted by human activity and rose with the increase of zooplankton diversity. Therefore, a complex macrophyte cover should be maintained to improve the ecological status of small water bodies.

Acknowledgments: This research work was financed by the Polish State Committee for Scientific Research in 2010-2014 as research project N N305 042739. The author is thankful B. Nagengast for identification of macrophyte species and biometric features measurements as well as T. Joniak for chemical analyses.

Conflicts of Interest: The author declares no conflict of interest.

References

1. Williams, P.; Whitfield, M.; Biggs, J.; Bray, S.; Fox, G.; Nicolet, P.; Sear, D. Comparative biodiversity of rivers, streams, ditches and ponds in an agricultural landscape in Southern England. *Biol. Conserv.* **2004**, *115*, 329–341. [CrossRef]
2. Biggs, J.; Von Fumetti, S.; Kelly-Quinn, M. The importance of small waterbodies for biodiversity and ecosystem services: Implications for policy makers. *Hydrobiologia* **2016**, *793*, 3–39. [CrossRef]
3. Berta, C.; Tóthmérész, B.; Wojewódka, M.; Augustyniuk, O.; Korponai, J.; Bertalan-Balázs, B.; Nagy, S.; Grigorszky, I.; Gyulai, I.; Balázs, B.; et al. Community Response of Cladocera to Trophic Stress by Biomanipulation in a Shallow Oxbow Lake. *Water* **2019**, *11*, 929. [CrossRef]
4. Basińska, A.; Kuczyńska-Kippen, N.; Świdnicki, K. The body size distribution of Filinia longiseta (Ehrenberg) in different types of small water bodies in the Wielkoposka region. *Limnetica* **2010**, *29*, 171–181.

5. Feniova, I.; Sakharova, E.; Karpowicz, M.; Gladyshev, M.I.; Sushchik, N.N.; Dawidowicz, P.; Gorelysheva, Z.; Andrzej, G.; Stroinov, Y.V.; Dzialowski, A. Direct and Indirect Impacts of Fish on Crustacean Zooplankton in Experimental Mesocosms. *Water* **2019**, *11*, 2090. [CrossRef]

6. Nowosad, P.; Kuczyńska-Kippen, N.; Słodkowicz-Kowalska, A.; Majewska, A.C.; Graczyk, T.K. The use of rotifers in detecting protozoan parasite infections in recreational lakes. *Aquat. Ecol.* **2006**, *41*, 47–54. [CrossRef]

7. Ejsmont-Karabin, J. The usefulness of zooplankton as lake ecosystem indicators: Rotifer trophic state index. *Pol. J. Ecol.* **2012**, *60*, 339–350.

8. Dodson, S.I.; Everhart, W.R.; Jandl, A.K.; Krauskopf, S.J. Effect of watershed land use and lake age on zooplankton species richness. *Hydrobiologia* **2006**, *579*, 393–399. [CrossRef]

9. Kuczyńska-Kippen, N.; Joniak, T. The impact of water chemistry on zooplankton occurrence in two types (field versus forest) of small water bodies. *Int. Rev. Hydrobiol.* **2010**, *95*, 130–141. [CrossRef]

10. Añorve, L.E.S.; Lepretre, A.; Davoult, D. Diversity of benthic macrofauna in the eastern English Channel: Comparison among and within communities. *Biodivers. Conserv.* **2002**, *11*, 265–282. [CrossRef]

11. Pakulnicka, J.; Górski, A.; Bielecki, A. Environmental factors associated with biodiversity and the occurrence of rare, threatened, thermophilous species of aquatic beetles in the anthropogenic ponds of the Masurian Lake District. *Biodivers. Conserv.* **2014**, *24*, 429–445. [CrossRef]

12. Visconti, A.; Caroni, R.; Rawcliffe, R.; Fadda, A.; Piscia, R.; Manca, M. Defining seasonal functional traits of a freshwater zooplankton community using δ13C and δ15N stable isotope analysis. *Water* **2018**, *10*, 108. [CrossRef]

13. Ejsmont-Karabin, J.; Gulati, R.D.; Rooth, J. Is food availability the main factor controlling the abundance of Euchlanis dilatata lucksiana Hauer in a shallow, hypertrophic lake? *Hydrobiologia* **1989**, *186*, 29–34. [CrossRef]

14. Kuczyńska-Kippen, N.; Klimaszyk, P. Diel microdistribution of physical and chemical parameters within the dense Chara bed and their impact on zooplankton. *Biol. Bratisl.* **2007**, *62*, 432–437. [CrossRef]

15. Joniak, T.; Kuczyńska-Kippen, N. Habitat features and zooplankton community structure of oxbows in the limnophase: Reference to transitional phase between flooding and stabilization. *Limnetica* **2016**, *35*, 37–48.

16. De Marco, P.; Nogueira, D.S.; Correa, C.C.; Vieira, T.B.; Dias-Silva, K.; Pinto, N.S.; Bichsel, D.; Hirota, A.S.V.; Vieira, R.R.S.; Carneiro, F.; et al. Patterns in the organization of Cerrado pond biodiversity in Brazilian pasture landscapes. *Hydrobiologia* **2013**, *723*, 87–101. [CrossRef]

17. Kuczyńska-Kippen, N. Seasonal changes of the rotifer community in the littoral of a polymictic lake. *SIL Proc.* **1998**, *27*, 2964–2967. [CrossRef]

18. Pinel-Alloul, B.; Ghadouani, A. Spatial Heterogeneity of Planktonic Microorganisms in Aquatic Systems. In *The Spatial Distribution of Microbes in the Environment*; Springer Science and Business Media LLC: Dorsdrecht, The Netherlands, 2007; pp. 203–310.

19. Celewicz, S.; Kuczyńska-Kippen, N. Ecological value of macrophyte cover in creating habitat for microalgae (diatoms) and zooplankton (rotifers and crustaceans) in small field and forest water bodies. *PLoS ONE* **2017**, *12*, e0177317.

20. Collinson, N.; Biggs, J.; Corfield, A.; Hodson, M.; Walker, D.; Whitfield, M.; Williams, P. Temporary and permanent ponds: An assessment of the effects of drying out on the conservation value of aquatic macroinvertebrate communities. *Biol. Conserv.* **1995**, *74*, 125–133. [CrossRef]

21. Tolonen, K.T.; Holopainen, I.J.; Rahkola-Sorsa, M.; Mikkonen, K.; Karjalainen, J.; Ylöstalo, P.; Hämäläinen, H. Littoral species diversity and biomass: Concordance among organismal groups and the effects of environmental variables. *Biodivers. Conserv.* **2005**, *14*, 961–980. [CrossRef]

22. Sahuquillo, M.; Miracle, M.R. Rotifer communities in Mediterranean ponds in eastern Iberian Peninsula: Abiotic and biotic factors defining pond types. *Limnetica* **2019**, *38*, 103–117.

23. Van Onsem, S.; De Backer, S.; Triest, L. Microhabitat–zooplankton relationship in extensive macrophyte vegetations of eutrophic clear-water ponds. *Hydrobiologia* **2010**, *656*, 67–81. [CrossRef]

24. Klimaszyk, P.; Piotrowicz, R.; Rzymski, P. Changes in physico-chemical conditions and macrophyte abundance in a shallow soft-water lake mediated by a Great Cormorant roosting colony. *J. Limnol.* **2014**, *73*, 114–122. [CrossRef]

25. Nagengast, B.; Kuczyńska-Kippen, N. Macrophyte biometric features as an indicator of the trophic status of small water bodies. *Oceanol. Hydrobiol. Stud.* **2015**, *44*, 38–50. [CrossRef]

26. Kuczyńska-Kippen, N. Spatial distribution of zooplankton communities between the Sphagnum mat and open water in a dystrophic lake. *Pol. J. Ecol.* **2008**, *56*, 57–64.

27. Joniak, T.; Nagengast, B.; Kuczyńska-Kippen, N. Can popular systems of trophic classification be used for small water bodies? *Oceanol. Hydrobiol. Stud.* **2009**, *38*, 145–151. [CrossRef]

28. Ejsmont-Karabin, J.; Karabin, A. The suitability of zooplankton as lake ecosystem indicators: Crustacean trophic state index. *Pol. J. Ecol.* **2013**, *61*, 561–573.

29. Špoljar, M.; Tomljanović, T.; Lalić, I. Eutrophication impact on zooplankton community: A shallow lake approach. *Holist. Approach Environ.* **2011**, *1*, 131–142.

30. Vijverberg, J.; Boersma, M. Long-term dynamics of small-bodied and large-bodied cladocerans during the eutrophication of a shallow reservoir, with special attention for Chydorus sphaericus. *Hydrobiologia* **1997**, *360*, 233–242. [CrossRef]

31. Joniak, T.; Kuczyńska-Kippen, N.; Gąbka, M. Effect of agricultural landscape characteristics on the hydrobiota structure in small water bodies. *Hydrobiologia* **2016**, *793*, 121–133. [CrossRef]

32. Radwan, S.; Bielańska-Grajner, I.; Ejsmont-Karabin, J.; Rotifera, W. Fauna słodkowodna Polski. In *Rotifers Rotifera. Freshwater fauna of Poland*; Oficyna Wydawnicza Tercja: Łódź, Poland, 2004.

33. Margalef, R. Information theory in ecology. *Gen. Syst.* **1957**, *3*, 36–71.

34. Ter Braack, C.J.F.; Šmilauer, P. *CANOCO Reference Manual and User's Guide to Canoco for Windows: Software for Canonical Community Ordination (version 4.5)*; Microcomputer Power: Ithaca, NY, USA, 2002.

35. Špoljar, M. Microaquatic communities as indicators of environmental changes in lake ecosystems. *J. Eng. Res.* **2013**, *1*, 29–42.

36. Cottenie, K.; Nuytten, N.; Michels, E.; De Meester, L. Zooplankton community structure and environmental conditions in a set of interconnected ponds. *Hydrobiologia* **2001**, *442*, 339–350. [CrossRef]

37. Kornijów, R. A quantitative sampler for collecting invertebrates associated with deep submerged vegetation. *Aquat. Ecol.* **2014**, *48*, 417–422. [CrossRef]

38. Yin, X.; Jin, W.; Zhou, Y.; Wang, P.; Zhao, W. Hidden defensive morphology in rotifers: Benefits, costs, and fitness consequences. *Sci. Rep.* **2017**, *7*, 4488. [CrossRef]

39. Mouillot, D.; Gaillard, S.; Aliaume, C.; Verlaque, M.; Belsher, T.; Troussellier, M.; Dochi, T. Ability of taxonomic diversity indices to discriminate coastal lagoon environments based on macrophyte communities. *Ecol. Indic.* **2005**, *5*, 1–17. [CrossRef]

40. Barnett, A.; Finlay, K.; Grossart, H.-P. Functional diversity of crustacean zooplankton communities: Towards a trait-based classification. *Freshw. Biol.* **2007**, *52*, 796–813. [CrossRef]

41. Hilt, S.; Henschke, I.; Rücker, J.; Nixdorf, B. Can Submerged Macrophytes Influence Turbidity and Trophic State in Deep Lakes? Suggestions from a Case Study. *J. Environ. Qual.* **2010**, *39*, 725–733. [CrossRef]

42. Larkin, D.J.; Monfils, A.K.; Boissezon, A.; Sleith, R.S.; Skawinski, P.M.; Welling, C.H.; Cahill, B.C.; Karol, K.G. Biology, ecology, and management of starry stonewort (Nitellopsis obtusa; Characeae): A Red-listed Eurasian green alga invasive in North America. *Aquat. Bot.* **2018**, *148*, 15–24. [CrossRef]

43. Pełechaty, M.; Pronin, E.; Pukacz, A. Charophyte occurrence in Ceratophyllum demersum stands. *Hydrobiologia* **2013**, *737*, 111–120. [CrossRef]

44. Nagengast, B.; Gąbka, M. Apparent niche partitioning of two congeneric submerged macrophytes in small water bodies: The case of *Ceratophyllum demersum* L. and *C. submersum* L. *Aquat. Bot.* **2017**, *137*, 1–8. [CrossRef]

45. Kuczyńska-Kippen, N.; Nagengast, B.; Celewicz, S.; Klimko, M. Zooplankton community structure within various macrophyte stands of a small water body in relation to seasonal changes in water level. *Oceanol. Hydrobiol. Stud.* **2009**, *38*, 125–133. [CrossRef]

46. Basińska, A.M.; Antczak, M.; Świdnicki, K.; Jassey, V.E.J.; Kuczyńska-Kippen, N. Habitat type as strongest predictor of the body size distribution of Chydorus sphaericus (O. F. Müller) in small water bodies. *Int. Rev. Hydrobiol.* **2014**, *99*, 1–11. [CrossRef]

47. Declerck, S.A.J.; De Bie, T.; Ercken, D.; Hampel, H.; Schrijvers, S.; Van Wichelen, J.; Gillard, V.; Mandiki, S.N.M.; Losson, B.J.; Bauwens, D.; et al. Ecological characteristics of small farmland ponds: Associations with land use practices at multiple spatial scales. *Biol. Conserv.* **2006**, *131*, 523–532. [CrossRef]

48. Knapp, S.; Kühn, I.; Mosbrugger, V.; Klotz, S. Do protected areas in urban and rural landscapes differ in species diversity? *Biodivers. Conserv.* **2008**, *17*, 1595–1612. [CrossRef]
49. Duggan, I.; Green, J.D.; Shiel, R.J. Distribution of rotifer assemblages in North Island, New Zealand, lakes: Relationships to environmental and historical factors. *Freshw. Biol.* **2002**, *47*, 195–206. [CrossRef]

Article

The Effect of Hydrological Connectivity on the Zooplankton Structure in Floodplain Lakes of a Regulated Large River (the Lower Vistula, Poland)

Paweł Napiórkowski *, Martyna Bąkowska, Natalia Mrozińska, Monika Szymańska, Nikola Kolarova and Krystian Obolewski

Department of Hydrobiology, Faculty of Natural Sciences, Kazimierz Wielki University, 85-064 Bydgoszcz, Poland; bakowska@ukw.edu.pl (M.B.); nattam@ukw.edu.pl (N.M.); m.szymanska9306@gmail.com (M.S.); n.kolar77@gmail.com (N.K.); obolewsk@ukw.edu.pl (K.O.)
* Correspondence: pnapiork@ukw.edu.pl; Tel.: +48-523-419-171

Received: 29 July 2019; Accepted: 11 September 2019; Published: 14 September 2019

Abstract: The zooplankton community structure and diversity were analysed against the gradient of floodplain lakes connectivity and water level under different flood-pulse dynamics in the Vistula River. The lakes differed in terms of hydrology, among others in the degree/type of their connection with the river (permanent, temporary and no connection). The study was conducted during the growing seasons in the years 2006–2013 and involved the lower Vistula River and three floodplain lakes: isolated, transitional and connected. Water samples were collected biweekly from April to September. Zooplankton was the most diverse and abundant in the transitional lake (the highest Shannon α-diversity index H′ and Pielou's evenness index J′). The gentle washing of the lakes might have stimulated the development of zooplankton in accordance with the Intermediate Disturbance Hypothesis. The diversity and density of zooplankton were higher in the connected lake compared to the isolated one. We confirmed the hypothesis that zooplankton should be more abundant and diverse in floodplain lakes connected with the river (or transitional) than in isolated ones. Zooplankton analyses indicated that hydrological conditions (flood-pulse regime) contributed most substantially to zooplankton diversity and density in the floodplain lakes of the lower Vistula valley.

Keywords: invertebrates; hydrological regime; diversity; water bodies

1. Introduction

Floodplain lakes, known for high biodiversity and ecological value, are important elements of landscapes with large rivers [1–4]. A floodplain lake is defined as a small water body located in a river valley, permanently/periodically connected to or isolated from the main river channel, formed when a meander is cut off naturally or separated from the river by a flood embankment. Parts of the old river channel connected periodically or permanently with the river bed are also considered as floodplain lakes [5]. Floodplain lakes are generally shallow, astatic water bodies with varying environmental conditions and macrophyte dominance. They can be seen as flowing-to-stagnant water transition zones [6–8].

Floodplain lakes can be divided into isolated from the river, temporarily connected to the river, or permanently connected to the river [5]. According to the flood-pulse concept [9] the functioning of floodplain lakes depends on periodic river flooding. The resulting connection between the river and the lake allows the exchange of water with nutrients and organisms between all elements of the river system [1,2,10,11]. Hydrological connectivity is observed on four levels of fluvial systems: longitudinal, lateral, vertical and temporal [2]. The lateral and temporal (long-term dynamics) connectivity were investigated.

The degree of hydrological connectivity, duration and frequency of flooding as well as lake water level depend on many factors such as lake location, its distance from the river, river level fluctuations and catchment area [2,12]. Increased hydrological connectivity or flood-pulse act as a homogenizing factor [13,14].

Dias et al. [15], Dittrich et al. [16], Anderson and Bonecker [17] and Schöll et al. [18] point out that plankton communities in floodplain lakes are determined primarily by these two factors: whether a lake is connected with the river and whether inundation causes any disturbances in a lake.

The intensity of flooding determines species diversity and density in floodplain lakes. According to the Intermediate Disturbance Hypothesis (IDH) [19], medium flooding can increase species diversity [20]. The IDH predicts low species diversity in ecosystems exposed to high and low levels of disturbance. Under these conditions the survival is guaranteed only for species which can easily adapt to changing and/or extreme conditions or can quickly recolonize a given ecosystem (e.g., floodplain lakes after floods).

There are two theories describing how the connection between a floodplain lake and a river affects the zooplankton community. Kobayashi et al. [21] and Lemke et al. [22] observed high zooplankton density in hydrologically isolated floodplain lakes. The inflow of river water to lakes connected with the main river channel can periodically destabilize their environmental conditions. Lower water temperature and transparency inhibit the growth of macrophytes, which normally provide a habitat for zooplankton (space and food). Conditions in isolated floodplain lakes will lead to greater diversity and abundance of planktonic crustaceans (more species and higher abundance). However, Hein et al. [23] and Kasten [24] hypothesize that zooplankton should be more abundant and diverse in lakes connected to the river than in the isolated ones.

The main objective of our study was to answer the question of how different hydrological connectivity between a large regulated river and its floodplain lakes can shape zooplankton communities. Before the investigation we put forward the following hypotheses: (1) The river zooplankton will be less diverse and less abundant than zooplankton in the studied floodplain lakes owing to specific environmental conditions in the river (turbulent water flow, lower temperature, etc.). (2) The degree of connectivity between a particular lake and the river will affect the zooplankton structure. The diversity and abundance of zooplankton will be lower in the lakes connected with the river than in the isolated one as a result of less stable environmental conditions in the former. (3) Flood-pulse dynamics will have an impact on a degree of connectivity between the lakes and the river and will affect the zooplankton structure. During high water level the predominance of small organisms (rotifers) will be observed. During low water level the abundance of crustaceans will increase.

2. Materials and Methods

The study involved three floodplain lakes lying in the lower Vistula valley. Over almost its entire length Vistula is a typical lowland river. The first floodplain terrace has many lakes which are the remnants of the Vistula backwaters and are periodically flooded. The investigated lakes were created after the construction of flood embankments during the river regulation in the 19th century [25]. The lakes are shallow and relatively young (ca. 150 years). Before regulation the Vistula was a braided river, so its old riverbeds tend to have an elongated shape and be half-open (semi-lotic) or closed (lenitic) (Figure 1).

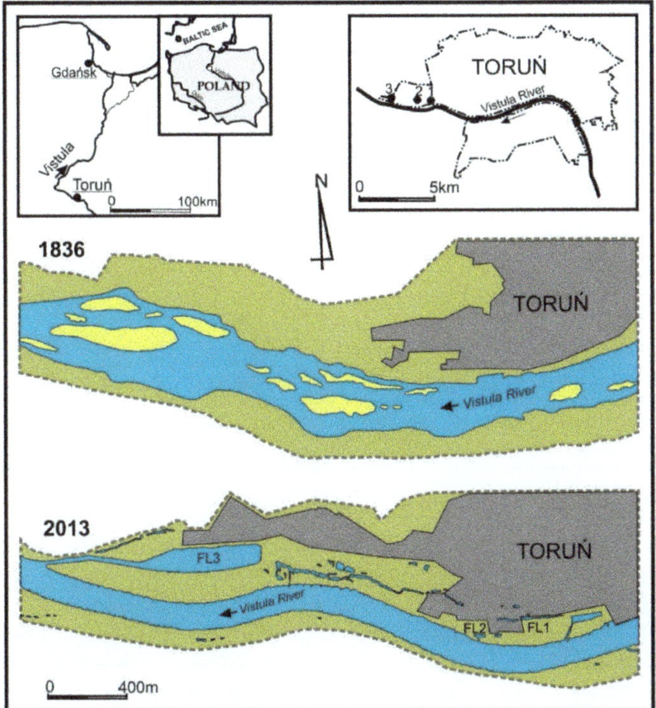

Figure 1. Hydrological system of the lower Vistula River in Toruń (Poland) before regulation (1836) and today (2013). FL1—floodplain lake isolated from the river, FL2—floodplain lake periodically connected to the river (transitional), FL3—floodplain lake permanently connected to the river.

The three studied floodplain lakes of the lower Vistula River are located in the city of Toruń: FL1 (53°00′ N, 18°34′ E)—isolated from the river (n = 34), FL2 (53°00′ N, 18°33′ E)—temporarily isolated from the river (n = 38: 19 with connection and 19 without connection) and FL3 (53°01′ N, 18°30′ E)— permanently connected with the river by a long channel (n = 39) (n—number of samples). FL1, located in Toruń city park at the 737th km of the river, is a small (2.5 ha) and shallow (2.0 m) water body without a direct surface contact with the Vistula River. It has a submerged macrophyte community and a rich littoral zone. The following macrophyte species inhabit the lake: yellow pond-lily (*Nuphar lutea*), rigid hornwort (*Ceratophyllum demersum*), star duckweed (*Lemna trisulca*). Different species of filamentous algae are also found in the lake.

FL2, located in Toruń at the 738th km of the river is small (1.0 ha). At the medium water level in the Vistula it is connected with the river through a wide channel (up to 30 m). At the low water level the lake is isolated. The area at the northern shore is covered by gardens. With a small surface and low exposure to wind, the lake is not very susceptible to water mixing. At the low water level in the Vistula elodeids (mainly *C. demersum* and *Potamogeton* spp.) could have developed however their density was relatively small (no reeds observed).

FL3 is located at the 745th km of the Vistula, (the largest) (Figure 1, Table 1). It is permanently connected with the river through a 1.2 km channel, constructed to make possible timber transportation. The direct catchment comprises woodlands, grasslands and agricultural areas. During growing seasons we observed rapid growth of elodeids on the lake bottom, resulting from the low water level in the Vistula. The dominant species included Canadian waterweed (*Elodea canadiensis*), hornwort (*C. demersum*), and potamogetons (*Potamogeton* spp.). At very low water level submerged vegetation filled almost the entire water column. Although floodplain lakes differed in size and morphology, the habitat complexity during the vegetation season seemed to be low.

Table 1. Lateral hydrological connectivity definitions for floodplain lakes of the lower Vistula River. Type corresponds to categories of hydrological connectivity (HC). Depth is at the deepest part of the floodplain lakes in 2006–2013. Age is the time since the cutoff of floodplain lakes, or construction, depending on origin.

Site	Depth (m)	Age (yr)	Area (ha)	Length (m)	Geographic Coordination	HC (%)	HC Definition
VR	-	-	-	-	N 53°00′ E 18°60′	-	Vistula River
FL1	2.5	155	2.8	640	N 53°00′ E 18°34′	0	Floodplain lake without connection with the Vistula River
FL2	0.6–1.6	155	1.0	160	N 53°00′ E 18°33′	50	Floodplain temporarily connected with the Vistula River
FL3	11.0	155	71	1800	N 53°01′ E 18°30′	100	Floodplain permanently connected with the Vistula River by one arm 1.2 km length and width 50 m

Our studies were conducted during the growing seasons in the years 2006–2013 and involved the lower Vistula River and three floodplain lakes. Water samples were collected biweekly from April to September. A total of 135 samples were collected. Water samples were collected with a 1 L Patalas bucket at the depth of ca. 0.5 m in the central part of each water body. Water was filtered through a plankton net, mesh size 25 μm. Ten litres of water were filtered to obtain one sample of zooplankton. All zooplankton samples were preserved in Lugol's solution [26,27]. The sample volume (10 L) was adjusted to 10 mL and a 1 mL aliquot of well mixed concentrate was pipetted into a Segdwick–Rafter chamber. The zooplankton was counted under a microscope in a Segdwick–Rafter chamber by the sub-sample method [26]. The abundance of zooplankton was calculated per volume of 1 L of water. The identification of zooplankton was performed with the use of a light microscope Nikon Alphaphot-2, a camera and MultiScan computer software for image analysis. The taxonomical identification of zooplankton was based on the commonly available studies and keys [26,28–33].

To characterize the abundance-dominance relationship the Shannon α-diversity index (H′) and Pielou evenness index (J′) were used. Sampling was accompanied by the measurements of physical and chemical parameters of water, such as: Secchi disk visibility (SD, m) (except the river), temperature (T_w, °C), dissolved oxygen (DO, mg L^{-1}), conductivity (EC, μS cm^{-1}), and pH. Measurements of physicochemical parameters were performed with the use of multimeter WTW Multi 3430SET F (Xylem Analytics, Weilheim, Germany) field probes. Data on the water level (WL, cm) in the Vistula River in Toruń city were obtained from the Meteorological and Hydrological Institute, the Regional Hydrological and Meteorological Station in Toruń.

For statistical analysis, the investigated floodplain lakes were classified based on the degree of their connectivity with the river. The final dataset contained basic environmental variables: dates of analysis combined with the degree of connectivity between each lake and the river (1—isolated, 2—transitional; 3—connected), water level, water physicochemical parameters, zooplankton richness and density.

Similarity analysis performed to classify abiotic data and confirm floodplain lake types was based on non-metric multidimensional scaling (nMDS) using Bray–Curtis dissimilarity indices and software Past v.3.01 [34].

To evaluate general differences in the zooplankton structure we investigated three water bodies (FL1, FL2 and FL3) on the given dates and performed analysis of variance (ANOVA) with the Kruskal–Wallis test (K–W) followed by post-hoc Dunn's test in Prism 5.01 software (GraphPad Software, San Diego, CA, USA).

The following variables were determined: water level, water physicochemical parameters (T_w, SD, pH, DO, EC), species richness (zooplankton species richness, rotifer species richness and crustacean species richness) and mean zooplankton density (total zooplankton density, rotifer, crustacean and predominant species densities), α-diversity and evenness indexes.

Redundancy analysis (RDA) was performed using the covariance method to determine the relative significance of environmental factors in explaining the variability of the tested samples. The dataset was log transformed (log (n + 1)) and centred on species, as this was obligatory for the constrained linear methods.

Consequently, the relationships between predictors (dates, sites, water level, water physicochemical properties) and zooplankton density were analysed by RDA [35,36]. The statistical significance of canonical axes was determined in the Monte Carlo permutation test [37]. A subset of independent variables representing the relationships between environmental factors and the taxa of planktonic fauna was identified by eliminating factors that were not significant for the model. The data were processed statistically in Canoco 5.0 software (Wageningen University & Research, Wageningen, Holland) at probability levels of * $p < 0.05$, ** $p < 0.01$ and *** $p < 0.001$ [38].

To explore significant positive and negative relationships between zooplankton and specified descriptors of physicochemical properties of lake water, t-value biplots (with van Dobben circles), which approximate the t-values of the regression coefficients of a weighted multiple regression, were generated. The t-value biplots indicated the zooplankton data that to a large extent reacted to the tested factor [38].

3. Results

3.1. Hydrological Conditions

From 2006 to 2013 water levels in the Vistula River at a gauging station in Toruń ranged from 1.45 m to 8.36 m above sea level (a.s.l.) and the average level was 3.0 m a.s.l. (Figure 2). In the research period water levels ranged from 1.45 m to 5.25 m a.s.l., with the average value of 2.84 m a.s.l.

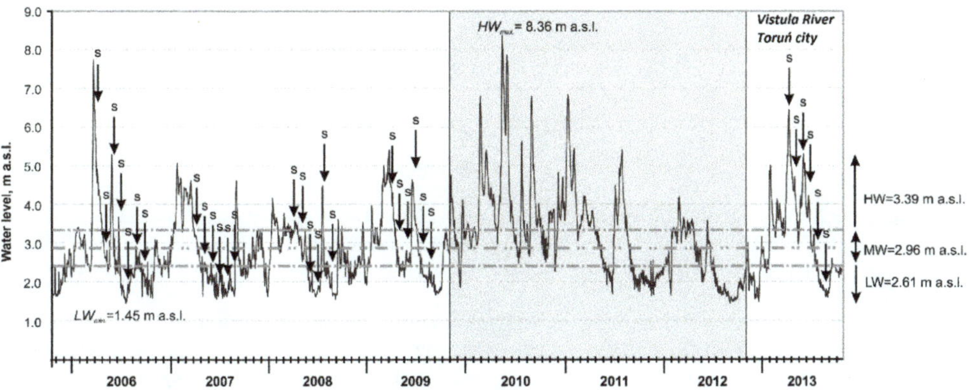

Figure 2. Hydrograph of the lower Vistula River from the gauge station in Toruń. HW_{max}—the maximum water level, LW_{min}—the lowest water level. HW—average of high water levels, MW—average of mean water levels, LW—average of low water levels have been calculated from the study period 2006–2009 and 2013 (grey area indicates a period without sampling). Arrows with "s" indicate sampling times.

The majority of samples (2006–2009) were collected during low and average water levels in the Vistula River. Only in 2013, the river level was high during sampling. In May 2010, a catastrophic flood that occurred on the Vistula inundated all the investigated floodplain lakes and changed their biotic and abiotic conditions. To eliminate the impact of the flood on the results, samples were not collected from 2010 to 2012. In 2013 we assumed that the river had returned to the hydrological state that was observed before the flood.

FL3 is directly connected with the river channel, therefore changes in lake levels and discharges are determined by the river. FL2 did not have permanent surface connectivity with the river during the research period. When the water level in the Vistula dropped below 2.31 m a.s.l. no connectivity was observed. For 38 observations conducted in the transitional floodplain lake, half were made during the isolation period. FL1, without direct contact with the Vistula River could be shaped indirectly by underground water movements when the water level in the river changed.

3.2. Physical and Chemical Parameters

The non-metric multidimensional scaling (nMDS) revealed remarkable differences in environmental conditions (SD, T_W, DO, EC and pH) of the studied floodplain lakes considering the degree of their connectivity with the river. It allowed us to determine the index of multivariate dispersion of data from individual lakes. Based on environmental conditions nMDS analysis divided the collected samples between the types of floodplain lakes: isolated, transitional and connected (Figure 3).

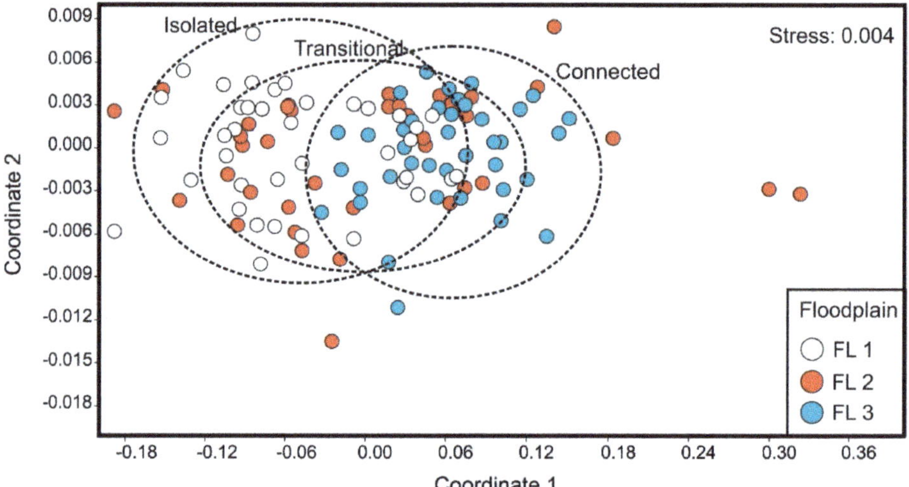

Figure 3. Non-metric multidimensional scaling analysis (nMDS) of the environmental conditions (Secchi disk visibility (SD), temperature (T_W), dissolved oxygen (DO), conductivity (EC) and pH) (Stress: 0.004) in three types of floodplain lakes (FL1—isolated; FL2—transitional; FL3—connected).

The highest average water transparency (1.45 m) was noted in the isolated lake, and the lowest (0.63 m), in the connected lake (K–W = 33.73, $p \leq 0.001$). In contrast, the differences in average temperatures (T_w) and conductivity (EC) among the studied floodplain lakes were not statistically significant. The highest average temperature was recorded in the connected lake (20.0 °C), and the lowest, in the transitional lake (19.1 °C). The highest average value of EC was recorded in the connected lake (625), and the lowest, in the isolated lake (576) (Table 2). The highest average concentration of dissolved oxygen in water (DO) was recorded in the connected lake (8.96 mg L^{-1}), and the lowest, in the isolated lake (7.87 mg L^{-1}) (K–W = 10.72, p = 0.01). The highest pH was recorded in the transitional lake (8.6), and the lowest, in the isolated lake (8.3), (K–W = 10.97, p = 0.01). Statistically significant differences among sites in visibility (SD), dissolved oxygen (DO) and pH value were observed.

Table 2. Water parameters of the investigated floodplain lakes and the Vistula River. Data present the mean (s.d.) for every month samples in 2006–2013. Bold shows significant differences (nonparametric Kruskal–Wallis test, * $p \leq 0.05$; *** $p \leq 0.001$) among water bodies.

	FL1 (n = 34)	FL2 (n = 38)	FL3 (n = 39)	RV (n = 24)
Water Temperature (°C)	19.2 (4.2)	19.1 (4.1)	20.0 (3.7)	19.3 (3.7)
Visibility (cm) *	1.45 (0.44)	0.75 (0.21)	0.63 (0.23)	0.54 (0.12)
pH *	8.32 (0.36)	8.60 (0.63)	8.51 (0.32)	8.26 (0.22)
Dissolved Oxygen (mg L^{-1}) *	7.87 (2.72)	8.29 (2.59)	8.96 (2.20)	7.14 (1.66)
EC (µS cm^{-1})	576 (95)	615 (114)	625 (84)	596 (80)

3.3. Taxonomic Richness and Abundances

Samples analysis revealed the presence of 97 zooplankton species in the investigated lakes, including 75 rotifer species (i.e., 77% of all species), 22 crustacean species (i.e., 33% of all species) and nauplii and copepodites, larval forms of Copepoda. The highest number of species (73) was recorded in the transitional lake. The lowest number of species (52) was recorded in the isolated lake (Table 3). There were fewer species (47) in the main channel of the Vistula River (Table 3). The results also indicated that the highest number of species (both rotifers and crustaceans) was recorded in the transitional lake (54 and 19, respectively). The lowest number of species (both rotifers and crustaceans) was recorded in the isolated lake (44 and 8, respectively). The statistically significant differences among the sites were observed in the number of crustacean species ($K–W = 20.30$, $p \leq 0.001$) (Table 3). The list of zooplankton taxa is presented in Table S1 (Supplementary Materials).

Table 3. Zooplankton structure of the investigated floodplain lakes and the Vistula River. Data present the mean (s.d.) for every samples between 2006 and 2013, including only dominant taxa (D > 10%). H'—α-diversity, J'—evenness. Significant differences (nonparametric Kruskal–Wallis test; ** $p \leq 0.01$; *** $p \leq 0.001$) among the investigated water bodies are shown in bold.

	FL1 (n = 34)	FL2 (n = 38)	FL3 (n = 39)	RV (n = 24)
Richness	52	73	63	47
No. of Crustacea species *	8	19	14	6
No. of Rotifera species	44	54	49	41
Total density (ind L^{-1})	1111 (1255)	1995 (3406)	1847 (2587)	344 (416)
H' index	1.58 (0.47)	1.80 (0.37)	1.70 (0.50)	1.97 (0.27)
J' index	0.581 (0.151)	0.656 (0.147)	0.633 (0.153)	0.745 (0.108)
Crustacea (ind L^{-1}) *	300 (298)	290 (329)	451 (697)	21 (22)
Bosmina longirostris	49	67	145	3
nauplii *	181	145	189	15
Rotifera (ind L^{-1}) *	811 (1001)	1705 (3285)	1396 (2359)	323 (407)
Keratella tecta **	21 (33)	517 (1859)	295 (632)	91 (146)
Keratella cochlearis	286 (577)	290 (775)	114 (190)	57 (71)
Keratella quadrata **	29 (34)	74 (173)	26 (38)	5 (9)
Polyarthra longiremis	308 (509)	207 (517)	359 (1584)	44 (81)
Brachionus angularis	5 (8)	178 (478)	43 (103)	25 (49)

The average zooplankton density in the studied floodplain lakes was 1651 ind L^{-1}. The highest average zooplankton density was recorded in the transitional lake (1995 ind L^{-1}), and the lowest, in the isolated lake (1111 ind L^{-1}) (Figure 4A). The average rotifer density was more than twice as high in the transitional lake as in the isolated one (Table 3, Figure 4A). The difference in the rotifer density among the studied sites was statistically significant ($K–W = 12.27$, $p = 0.007$). The average crustacean density was the highest in the connected lake (451 ind L^{-1}) and the lowest in the transitional lake (290 ind L^{-1}) (Table 3, Figure 4A). The difference in the crustacean density between the sites was also statistically significant ($K-W = 46.56$, $p \leq 0.0001$). The density of dominant species was the highest in the transitional lake, e.g., *Keratella tecta*, 517 ind L^{-1}, *Keratella quadrata*, 74 ind L^{-1} (Table 3). The lowest

K. tecta density was noted in the isolated lake, and the lowest *K. quadrata* density, in the connected lake (Table 3, Figure 4B). The differences in the density of *K. tecta* and *K. quadrata* between the studied sites were statistically significant (K–W = 12.78 and 15.80, $p \leq 0.001$, respectively).

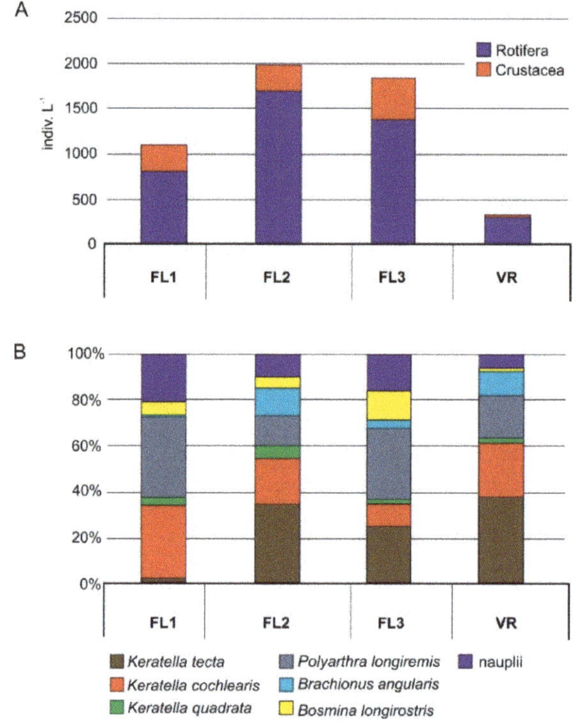

Figure 4. Zooplankton density (ind L^{-1}) in the investigated floodplain lakes (FL1—isolated, FL2—transitional, FL3—connected) and in the Vistula River (VR) (**A**); Percentage share of dominant taxa in zooplankton density in the studied lakes and in the Vistula River (**B**).

In addition, the average density of *Keratella cochlearis* and *Brachionus angularis* were also the highest in the transitional lake (Table 3). However, for these two species, the differences in the average density among the studied sites were not statistically significant.

The highest density of dominant crustacean forms (e.g., copepod larval forms nauplii) and species was recorded in the connected lake. The lowest density of nauplii was recorded in the transitional lake (K–W = 34.47, $p \leq 0.0001$). *Bosmina longirostris* (Cladocera) was almost three times more abundant in the connected lake than in the isolated one (Table 3). *K. tecta* had the largest share in density among species in the river (VR) and in the transitional lake (Figure 4). *Polyarthra longiremis* and *K. cochlearis* had the biggest share among the dominant species in the isolated lake, while *P. longiremis* and *K. tecta*, in the connected lake. Based on the results it can be concluded that the Vistula River had the highest impact on the transitional and connected floodplain lakes (e.g., *K. tecta*—Dunn's test, $p \leq 0.01$) (Table 3, Figure 4B). The highest value of α-diversity (H′ = 1.80 ± 0.37) and evenness index (J′ = 0.656 ± 0.147) were noted in the transitional lake (Table 3) while the lowest, in the isolated lake (α-diversity H′ = 1.58 ± 0.47), evenness index (J′ = 0.581 ± 0.151).

3.4. Influence of Environmental Factors on Zooplankton Communities

The RDA revealed a relationship between zooplankton species composition and environmental variables (Figure 5A). The results of the ordination showed that the eigenvalues of the first ($\lambda_{RDA1} = 0.407$) and second ($\lambda_{RDA2} = 0.169$) RDA axes accounted for 57.6% of the variation in the environmental data. All canonical axes were significant (Monte Carlo test, $p = 0.002$).

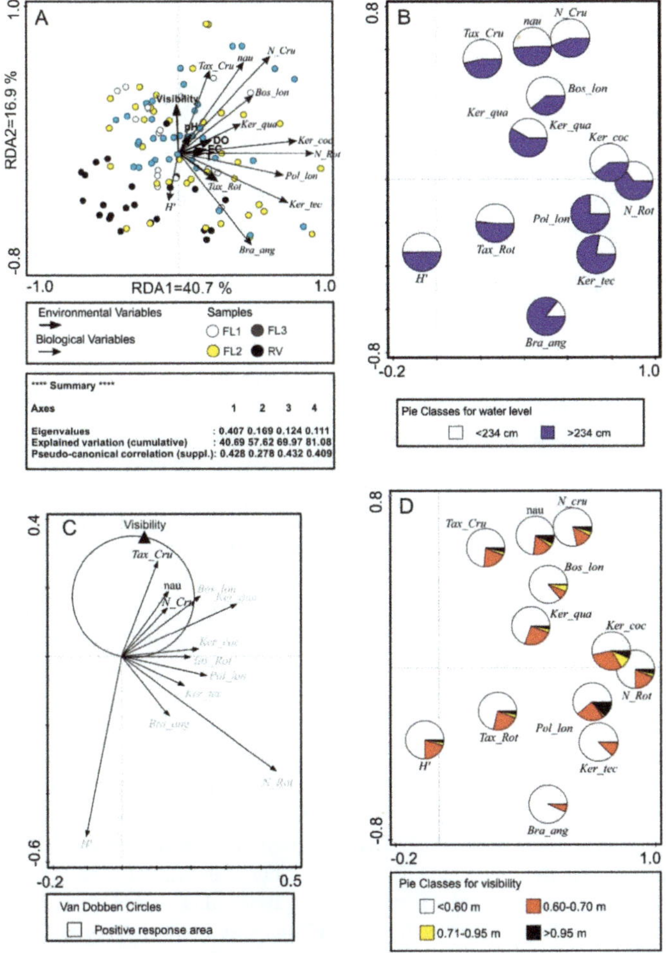

Figure 5. Results of redundancy analysis (RDA) performed on zooplankton and environmental data for the three types of floodplain lakes using forward selection of variables ($p < 0.05$). (**A**) Triplot of significant environmental variabilities, zooplankton and samples; (**B**) relative values of zooplankton communities (pies charts) in relation to the water level in river channel and floodplain lakes (<234 cm—limnophase, >234 cm—potamophase); (**C**) Van Dobben circles analysis (visibility vs. zooplankton structure). A circle indicates positive responses; (**D**) relative values of zooplankton communities in pie charts in relation to visibility. Codes: H'—α-diversity index, Tax_Rot—number of rotifer species, Tax_Crus—number of crustacean species; N_Rot—density rotifers, N_Cru—crustaceans, nau—nauplii, Bos_lon—*Bosmina longirostris*, Ker_qua—*Keratella quadrata*, Ker_coc—*Keratella cochlearis*, Pol_lon—*Polyarthra longiremis*, Ker_tec—*Keratella tecta*, Bra_ang—*Brachionus angularis*; Tw—Temperature, EC—Electrolytic Conductivity, DO—Dissolved Oxygen.

The longest vector describing the visibility is closely correlated with the second axis of ordination. In contrast, the temperature, conductivity and oxygen concentration vectors were well correlated with the first axis of ordination, but were short, which means that these parameters are of less importance. Similarly, the pH vector, which was more closely correlated with the second axis of ordination, was short so that pH was also less important in this analysis.

The RDA indicated that zooplankton did not show a preference for river habitat. Crustacean zooplankton preferred the lake connected with the river, while rotifers the transitional lake (Figure 5A). According to pie charts analysis for water level (Figure 5B) we noted the following relationships: during higher river level (>234 cm) rotifers developed more efficiently, as manifested in higher average density of rotifers, higher average density of *K. tecta*, *B. angularis* and *P. longiremis*; during lower river level (<234 cm) crustaceans developed more efficiently, as manifested in the higher average density of Crustacea as well as in the higher average density of *B. longirostris*. In addition, rapid development of *K. cochlearis* (rotifers) was also observed.

Based on the analysis of RDA and Van Dobben circles (Figure 5C) the visibility was positively correlated with the average number of crustacean species, average density of crustaceans and average density of nauplii, copepod larval forms. Low water transparency was preferred by *K. tecta* and *B. angularis* (indicator species of high trophy) (Figure 5D).

4. Discussion

We investigated zooplankton in three floodplain lakes, created as a result of the regulation of the lower Vistula in the mid-19th century. The lakes differed in terms of hydrology, among others in the degree of their connection with the river (isolated, transitional and connected lake). Lateral connectivity of the lakes was responsible for nutrient cycling and biodiversity in the river–floodplain system [1,2,9,15]. The type of connection between the floodplain lake and the river is also important. According to Paira and Drago [5] direct connection occurs when the channel between a floodplain lake and a river is shorter than 1 km. Periodic floods in the Vistula River cause the inclusion of the floodplain lakes into the river system (Figure 2). Massive floods destabilize environmental conditions in floodplain lakes by reducing water transparency, lowering water temperature and inhibiting macrophyte development [39–42]. All these changes have a negative impact on the zooplanktonic population. This observation is true only for intensive flooding. However, on the majority of dates (Figure 2) we recorded medium or small flooding, which did not significantly affect environmental conditions in the studied lakes [20].

The diversity of habitat conditions in floodplain lakes results from several factors including the following: lake–river distance, permanent versus temporary connection between the lake and the river, the size and shape of the lake [2]. The results of our studies, based on environmental conditions, divided floodplain lakes into three types: isolated (FL1), transitional (FL2) and connected (FL3) (Figure 3). The abundance and diversity of invertebrates is generally higher in floodplain lakes compared to the main channel of the river while the taxonomic structure of individual clusters is usually similar [11,43,44]. Zooplankton in the Vistula River was less diverse and less abundant than zooplankton in the studied floodplain lakes because of specific environmental conditions in the river e.g., turbulent water flowing (Figure 4). The taxonomic composition of zooplankton in the Vistula river has impact on the structure of zooplankton communities in the studied floodplain lakes (Figure 4B). A similar relationship has also been observed by other authors [45].

Rotifers predominated in zooplankton diversity and density in the river and floodplain lakes but their predominance varied depending on hydrological conditions. They constituted approximately 90% of zooplankton density in the river, and 60%–80% in the floodplain lakes (depending on the connectivity) (Table 3, Figure 4A). In the lakes, rotifers were (partially) replaced by crustaceans (Figure 4A). According to many authors [18,46,47], Rotifera predominate in both standing and flowing waters owing to their tolerance to changing environmental conditions. Their dominance is believed to be connected with their small size and relatively short development time compared to crustaceans [48–50]. Rotifera display life history r-strategies and are more adaptable to environmental disturbances of lotic and semi-lotic habitats [18,50].

The greatest mean density of zooplankton and rotifers was recorded in the transitional floodplain lake (Table 3). In this water body rotifers also had the largest percentage share among the dominant species (over 80%) (Figure 4B).

The transitional floodplain lake was temporarily washed out with water from the Vistula: on 19 out of 38 sampling dates the lake was connected to the river. Presumably because of temporary inundation it had the highest number of zooplankton species (both rotifers and crustaceans) (Table 3). The highest mean density of the most dominant species among rotifers and the highest values of H′ and J′ indices were also observed here. The results indicated that it had more diverse zooplankton compared to the isolated and connected lakes.

The gentle washing of the lake at medium water level in the Vistula River might have stimulated the development of zooplankton in accordance with the Intermediate Disturbance Hypothesis [19]: the density and diversity of zooplankton was greater in the lakes connected even temporarily with the main river channel compared to the isolated one [23,24]. At intermediate levels of disturbance (medium flooding during potamophase), species diversity should be greatest because many taxa tolerate existing environmental conditions but none can dominate in the population [51]. Our observations confirmed that the best conditions for zooplankton development were found in the lake periodically connected with the river. According to Mitrovic et al. [52] organic matter supply from inundation could stimulate heterotrophic bacterioplankton and affect zooplankton density and structure. High food availability is related to greater density of zooplankton, which is dominated by small-sized rotifers [53], e.g., *K. tecta* (Table 3, Figure 4B). Rotifers preferred the transitional floodplain lake (Figure 5A).

Based on the percentage share of dominant species (e.g., *K. tecta)* it could be concluded that the river had the greatest impact on zooplankton density in the transitional and connected floodplain lakes (Figure 4B). The differences between the sites were statistically significant. A river can shape the zooplankton community structure in floodplain lakes by periodical flooding [54,55].

The greatest mean density of crustacean zooplankton was observed in the connected floodplain lake, where it constituted approximately 30% of the predominant species density (Figure 4B). The isolated floodplain lake (FL1) did not have the best conditions for the development of crustacean zooplankton: neither its density nor the total number of crustacean species was the highest at this site. On the contrary, the number of crustacean species was the lowest in this lake. A small representation of crustaceans in the isolated floodplain lakes may result from fish pressure [45,46].

On the other hand, it was surprising that the highest density of crustacean zooplankton was recorded in the floodplain lake permanently connected to the river. Presumably, the development of crustaceans was determined by the way in which the lake was connected with the river, i.e., through a narrow channel with the length of 1.2 km and the width of 50 m. With this length the channel could limit or prevent the connection. Moreover, the large size of the lake could stabilize the environmental conditions, which promoted the development of macrophytes (e.g., *Elodeanuttalli)*. Owing to that, living conditions for large-sized zooplankton improved significantly. Our study demonstrated that crustacean zooplankton preferred the lake connected with the river (the highest density of crustacean) (Figure 5A). Similar results are obtained by Vadadi-Fülöp et al. [56] for lakes in the Danube valley. A lower degree of connectivity with the river ensures better (more stable) conditions for zooplankton, especially Crustacea. The isolated lake has a well-developed macrophyte community, which serves as a hiding place for zooplankton [57,58]. Cladocera groups, very sensitive to periodic flooding [56], develop faster in isolated or semi-isolated floodplains. Several authors observe that the highest density of zooplankton is recorded in lakes permanently connected to the river by a long channel [17,39,59].

Conditions in floodplain lakes permanently (FL3) or temporarily (FL2) connected to the river depend on flood pulse dynamics (intensity) [47,60–62]. During higher river levels (>234 cm, potamphase), rotifers developed better in the lakes, which was manifested in the higher average density of rotifers and higher density of dominant species, such as *K. tecta*, *B. angularis* and *P. longiremis*

(Figure 5A,B). Higher water level in the Vistula caused an inflow of river water to the lakes (transitional and connected) and facilitated the higher entry of species such as *K. tecta* or *B. angularis* (Figure 4B). The inflow of river water enriched the lakes with river species (mainly rotifers) and this resulted in an increase of the α-diversity index (H′). Such a regularity is also observed by Simões et al. [63]. The river water carried mineral and organic suspension into the connected lakes, thus providing food for detritivorous rotifera, such as *K. tecta* or *B. angularis*. Similar phenomena are observed by Bomfim et al. [53].

Flood pulses also cause nutrient inputs [64], which stimulate phytoplankton production [65,66]. Rapid growth of small edible phytoplankton in floodplain lakes provides good quality food for zooplankton [67–70]. During lower river levels (<234 cm; limnophase), crustacean zooplankton developed better, as noted in the crustacean density and *B. longirostris* density. Lower river levels stabilized environmental conditions not only in the transitional lake (limnophase), but also in the connected lake due to the specific character of the connection (Figure 5B). Periodic flooding causes the succession of two phases in the floodplain lake life cycle: limnophase (isolation) and potamophase (connection) [71]. These two phases differ in water transparency. Water transparency expressed as visibility (SD) was correlated with the average number of crustacean species, average density of crustaceans and average density of nauplii (Figure 5C). Lake isolation promotes the development of Crustacea. Burdis and Hoxmeier [72] note that slower current velocity, lower turbulence and longer residence time are important mechanisms favouring crustacean development in floodplain habitats. Copepods predominated in the crustacean zooplankton of the studied floodplain lakes of the lower Vistula. Their most common forms included larval nauplii (isolated and transitional lake during limnophase), similarly to what is observed in other investigated floodplain lakes [18,65]. During periods with low water transparency rotifer species (*K. tecta* and *B. angularis*) which are indicators of high trophy developed faster [73] (Figure 5D).

It is not easy to answer how, directly, connectivity (dispersal) matters for the persistence and performance of metacommunities [10]. Unfortunately, it is difficult to distinguish between organisms belonging to the adapted vs. dispersed group in zooplankton of floodplain lakes. All pelagic organisms (both alive forms and resting eggs) could be dispersed from river to local communities so it is a very important process. However, local factors such as habitat heterogeneity, water quality, and community interactions can affect the survival and reproduction of individuals [4,13,21,58]. This issue can be explored by studying resting eggs in bottom sediments and interstitial waters. This kind of investigation would help answer the question about the origin of zooplankton in floodplain lakes and dispersion possibilities. We intend to conduct this type of research in future.

Also, it is difficult to compare dispersal probability of zooplankton and settled macroinvertebrates [11] in floodplain-river systems. Based on literature and on our studies Rotifera of the Brachionidae family are best adapted to unstable conditions in floodplain lakes and could be easily dispersed in different water bodies [17,18,39].

5. Conclusions

The degree of connection between floodplain lakes and the river affected the zooplankton structure. However, contrary to the initial assumption, the diversity and density of zooplankton were higher in the lake connected with the river (FL3) than in the isolated one (FL1). Zooplankton was the most diverse and abundant in the transitional lake (FL2). The number of crustacean species was also the highest at this site. Regardless of the type of connection, the zooplankton community was less diverse and less abundant in the river than in its floodplain lakes.

The gentle washing of the lake might have stimulated the development of zooplankton in accordance with the IDH theory. The intensity of flood-pulse (inundation) determined a degree of connection between the floodplain lakes and the river and shaped the zooplankton structure (FL2). Higher river level (potamophase) increased zooplankton diversity (higher α-diversity and rotifer density). Lower river level (limnophase) increased crustacean density.

Supplementary Materials: The following are available online at http://www.mdpi.com/2073-4441/11/9/1924/s1, Table S1: List of zooplankton taxa at studied floodplain lakes and at the Vistula River.

Author Contributions: P.N. conceived and designed the study; P.N., M.B., N.M., M.S. and K.O. performed the field sampling, measurements and data analysis; P.N., K.O. and N.K. wrote and revised the manuscript; P.N. and K.O. provided conceptual overview of the manuscript preparation, writing

Funding: This research was funded by the Polish Minister of Science and Higher Education grant number 008/RID/2018/19 by, under the program "Regional Initiative of Excellence" in 2019–2022.

Conflicts of Interest: The authors have declared no conflict of interest.

References

1. Tockner, K.; Schimer, F.; Baumgartner, C.; Kum, G.; Weigand, E.; Zweimuller, I.; Ward, J.W. The Danube restoration project: Species diversity patterns across connectivity gradients in the floodplain system. *Regul. River* **1999**, *15*, 245–255. [CrossRef]

2. Amoros, C.; Bornette, G. Connectivity and biocomplexity in water bodies of riverine floodplains. *Freshw. Biol.* **2002**, *47*, 761–776. [CrossRef]

3. Funk, A.; Reckendorfer, W.; Kucera-Hirzinger, V.; Raab, R.; Schiemer, F. Aquatic diversity in a former floodplain: Remediation in an urban context. *Ecol. Eng.* **2009**, *35*, 1476–1484. [CrossRef]

4. Górski, K.; Collier, K.J.; Duggan, I.C.; Taylor, C.M.; Hamilton, D.P. Connectivity and complexity of floodplain habitats govern zooplankton dynamics in a large temperate river system. *Freshw. Biol.* **2013**, *58*, 1458–1470. [CrossRef]

5. Paira, A.R.; Drago, E.C. Origin, evolution and types of floodplain water bodies. In *The Middle Paraná River: Limnology of Subtropical Wetlands*; Iriondo, M.H., Paggi, J.C., Parma, M.J., Eds.; Springer: Berlin/Heidelberg, Germany, 2007; pp. 53–80.

6. Baranyi, C.; Hein, T.; Holarek, C.; Keckeis, S.; Schiemer, F. Zooplankton biomass and community structure in a Danube River floodplain system: Effects of hydrology. *Freshw. Biol.* **2002**, *47*, 473–482. [CrossRef]

7. Dembowska, E.; Napiórkowski, P. A Case Study of the Planktonic Communities in Two Hydrologically Different Oxbow Lakes (Vistula River, Central Poland). *J. Limnol.* **2015**, *74*, 346–357. [CrossRef]

8. Obolewski, K.; Glińska-Lewczuk, K.; Ożgo, M.; Astel, A. Connectivity restoration of floodplain lakes: An assessment based on macroinvertebrate communities. *Hydrobiologia* **2016**, *774*, 23–37. [CrossRef]

9. Junk, W.J.; Bayley, P.B.; Sparks, R.E. The flood pulse concept in river-floodplain systems. *Can. Spec. Publ. Fish Aquat. Sci.* **1989**, *106*, 110–127.

10. Chaparro, G.; Horvath, Z.; O'Farrel, I.; Ptacnik, R.; Hein, T. Plankton metacommunities in floodplain wetlands under contrasting hydrological conditions. *Freshw. Biol.* **2018**, *63*, 380–391. [CrossRef]

11. Obolewski, K.; Glińska-Lewczuk, K.; Bąkowska, M. From isolation to connectivity: The effect of floodplain lake restoration on sediments as habitats for macroinvertebrate communities. *Aquat. Sci.* **2018**, *80*, 4. [CrossRef]

12. Chaparro, G.; Kandus, P.; O'Farrel, I. Effect of spatial heterogeneity on zooplankton diversity: A multiscale habitat approximation in a floodplain lake. *River Res. Appl.* **2015**, *31*, 85–97. [CrossRef]

13. Thomaz, S.M.; Bini, L.M.; Bozelli, R.L. Floods increase similarity among aquatic habitats in River-floodplain systems. *Hydrobiologia* **2007**, *579*, 1–13. [CrossRef]

14. Paidere, J. Influence of flooding frequency on zooplankton in the floodplains of the Daugava River (Latvia). *Acta Zool. Litu.* **2009**, *19*, 306–313. [CrossRef]

15. Dias, J.D.; Simões, N.R.; Meerhoff, M.; Lansac-Tôha, F.A.; Velho, L.F.M.; Bonecker, C.C. Hydrological dynamics drives zooplankton matacommunity structure in a Neotropical floodplain. *Hydrobiologia* **2016**, *781*, 109–125. [CrossRef]

16. Dittrich, J.; Dias, J.D.; Bonecker, C.C.; Lansac-Tôha, F.A.; Padial, A.A. Importance of temporal variability at different spatial scales for diversity of floodplain aquatic communities. *Freshw. Biol.* **2016**, *61*, 316–327. [CrossRef]

17. Anderson, S.M.A.; Bonecker, C.C. Rotifers in different environments of the Upper Parana River floodplain (Brazil): Richness, abundance and the relationship with connectivity. *Hydrobiologia* **2004**, *522*, 281–290.

18. Schöll, K.; Kiss, A.; Dinka, M.; Berczik, A. Flood-Pulse effects on zooplankton assemblages in a river-floodplain system (Gemenc Floodplain of the Danube, Hungary). *Int. Rev. Hydrobiol.* **2012**, *97*, 41–54. [CrossRef]

19. Connel, J.H. Diversity in tropical rain forest and coral reefs. *Science* **1978**, *199*, 1302–1310. [CrossRef]

20. Dembowska, E.A.; Kubiak-Wójcicka, K. Influence of water level fluctuations on phytoplankton communities in an oxbow lake. *Fundam. Appl. Limnol.* **2017**, *190*, 221–233. [CrossRef]

21. Kobayashi, T.; Ralph, T.J.; Ryder, D.S.; Hunter, S.J.; Shiel, R.J.; Segers, H. Spatial dissimilarities in plankton structure and function during flood pulses in a semi-arid floodplain wetland system. *Hydrobiologia* **2015**, *747*, 19–31. [CrossRef]

22. Lemke, M.J.; Paver, S.F.; Dungey, K.E.; Velho, L.F.M.; Kent, A.D.; Rodrigues, L.C.; Kellerhals, D.M.; Randle, M.R. Diversity and succession of pelagic microorganisms communities in a newly restored Illinois River floodplain lake. *Hydrobiologia* **2017**, *804*, 35–58. [CrossRef]

23. Hein, T.; Baranyi, C.; Reckendorfer, W.; Schimer, F. The impact of surface water exchange on the nutrient and particle dynamics in side-arms along the River Danube, Austria. *Sci. Total Environ.* **2004**, *328*, 207–218. [CrossRef] [PubMed]

24. Kasten, J. Innundation and isolation: Dynamics of phytoplankton communities in seasonal inundated flood plain waters of the Lower Odra Valley National Park – Northeast Germany. *Limnologica* **2003**, *33*, 99–111. [CrossRef]

25. Makowski, J. Second Part: Section from Toruń to Biała Góra. In *The Floodbanks of the Lower Vistula River, Historical Shaping, Current State and Behavior during the Major Floods*; Publisher of the Institute of Hydraulic Construction PAS Scientific Library of Hydrotechnics: Gdańsk, Poland, 1998; Volume 27, pp. 1–78. (In Polish)

26. Nogrady, T.; Wallance, R.L.; Snell, T.W. Rotifera, biology, ecology and systematic. In *Guides to the Identification of the Microinvertebrates of the Continental Waters of the World*; Dumont, H.J., Ed.; SPB Academy Publisher: The Hague, The Netherlands, 1993; Volume 4, pp. 1–142.

27. Harris, R.P.; Wiebe, P.H.; Lenz, J.; Skjoldal, H.R.; Huntley, M. *Zooplankton Methodology Manual*; ICES Academic Press: Cambridge, MA, USA, 2000; pp. 147–173.

28. Flössner, D. *Krebstiere, Crustacea. Kiemen- und Blattfüsser, Branchiopoda. Fischläuse, Branchiura. Die Tierwelt Deutschlands. 60 Teil*; VEB Gustav Fischer Verlag: Jena, Germany, 1972; p. 501. (In German)

29. Kiefer, F. Freilebende Copepoda. In *Das Zooplankton der Binnengewässer, 2. Teil, Die Binnengewässer, Band 26*; Kiefer, F., Fryer, G.E., Eds.; Schweizerbartsche Verlagsbuchhandlung: Stuttgart, Germany, 1978; p. 343. (In German)

30. Einsle, U. Copepoda: Cyclopoida—genera Cyclops, Megacyclops, Acanthocyclops. In *Guides to the Identification of the Microinvertebrates of the Continental Waters of the World*; Dumont, H.J., Ed.; SPB Academy Publisher: The Hague, The Netherlands, 1996; p. 82.

31. Smirnov, N.N. Cladocera: The Chydoridae and Sayciinae (Chydoridae) of the World. In *Guides to the Identification of the Microinvertebrates of the Continental Waters of the World*; Dumont, H.J., Ed.; SPB Academy Publisher: The Hague, The Netherlands, 1996; 197p.

32. Radwan, S.; Bielańska-Grajner, I.; Ejsmont-Karabin, J. *Rotifers (Rotifera, Monogononta), Freshwater Fauna of Poland*; University of Lodz Press: Łódź, Poland, 2004; p. 579.

33. Rybak, J.I.; Błędzki, L.A. *Freshwater planktonic Crustacea*; WUW: Warsaw, Poland, 2010; p. 365. (In Polish)

34. Hammer, Ø.; Harper, D.A.T.; Ryan, P.D. Past: Paleontological statistics software package for education and data analysis. *Palaeontol. Electron.* **2001**, *4*, 9.

35. Ter Braak, C.J.F. Canonical correspondence analysis: A new eigenvector technique for multivariate direct gradient analysis. *Ecology* **1986**, *67*, 1167–1179. [CrossRef]

36. Ter Braak, C.J.F. Non-linear methods for multivariate statistical calibration and their use in palaeoecology: A comparison of inverse (k-Nearest Neighbours, PLS and WA-PLS) and classical approaches. *Chemom. Intell. Lab. Syst.* **1995**, *28*, 165–180. [CrossRef]

37. Legendre, P.; Legendre, L. *Numerical Ecology*, 3rd ed.; Elsevier: Amsterdam, The Netherlands, 2012; p. 1006.

38. ter Braak, C.J.F.; Smilauer, P. *Manual CANOCO and CanoDraw for Windows: Software for Canonical Community Ordination 4.5*; Microcomputer Power: Ithaca, NY, USA, 2002.

39. de Paggi, S.B.J.; Paggi, J.C. Hydrological connectivity as shaping force in the zooplankton community of two lakes in the Parana River Floodplain. *Int. Rev. Hydrobiol.* **2008**, *93*, 659–678. [CrossRef]

40. Bozelli, R.L.; Thomaz, S.M.; Padial, A.A.; Lopes, P.M.; Bini, L.M. Floods decrease zooplankton beta diversity and environmental heterogeneity in an Amazonian floodplain system. *Hydrobiologia* **2015**, *753*, 233–241. [CrossRef]

41. Paillex, A.; Castella, E.; zu Ermagassen, P.S.E.; Gallardo, B.; Aldridge, D.C. Large river floodplain as a natural laboratory: Non-native macroinvertebrates benefit from elevated temperatures. *Ecosphere* **2017**, *8*, e01972. [CrossRef]

42. Zhang, K.; Xu, M.; Wu, Q.; Lin, Z.; Jiang, F.; Chen, H.; Zhou, Z. The response of zooplankton communities to the 2016 extreme hydrological cycle in floodplain lakes connected to the Yangtze River in China. *Environ. Sci. Pollut. Res.* **2018**, *25*, 23286–23293. [CrossRef]

43. Gallardo, B.; García, M.; Cabezas, A.; Gonzáles, E.; Gonzáles, M.; Ciancarelli, C.; Comín, F.A. Macroinvertebrate patterns along environmental gradients and hydrological connectivity within a regulated river-floodplain. *Aquat. Sci.* **2008**, *70*, 248–258. [CrossRef]

44. Obolewski, K. Macrozoobenthos patterns along environmental gradients and hydrological connectivity of oxbow lakes. *Ecol. Eng.* **2011**, *37*, 796–805. [CrossRef]

45. Goździejewska, A.; Glińska-Lewczuk, K.; Obolewski, K.; Grzybowski, M.; Kujawa, R.; Lew, S.; Grabowska, M. Effects of lateral connectivity on zooplankton community structure in floodplain lakes. *Hydrobiologia* **2016**, *774*, 7–21. [CrossRef]

46. Pithart, D.; Pichlová, R.; Bílý, M.; Hrbáček, J.; Novotná, K.; Pechar, L. Spatial and temporal diversity of small shallow waters in river Lužnice floodplain. *Hydrobiologia* **2007**, *584*, 265–275. [CrossRef]

47. Galir Balkić, A.G.; Ternjej, I.; Špoljar, M. Hydrology driven changes in the rotifer trophic structure and implications for food web interactions. *Ecohydrology* **2017**, *11*, 1–12. [CrossRef]

48. Špoljar, M.; Tomljanović, T.; Dražina, T.; Lajtner, J.; Štulec, H.; Matulić, D.; Fressl, J. Zooplankton structure in two interconnected ponds: Similarities and differences. *Croat. J. Fish.* **2016**, *74*, 29–42. [CrossRef]

49. Branko, C.W.C.; de Moraes Lima Silveira, R.; Morinho, M.M. Flood pulse acting on a zooplankton community in a tropical river (Upper Paraguay River, Northern Pantanal, Brazil). *Fundam. Appl. Limnol.* **2018**, *192*, 23–42. [CrossRef]

50. Galir Balkić, A.; Ternjej, I. Assessing Cladocera and Copepoda individual disturbance levels in hydrologically dynamics environment. *Wetl. Ecol. Manag.* **2018**, *26*, 733–749. [CrossRef]

51. Ward, J.V.; Tockner, K. Biodiversity: Towards a unifying theme for river ecology. *Freshw. Biol.* **2001**, *46*, 807–819. [CrossRef]

52. Mitrovic, S.M.; Westhorpe, D.P.; Kobayashi, T.; Baldwin, D.S.; Ryan, D.; Hitchcock, J.N. Short-term changes in zooplankton density and community structure in response to different sources of dissolved organic carbon in an unconstrained lowland river: Evidence for food web support. *J. Plankton Res.* **2014**, *36*, 1488–1500. [CrossRef]

53. Bomfim, F.F.; Braghin, L.S.M.; Bonecker, C.C.; Lansac-Tôha, F.A. High food availability linked to dominance of small zooplankton in a subtropical floodplain. *Int. Rev. Hydrobiol.* **2018**, *103*, 26–34. [CrossRef]

54. Medley, K.A.; Havel, J.E. Hydrology and local environmental factors influencing zooplankton communities in floodplain ponds. *Wetlands* **2007**, *27*, 864–872. [CrossRef]

55. Napiórkowski, P.; Napiórkowska, T. Limnophase versus potamophase: How hydrological connectivity affects the zooplankton community in an oxbow lake (Vistula River, Poland). *Ann. Limnol. Int. J. Limnol.* **2017**, *53*, 143–151. [CrossRef]

56. Vadadi-Fülöp, C.; Hufnagel, L.; Jablonszky, G.; Zsuga, K. Crustacean plankton abundance in the Danube River and its side arms in Hungary. *Biologia* **2009**, *64*, 1184–1195. [CrossRef]

57. Kuczyńska-Kippen, N.; Nagengast, B.; Celewicz-Gołdyn, S.; Klimko, M. Zooplankton community structure within various macrophyte stands of a small water body in relation to seasonal changes in water level. *Oceanol. Hydrobiol. Stud.* **2009**, *38*, 125–133. [CrossRef]

58. Karpowicz, M.; Ejsmont-Karabin, J.; Strzałek, M. Biodiversity of zooplankton (Rotifera and Crustacea) in water soldier (Stratiotes aloides) habitats. *Biologia* **2016**, *71*, 563–573. [CrossRef]

59. Alves, G.M.; Velho, F.L.M.; Lansac-Tôha, F.A.; Robertson, B.; Bonecker, C. Effect of the connectivity on the diversity and abundance of cladoceran assemblages in lagoons of the Upper Paraná river floodplain. *Acta Limnol. Bras.* **2005**, *17*, 317–327.

60. Paillex, A.; Doledec, S.; Castella, E.; Merigoux, S.; Aldridge, D.C. Functional diversity in a large river floodplain: Anticipating the response of native and alien macroinvertebrates to the restoration of hydrological connectivity. *J. Appl. Ecol.* **2013**, *50*, 97–106. [CrossRef]

61. Obolewski, K.; Glińska-Lewczuk, K.; Strzelczak, A. Does hydrological connectivity determine the benthic macroinvertebrate structure in oxbow lakes? *Ecohydrology* **2015**, *8*, 1488–1502. [CrossRef]

62. Batzer, D.; Gallardo, B.; Boulton, A.; Whiles, M. Invertebrates of Temperate-Zone River Floodplains. In *Invertebrates in Freshwater Wetlands: An International Perspective on their Ecology*; Batzer, D., Boix, D., Eds.; Springer International Publishing: Cham, Switzerland, 2016; pp. 451–492.

63. Simões, N.R.; Lansac-Tôha, F.A.; Bonecker, C. Drought disturbance increase temporal variability of zooplankton community structure in floodplains. *Int. Rev. Hydrobiol.* **2013**, *98*, 24–33. [CrossRef]

64. de Melo, T.X.; Dias, J.D.; Simoés, N.R.; Bonecker, C.C. Effects of nutrient enrichment on primary and secondary productivity in a subtropical floodplain system: An experimental approach. *Hydrobiologia* **2019**, *827*, 171–181. [CrossRef]

65. Keckeis, S.; Baranyi, C.; Hein, T.; Holarek, C.; Riedler, P.; Schiemer, F. The significance of zooplankton grazing in a floodplain system of the River Danube. *J. Plankton Res.* **2003**, *25*, 243–253. [CrossRef]

66. Cardoso, S.J.; Nabout, J.C.; Farjalla, V.F.; Lopes, P.M.; Bozelli, R.L.; Huszar, V.L.M.; Roland, F. Environmental factors driving phytoplankton taxonomic and functional diversity in Amazonian floodplain lakes. *Hydrobiologia* **2017**, *802*, 115–130. [CrossRef]

67. Mihaljević, M.; Stević, F.; Horvatić, J.; Hackenberger-Kutuzović, B. Dual impact of the flood pulses on the phytoplankton assemblages in a Danubian floodplain lake (Kopački Rit Nature Park, Croatia). *Hydrobiologia* **2009**, *618*, 77–88. [CrossRef]

68. Stević, F.; Mihajlević, M.; Špojlarić, D. Changes of phytoplankton functional groups in a floodplain lake associated with hydrological perturbations. *Hydrobiologia* **2013**, *709*, 143–158. [CrossRef]

69. Chaparro, G.; Fontanarrosa, M.; O'Farrel, I. Colonization and succession of zooplankton after a drought: Influence of hydrology and free-floating plant dynamics in a floodplain lake. *Wetlands* **2015**, *36*, 85–100. [CrossRef]

70. Dembowska, E.A. Seasonal variation in phytoplankton and aquatic plants in floodplain lakes (lower Vistula River, Poland). *Wetl. Ecol. Manag.* **2015**, *23*, 535–549. [CrossRef]

71. Neiff, J.J. Large rivers of South America: Toward the new approach. *Int. Ver. Theor. Angew. Limnol. Verh.* **1996**, *26*, 167–180. [CrossRef]

72. Burdis, R.M.; Hoxmeier, R.J.H. Seasonal zooplankton dynamics in main channel and backwater habitats of the Upper Mississippi River. *Hydrobiologia* **2011**, *667*, 69–87. [CrossRef]

73. Ejsmont-Karabin, J. The usefulness of zooplankton as lake ecosystem indicators: Rotifer Trophic State Index. *Pol. J. Ecol.* **2012**, *60*, 339–350.

Article

Rainfall Variability and Trend Analysis of Rainfall in West Africa (Senegal, Mauritania, Burkina Faso)

Zeineddine Nouaceur [1] and Ovidiu Murarescu [2,*]

[1]   UMR IDÉES CNRS 6266, Rouen University, 76821 Mont Saint Aignan CEDEX, France; zeineddine.nouaceur@univ-rouen.fr
[2]   Department of Geography, Valahia University, 130001 Târgovişte, Romania
*   Correspondence: ovidiu.murarescu@valahia.ro

Received: 13 May 2020; Accepted: 12 June 2020; Published: 19 June 2020

Abstract: This study concerns the West African Sahel. The Sahelian climate is characterized by a long dry season and a rainy season which starts in June and ends in September–October. This latter season is associated with the process of oceanic moisture transfer to the mainland (the West African Monsoon). This movement is governed by an overall moving of the meteorological equator and its low-pressure corridor (Intertropical Convergence Zone, ITCZ) towards the north, under the effect of the attraction of the Saharan thermal depressions and a greater vigor of the anticyclonic nuclei. This study was conducted on 27 Sahelian climatic stations in three countries (Burkina Faso, Mauritania, and Senegal). The method used to determine the modes of this variability and the trends of rainfall is the chronological graphic method of information processing (MGCTI) of the "Bertin Matrix" and continuous wavelets transform (CWT). Results show a rain resumption observed in the recent years over the Sahelian region and a convincing link with the surface temperature of the Atlantic Ocean.

Keywords: West Africa; climate change; rainfall variability

---

1. Introduction

If, on a global scale, the rise in temperatures is a certainty, the evolution of global rainfall is much more contrasting, as it is subject to a strong spatiotemporal variability [1].

Increased power of evaporation will lead to a greater availability of water vapors. An increase in the quantity of the humidity in the lower layers of the atmosphere might be the cause of an intensification of rainfall in a torrential form [2]. Using a climate simulation model, it was proved that a 22% rise in air humidity may be brought towards continents by maritime flows [3]. The results of various studies regarding rainfall evolution also show that climate changes have entailed an intensification of precipitation and a repetition of extreme events [4–8].

Given this climate change, a likely increase in extreme events, particularly floods, is to be expected. According to the World Meteorological Organization, floods are the most frequent extreme phenomenon that occurred in the 2001–2010 decade [9]. This phenomenon affected several regions of the World, producing hundreds or even thousands of victims. In 2016, almost 23.5 million people were dislocated because of natural disasters related to extreme meteorological events—mainly storms and floods in the Asia-Pacific region [10]. In Europe, a significant change in the flood calendar due to current climate changes [11] was highlighted. Some researchers have pointed out increasing trends of extreme rainfall in more than 8326 weather stations worldwide [12]. Furthermore, they have been able to prove a significant statistical association between the average temperature and the meridian variation (the highest sensitivity occurs in the tropics and at high elevations, whereas the highest uncertainty is near the Equator, owing to a limited number of sufficiently long recordings on precipitation). The Sahelian climate is characterized by a long dry season and a rainy season which starts in June and ends in September–October. The latter season is associated with the oceanic moisture transfer to the mainland (the West African monsoon). This movement is governed by an overall translation of the meteorological Equator and its low-pressure corridor (Intertropical Convergence Zone—ITCZ) towards the north, under the effect of the attraction of the Saharan thermal depressions and the intensification of the anticyclonic nuclei in the southern hemisphere in winter. The Sahelian climate is subject to a very high rainfall variability. In the 1970s–1980s, the scientific world mobilized to conduct studies on climatic droughts, which was one of the most inter-decade powerful signals ever observed on the planet. Researchers are currently focusing on the African monsoon in order to elucidate the complexity of mechanisms related to its spatiotemporal variability. The international program, African Monsoon Multidisciplinary Analysis (AMMA) illustrates this approach [13].

The Eastern Equatorial Atlantic is the region of the tropical Atlantic basin where the seasonal cycle of ocean surface temperatures is most marked (a drop of 5 to 7 °C is observed in spring). This anomaly called "cold water tongue" [14] appears around 10° W and extends to the area south of the equator between the African coasts and 20° W. Under the effect of this contrast, the southeast trade winds cross the equator, pivot in a southwest wind, and penetrate the continent where they form the monsoon flow in the lower layers of the atmosphere.

It was the presence of abnormally cold waters in the North Atlantic and abnormally warm waters in the South tropical Atlantic that was first associated with Sahelian rainfall deficits, leading to a reduced rise in rainfall to the north [15]. Another research suggests that this Atlantic dipole structure was part of a global inter-hemispheric structure linked to the strongest warming of the tropical ocean recorded in the last century [16].

Regarding the research of the role of sea surface temperature (SST), Tropical Southern Atlantic (TSA), and Tropical Northern Atlantic (TNA), this is frequently highlighted as a source for decadal variability of rainfall in West Africa [17,18]. Atlantic multidecadal variability has been given much attention, particularly regarding the interhemispheric pattern of SST and Atlantic multidecadal oscillation (AMO) [19–24].

In terms of the surface temperature of the oceans, predicting the onset of this climatic phenomenon and knowing its intra-seasonal and annual variability is crucial for local populations. It is known this region is vulnerable to drought on a large scale, because some of its economy is based on an agricultural system dependent on precipitation. The Sahelian economy is in its infancy with a human development index (HDI, this index varies from the highest value 0.954 for Norway to the lowest value 0.354 for Niger) calculated for all countries in the studied region of 0.43 (Mauritania 0.52; Senegal 0.50; Mali and Burkina Faso 0.42; Chad 0.40; and Niger 0.35).

2. Data and Working Methods

This study relies on data from 27 Sahelian weather stations in three countries (Burkina Faso, Mauritania, and Senegal). Annual data for the period between 1947 and 2014 has been used for the chronological graphic method of information processing (MGCTI) of "Bertin matrix" (this choice was prompted by data availability—Figure 1). The mean monthly values were available only for stations in Senegal and Burkina Faso. Available monthly rainfall data were extracted between 1948 and 2017 for each country for the continuous wavelets transform (CWT) analysis. The data used have not been interpolated but come from measurements from observation stations that appear on the map in Figure 1.

The data were collected from the weather services in Senegal (ANACIM) (http://www.anacim.sn/meteorologie/), Mauritania (http://www.onm.mr/). Rainfall data from Burkina Faso were collected from the KNMI Climate Explorer (http://climexp.knmi.nl/selectstation.cgi?id=someone @somewhere). Some missing data were found on "TuTiempo.net" (https://fr.tutiempo.net/climat/afrique.html). This site uses the National Climatic Data Center's Global Database (NCDC—https://www.ncdc.noaa.gov/data-access/quick-links).

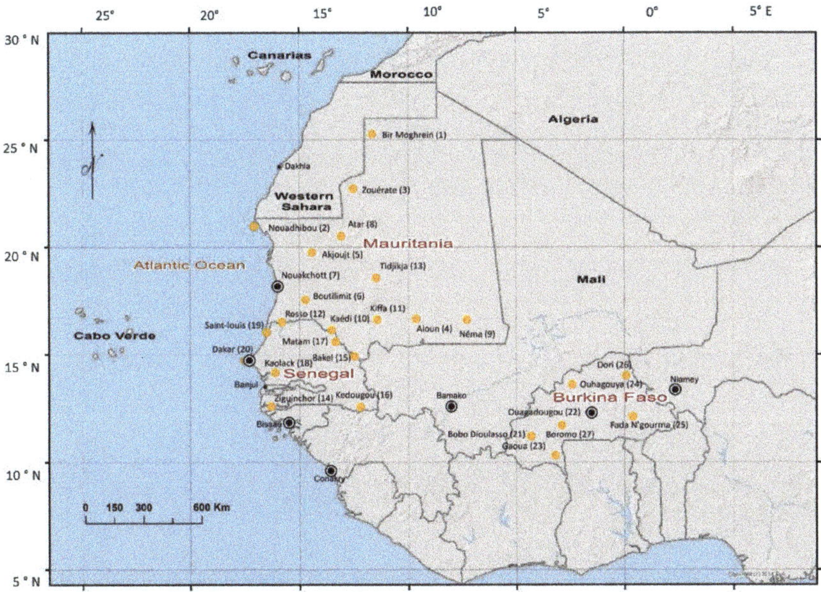

Figure 1. Geographical position of the 27 weather stations cited. The number between the brackets indicates the position of stations in the chronological graphic method of information processing (MGCTI) of "Bertin matrix" type (Figure 2).

Three regionalized indices of surface temperatures of the North Atlantic and South Atlantic Tropical, freely accessible on the site (https://www.esrl.noaa.gov/psd/data/climateindices), were used: The AMO index (25–60° N, 7–75° W), the TNA anomaly of the average of the monthly SST from 5.5° N to 23.5° N and 15° W to 57.5° W), and the TSA anomaly of the average of the monthly SST from Eq-20° S and 10° E–30° W).

The analysis of rainfall variability employs two working methods:

- The chronological graphic method of information processing of "Bertin matrix" type. The MGCTI is an analytical method based on a statistical analysis and on a graphical representation of results. This method was used for the first time in 2013 [25,26]. The MGCTI is developed to facilitate the interpretation of the statistical results for the Mediterranean rainfall analysis, due to the high variability affecting this parameter. The Bertin matrix was introduced to harmonize and consolidate information after the statistical treatment. The Bertin matrix is a manual and visual method of classification of information based on data. This matrix is used to group the data that have similarities according to all the criteria studied. It provides a simple and effective way to establish a multivariate typology based on observations of the user. This provides the concordant results (even rainfall character for all stations studied in the same year) but also identifies conflicting information (different characters between stations for the same year). The MGCTI and its graphic representation allow a chronological reading and a spatial analysis of the phenomenon. This method has been successfully tested in many North African regions [27–30], and in the Sahel [31]. A comparative study with the Standard Precipitation Index (SPI) method for detecting climate drought was conducted in 2015 [32]. This study showed the simplicity and clarity of the results obtained with the MGCTI method. One of the aims of this article is to show the trend of rainfall over nearly half a century and to detect the date of changes of cycles.

The First Stage. An annual data precipitation (cumulative rainfall over a calendar year) hierarchy in terms of limit values (Q1, Q2, Median, Q3, and Q4) is done for all stations and for the entire series

(Table 1). Depending on data position in relation to limit values, the years are considered as very dry, dry, normal, rainy, very rainy (Table 2):

(i)   very dry, below the first quintile;
(ii)   dry, between the first and the second quintile;
(iii)   normal with trends towards drought, between the second quintile and the third quintile;
(iv)   rainy, between the third and the fourth quintile;
(v)   very rainy, above the fourth quintile.

Table 1. Distribution and hierarchization of annual rainfall according to the quintiles.

Mauritania	Q1	Q2	ME	Q3	Q4
Aioun	146	209.4	221.1	256.6	325.6
Akjoujt	32.6	54.7	72.8	84.8	137.5
Atar	38.9	68.3	76.9	90.4	127.8
Bir Moghrein	8.8	21.7	27	37.1	70.5
Boutilimit	96	135.2	159.5	170.8	227.4
Kaédi	218.8	251.1	288.9	319.9	415
Kiffa	176.1	264.1	281	319.2	422.9
Néma	164.5	220.4	231.5	258.2	331.5
Nouadhibou	4.9	12.5	19.9	24.3	45.6
Nouakchott	45.4	81.7	95.4	119	189.7
Rosso	163.3	239.2	258.5	297.3	338.9
Tidjikja	53	95.5	110.2	130.2	171.7
Zouérate	21.3	38	47.2	56.7	85
Senegal					
Bakel	397.1	468.7	503.4	542.3	670
Dakar	274.3	382.8	421.6	477.2	624.3
Ziguinchor	1105.3	1281.5	1383.2	1531.4	1677.3
Saint Louis	188.7	240.6	278.9	300	373.5
Kedougou	1062	1154.5	1186.3	1264	1375.8
Kaolack	510.5	590.2	636.6	716.5	849
Matam	292.7	368.9	412.2	452.4	522.5
Burkina Faso					
Bobo Dioulas	888.8	972	1037.7	1085	1246
Boromo	773	884	930	963	1051
Dori	397	455	476	531	625
Fada N'gourma	699.5	789	824.5	907	999.5
Gaoua	954	1024	1059	1096	1207
Ouagadougou	675	738.87	765	803	928
Ouhagouya	536	612	649	709	767

The Second Stage. A recoding of values is made by means of a range of colors (the color varying in terms of the annual cumulative rainfall position in relation to limit values). This first processing is followed by a reordering procedure (permutations of columns) in order to get a ranking that allows the visualization of a homogenous colored structure (Bertin matrix) (Figure 2). This procedure allows for the visualization of the climate parameter evolution in terms of two dimensions (time and space).

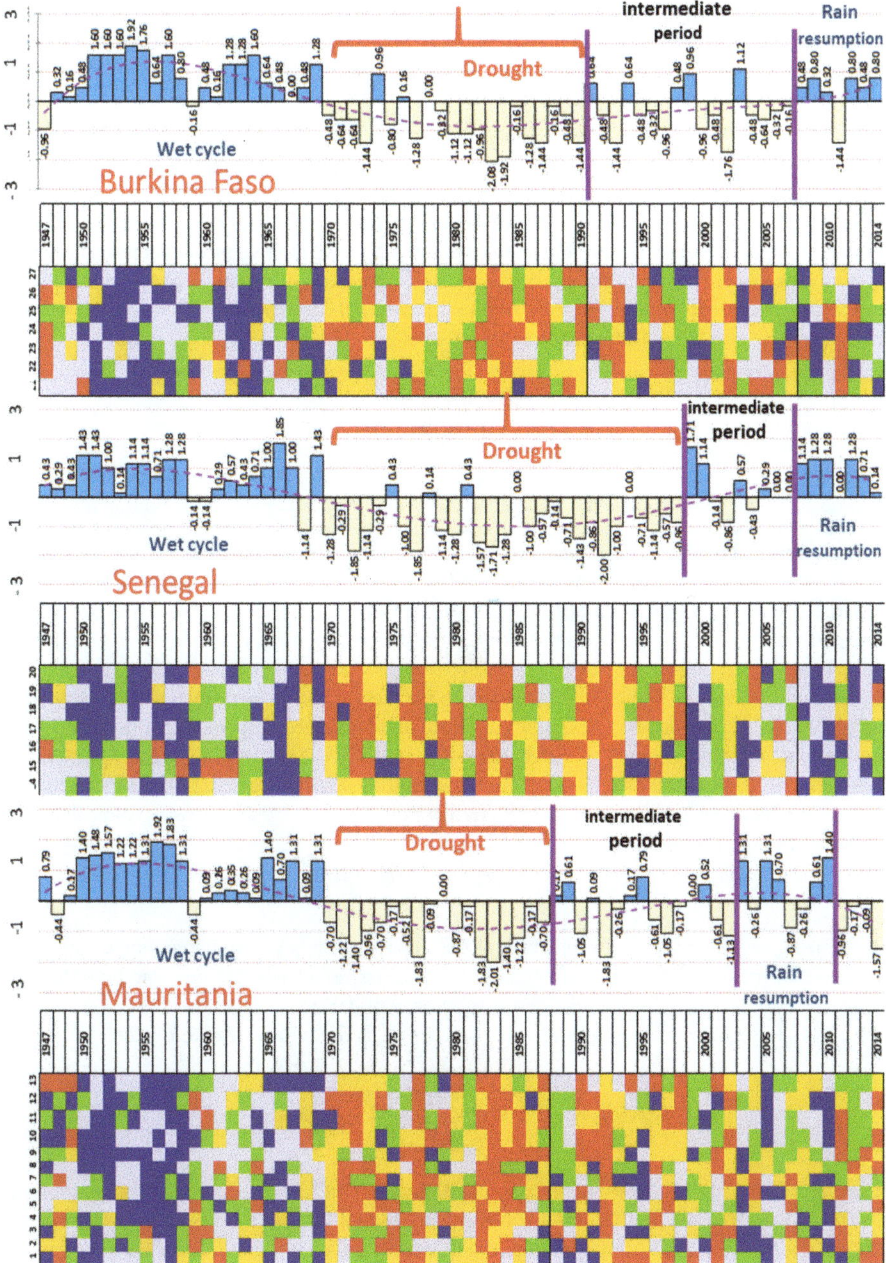

Figure 2. Chronological graphical matrix of data processing applied to rainfall (1947–2014).

Table 2. Trend of annual rainfall according to the quintiles.

Thresholds	Q1	Q2	Median	Q3	Q4
Distribution of values (%)	0–Q1 (0–20%)	Q1–Q2 (20–40%)	Q2–Median (40–50%) Median –Q3 (50–60%)	Q3–Q4 (60–80%)	>Q4 (80–100%)
Annual rainfall	Very dry	Dry	Median	Wet	Very wet
Trend at station					
Regional trend			Not expressed		

The Third Stage. To determine the typical breaks and periods, a second procedure is conducted. It consists in assigning a number ranging from one (very dry year) to five (very wet year) according to the already determined features assigned to each year. The sum of numbers of all stations for each year is centered and reduced, thus getting a regional index (*RI*) varying from $+\infty$ for a very wet year to $-\infty$ for a very dry year. The "*RI*" is calculated as follows:

$$RI = (X_i - X)/S \tag{1}$$

where X_i is yearly value. X is the series average, and S is standard deviation. The projection of the result on a graph allows for the visualization of the evolution of the phenomenon on a regional scale in a first stage and. in a second stage for the determination of data on breaks and trend change.

- The continuous wavelets analyses allow a temporal location of the variability of a given signal. It breaks down the signal, both, in time and in frequency, which can correctly describe these hydrological or climatic fluctuations, periodic or not [33] introducing the transformation into wavelets that, unlike the Fourier's transformation, breaks down the signal into a sum of finite-sized functions located over time for each frequency detected in the signal. For this, a mother wavelet is dissociated into girl waves to find the given frequency and is then translatable to analyze the neighboring frequencies. So, these analyses were developed to compensate for the disadvantages of conventional Fourier analysis. The girl wavelets have the result of the decomposition of the reference wavelet (mother wavelet). Each wavelet has a finite length (a ladder) and is highly localized over time. The mother wave has two parameters for time-frequency exploration: A scale a and a time location b:

$$\psi_{a,b}(t) = \frac{1}{\sqrt{a}}\psi\left(\frac{t-b}{a}\right) \tag{2}$$

The setting in scales and the translation of the girl wavelets allow the detection of the different frequencies that make up the signal. In addition, these frequency components can be detected and studied over time, allowing for a better description of non-stationary processes [34]. The continuous wavelet of an *S(t)* signal produces a local wave spectrum, as defined by (Equation (2)):

$$S(a,b) = \int_{-\infty}^{+\infty} s(t) \times \frac{1}{\sqrt{a}} \times \psi\left(\frac{t-b}{a}\right) \times dt \tag{3}$$

The convolution of the filtered monthly signal (Rainfall, AMO, TNA, and TSA) by a non-orthogonal wavelet basis was applied to define the continuous wavelet time-scale spectrum, which is able to identify the spectral components assigned to the dominant mode of variability of the total signal.

The rainfall signal was analyzed using CWT to identify the dominant modes of variability characterizing the Sahelian rainfall (Figure 3). Furthermore, the convolution of this signal with the wavelet basis generates a contour diagram having three variables: (1) the time graduation on the x-axis, (2) the time scale of wavelet on the y-axis, and (3) the power or variance of variability on the z-axis which can also can be explained by the correlation degrees between the signal and the wavelet basis. Such representation illustrates eventual changes in the variance based on the non-stationary signals. Distribution of the power (expressed as normalized decibels, i.e., maximum power = 0 bd for each scale) in the wavelet contour diagram is assigned by a variation of color from dark blue, that represents the low power (variance), to dark red, which represents the increasing power observed.

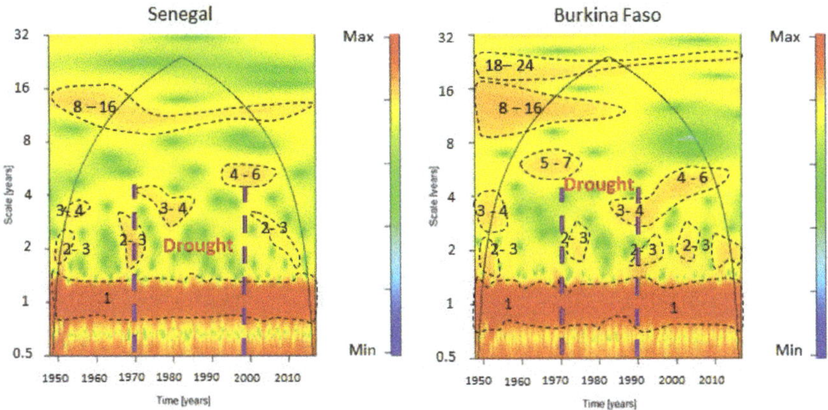

Figure 3. Continuous wavelet of precipitation regime in Senegal and Burkina Faso.

Then, correlation between the index of climate AMO, TNA, TSA, and rainfall signal was computed by the coherence diagram (WCO) for different modes of variability.

To compare the time series with each other, cross-correlation analysis is used.

By analogy to the spectra crossed by Fourier transform, the spectrum in crossed wavelets is a method which makes it possible to evaluate the correlation between two signals according to the various scales (frequencies) during time [35,36].

The spectrum by crossed wavelets $W_{xy}(a, \tau)$ between two signals $x(t)$ and $y(t)$ is calculated according to the equation below, where $C_X(a, \tau)$ and $C^*_Y(a, \tau)$ are the wavelet coefficient of the continuous signal $x(t)$ and the conjugate of the wavelet coefficient of $y(t)$, respectively:

$$W_{xy}(a, \tau) = c_x(a, \tau) \times C^*_y(a, \tau) \tag{4}$$

Continuous wavelet coherence can be defined as the estimate of the temporal evolution of linearity and of the relationship between two signals on a given scale [34,37]. Wavelet consistency is calculated using smoothed wavelet spectra of the $SW_{xx}(a, \tau)$ and $SW_{yy}(a, \tau)$ series and a smoothed crossed wavelet spectrum $SW_{xy}(a, \tau)$ [38].

$$WC(a, T) = \frac{|SW_{xy}(a, \tau)|}{\sqrt{[|SW_{xy}(a, \tau)| . |SW_{yy}(a, \tau)|]}} \tag{5}$$

Consistency is defined as being the modulus of the crossed spectrum, normalized of the same spectrum, having values between zero and one, and represents the degree of linear between two processes. A value of 1 means a linear correlation between the two signals at a time T on the scale a and a value of 0 that indicates a zero correlation [34,39].

This method therefore makes it possible, in our case, to be able to assess and describe the links existing between rainfall variability in the Sahel and climatic (represented by the AMO, TNA, and TSA indices), both for the different scales (frequencies) and according to the evolution of relationships over time.

The statistical significance of the fluctuations observed by the wavelet transform is evaluated by comparing the local wavelet spectra against randomly distributed theoretical spectra. More details on these statistical tests are indicated and discussed in the associated literature [38,39].

In this article we consider that the percentages of consistency express a weak relationship when this percentage is less than 50%. The relationship is average when the percentages are greater than 50% and less than 70%. The relationship becomes strong when the percentages are between 70% and 90% and very strong if the percentage exceeds 90%.

3. Results

3.1. Sahelian Rainfall Trend

The results of statistical processing show an evolution marked by various cycles structured into three main periods (Figure 2).

The period of the fat cow, called as such in reference to the abundance of pastures, is visible between 1947 and 1969. This period is considered wet, as shown by the matrix carried out for the three countries. The number of rainfall-challenged years, with a negative index, is very small. Two years of deficit were observed in Mauritania (1948 and 1959) and in Burkina Faso (1947, 1959). Three years were observed for Senegal (1959, 1960, and 1968).

Starting with the 1970s, it was noted that a period of severe drought occurred throughout the entire Sahelian region (1970–1987—Mauritania; 1970–1990—Burkina Faso; and 1970–1998—Senegal):

In Mauritania, the succession of dry years was interrupted between 1970 and 1987. The drought was severe in 1977 and 1982–1983. During these years, the regional index exceeded −1.5.

For Burkina Faso, this long period is marked by a succession of years of deficit for two decades (except for 1974 [0.96] and 1976 [0.16]). Drought was severe during 1982, 1983, 1987, and 1997 when regional indices reached −2.08, −1.92, −1.44, and −1.44, respectively.

As for Senegal, the dry climatic phase extended over almost 30 years.

The period of the return of precipitations is marked by two clearly identified cycles.

The first period is characterized by a break in the dry conditions of the past (we observe an alternation of wet and dry years on the matrix). This trend may be interpreted as an intermediate period which started in Mauritania in 1988, in Burkina Faso in 1990, and in Senegal in 1999. This cycle, which overlaps the climatic conditions of the previous years, announces the major climate changes currently noticed in the West African region.

The second period is identified by a higher frequency of wet years (positive indices). As regards Mauritania, it was visible starting with 2003, with several wet years exceeding the +1 threshold (2003, 2005, and 2010). In Senegal, this trend was noted as early as 2008 with a succession of wet years that exceeded the index +1 (2008–2011), interrupted in 2011, which is considered to be a "normal" climatologic year. In Burkina Faso, the conditions are similar to those in Senegal. The last years of the pluviometric series (starting with 2008) marked the return to favorable conditions for rainfall. Except 2011, just like in Senegal, a negative index of −1.4 was recorded. The situation was the same in Mauritania as well, with an index of −0.96, which points to the aridity conditions of this Sahelian area in West Africa. This rainfall variability, which marked the entire region, was more pronounced in Mauritania, due to the geographical position of this country in relation to the West African Monsoon movement to the northern areas of Sahelian Africa.

We can note two normal years with a dryness trend in 2012 and 2013, for Mauritania, whereas throughout the rest of the area studied this interval is considered to be wet, with an index of +1.28 in Senegal and +0.80 in Burkina Faso. In 2013, rainfall quantities in these countries were relatively small, falling under moisture conditions.

Thus, after almost three decades of drought, a large part of Sahelian West Africa recorded a return of rains. This trend is subject to a differentiated spatial distribution. The intermediate period (the succession of wet and dry years) started in Senegal later. The wet cycle (higher frequency of wet years) is less significant in Mauritania, even absent in the last years, namely 2012 and 2013 (this last year recorded a drought tendency). In 2014, there was a −1.57 index in Mauritania, whereas in Senegal and Burkina Faso there were wet conditions of +0.14 and +0.80, respectively.

3.2. Rainfall Variability in Senegal and Burkina Faso

The continuous wavelets transform is a good method to study the relation between rainfall and the climate index AMO, TNA, and TSA. This method was used by some authors to identify the non-stationary behavior of North Atlantic Oscillation (NAO) evolution. Similar researches were carried out by [34,40–46].

In this research, CWT was performed on monthly dataset aiming to identify the spectral components of the signals which can be assigned to different modes of variability characterizing the geophysical signal and, eventually, the time scales involved. The mean monthly values were available only for stations of Senegal and Burkina Faso used in the first part of this research (Rainfall, 1948–2017). Available monthly rainfall data were extracted between 1948 and 2017 for each country. These data vary from 697.27 mm for Senegal and 839.84 mm for Burkina Faso. We do not have monthly data for Mauritania and it is for this reason that this country is not associated with this analysis.

The CWT of the monthly data of precipitation was simulated (Figure 3). This analysis shows several variability scales for the two countries: Seasonal (6 months–1 year), interannual (2–3, 3–4, 4–6 and 5–7 years), quasi-decadal (QDO: 8–16), and multi-decadal (MDO: 18–24 only for Burkina Faso), as shown by Figure 3.

As for the annual band, it is of strong power for the two countries. The signal has no visible discontinuity. Seasonal variability is linked to the contrast between the dry seasons (November–March) and the wet seasons (April–October). Seasonal patterns normally weaken in arid years and only get stronger in wet years. This is not visible in Figure 3 and probably comes from the fact that we have used the monthly averages of the different stations in each country (a test carried out on the data of the Dakar station confirm this particularity).

The 2–3-year inter-annual band was present in both countries at specific dates. At the start of the series, we find it in 1950 for both countries. It was also present in 1969 in Senegal and in 1975 in Burkina Faso. In the 1990s, this band disappeared in Senegal and reappeared in the early 2010s, spreading to the 3–4 year mode. In Burkina Faso, this band reappeared between 1988 and 1995 then between 1998 and 2003 and finally, between 2008 and 2017.

The 3–4 year band is visible between 1950 and 1955 in Senegal and from 1948 to 1955 in Burkina Faso; it was also found between 1985 and 1995.

The 4–6 year band is present at the end of the series between 1995 and 2005 in Senegal and from 1995 until 2010.

The 5–7 year band was only present in Burkina Faso, between 1962 and 1972.

The quasi-decadal 8–16-year mode was present for the two countries studied. It was more spread out for Senegal but from 1975, it weakens and turns into 10–14-year-old fashion. In Burkina Faso, it was more powerful and concerns the period 1947–1985. There is a loss of energy after this last date.

The multi-decade mode does not appear on the wavelet of Senegal. It was present in Burkina Faso (18–24-years) from 1947 and until 1995 with a weakening since this last year.

4. Discussion

Potential Relations between the Global Variability of AMO, TNA, TSA, and the Rainfall

The West African Monsoon is a coupled atmosphere-ocean-land system and the major phenomenon of interest in the Sahelian zone in winter. Since the beginning of spring, temperatures increase, and a cold zone is formed in the Gulf of Guinea. This first thermal contrast explains the oceanic moisture transfer to the mainland in accordance with the trans-equatorial movement of trade winds in the southern hemisphere [34,41]. In West Africa, this transfer is governed by an overall movement of the meteorological Equator and its low-pressure corridor (ITCZ, Intertropical Convergence Zone) towards the north, under the effect of the attraction of the Saharan thermal depressions and a greater vigor of the anticyclonic nuclei in the southern hemisphere in winter. Rainfall variability in the Sahelian area is a climate feature in this region. In the last years, real scientific progress has been made in understanding climatic mechanisms in this region and on large part of the planet as well [14,42]. Today, this knowledge allows one to state that climate is subject to natural fluctuations overlapped by some anthropic signals (global changes). The role of oceans in regulating convective flows has been extensively studied by several specialists [14,47–49]. Thus, due to studies on pressure fields and oceanic temperature, two natural signals known as; multi-decadal signal (a signal occurring over a period larger than 40 years—low frequency variability) and quasi-decadal signal (signal with a shorter period, of 8–14 years. low-frequency variability) have been identified [16,50–52].

According to these researches, the possible links between SST and rainfall conditions are very complex and should be investigated separately for each frequency by the use of high statistical methods as the wavelets. The correlation between the index climate AMO, TNA, TSA, and rainfall signal was computed by the coherence diagram for the different modes of variability. Wavelet consistency analyses identify significant common oscillations between two signals (precipitation/Atlantic TSM) at certain variability scales for certain time intervals [53]. The result of coherence wavelet with the global mode of variability (1, 2–4, 4–8, 8–16, 18–24, and 20–24-year are shown in the Figure 4 and Tables 3–5).

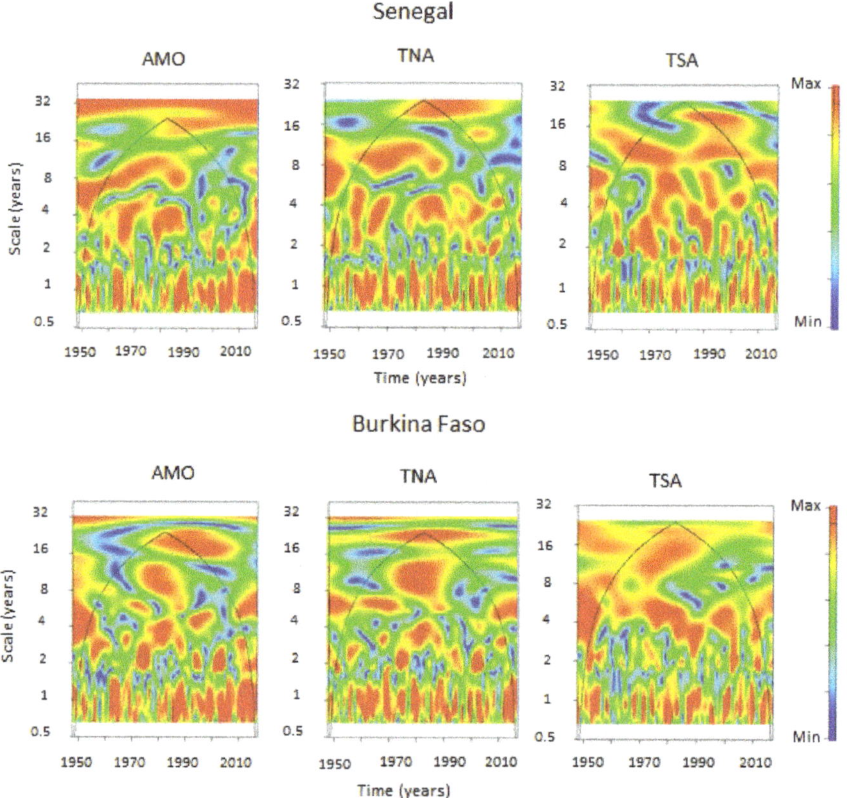

Figure 4. Wavelet coherence analysis between average monthly rain in Senegal and Burkina Faso (average of the different stations in each country) and climate indices: Atlantic multidecadal oscillation (AMO), Tropical Northern Atlantic Index (TNA), and Tropical Southern Atlantic Index (TSA).

The AMO shows moderate consistency with the monthly rainfall averages for the two countries studied.

On an interannual scale, the contribution of AMO is variable, we observe however a loss of coherence for the two countries between 1970 and 1980 (Table 3).

Table 3. Percentage of consistency between precipitation and the AMO index (%).

Variability Mode	1	2–4	4–8	8–12	8–16	18–24	20–24
Senegal	68.36	65.74	61.26	62.45	60.05	63.60	69.20
Burkina Faso	67.83	65.32	62.73	62.10	62.21	61.21	59.34

The contribution of the TNA on rainfall shows moderate consistency if we consider the different modes of variability studied (Table 4). The highest percentages are observed according to the modes of multi-decadal variability (20–24-years).

Table 4. Percentage of consistency between precipitation and the TNA index (%).

Variability Mode	1	2–4	4–8	8–12	8–16	18–24	20–24
Senegal	67.69	67.68	61.26	63.12	57.51	68.53	69.99
Burkina Faso	72	59.08	64.25	55.72	57.80	72	81.62

The consistency is well distributed in the multi-annual mode during the dry period for the two countries. In Burkina Faso, this strengthening was visible from the late 1960s to the mid-2000s, according to the 18–24 age group. For the 20–24 age group, this consistency is strong and reaches almost 82% (Table 4). Slightly offset in time, this reinforcement is also visible on the wavelet coherence spectrum of Senegal. We can see it between 1970 and 2005 but according to a lower consistency percentage of 72% for the 18–24-year mode (Table 4).

On a quasi-decadal scale (8–12 and 8–16 years), strong consistency is noted during the dry period. It is located between 1970 and 1990 for Burkina Faso with a loss of consistency before and after this date. In Senegal, this strengthening of coherence with TNA was visible between 1960 and 1985 and the loss of coherence appeared after this last date.

TNA's contribution to the interannual scale is variable. In Burkina Faso, the 4–8-year age group was reinforced between 1947 and 1975 then between 2004 and 2017. Between these two periods, a loss of power occurred. In Senegal, coherence is reinforced for this frequency band only between 1948 and 1952, there is an overall loss of power after this date. Despite the significant variability of the 2–4-year mode, we can nevertheless observe a structuring of this frequency band during the dry period. A strengthening was noted between the early 1970s and the mid-1990s for Senegal. In Burkina Faso, it was noted between the end of the 1970s and the mid-2000s. Reinforcement of consistency with the TNA was noted punctually according to different years for Burkina Faso (mid-1950s and 1960s) and for Senegal (early 1950s, mid 1960s, and early 2000s).

The consistency of the TNA on a 1-year interannual scale is very variable, however, there was a loss of power for the two countries in 2010.

The contribution of TSA to Sahelian rainfall is moderate according to the different modes of variability studied (Table 5). On a multiannual scale, this relationship is reinforced for the 18–24-year mode in Burkina Faso between 1947 and the mid-1990s and corresponds to a consistency of almost 79% for the 18–24-year mode. In Senegal, this link is not convincing, we note it by the loss of the signal observed on the wavelet graphs (Figure 4) but we find it quite intense according to the 16–20-year mode with 60. 14% of coherence between the late 1970s and early 2000s.

Table 5. Percentage of consistency between precipitation and the TSA index (%).

Variability Mode	1	2–4	4–8	8–12	8–16	18–24	20–24
Sénégal	66.16	67.79	69.70	75.82	57.51	53	48.51
Burkina Faso	63.62	63.81	73.84	55.72	67.21	78.91	75.24

On a quasi–decennial scale, the TSA contributes to the rainfall of Burkina Faso with a little more than 67% for the 8–16-year-old mode (Table 5). This frequency band is spread out over the entire period studied. However, we note that this relationship is more present during the dry period (from the late 1960s to the early 1990s). In Senegal the consistency of TSA is very low with this frequency band. However, there has been a strengthening of the 8–12-year-old mode (75.82%) (Table 5). This mode is generally present throughout the period studied.

Consistency with TSA on an interannual scale for mode 4–8 reached in Burkina Faso almost 74% (Table 5). It is reinforced at the start of the series and during the drought period (1970–2004) with a loss of power between 1985 and 1990 and a spread over the 2–4-year mode. In the 2–4-year mode, the consistency with TSA is close to 68% in Senegal and reaches 63.81% in Burkina Faso. In the latter country, this link is variable and is only reinforced at the start of the series, in the mid-1980s, in 2000 and in 2010. In Senegal, the consistency linked to this frequency is not well structured. However, some reinforcements occurred in the late 1960s, between 1960 and 1970, in the late 1970s, and late 1990 and early 2000s.

Consistency on an annual scale is very variable in the two countries and seems to present the same fluctuations.

A convincing link with the surface temperature of the oceans but difficult to define over time.

According to Caminade et al. [54] rainfall in the Sahel is characterized by a variability between 2 and 4 years in the Sahel, superimposed by slower oscillations (over 8–16 years) as well as a multi-decade evolution [55]. The results obtained thanks to wavelet coherence show convincing but different relationships according to the mode of variability and the geographic areas. We shall further summarize the main significant results of this analysis.

In Senegal AMO and TNA contribute more to rainfall according to the multiannual mode of variability (more than 69% for mode 20–24) while consistency with TSA is stronger according to the quasi–decennial and interannual modes (69.70% on the scale of 4–8 years and 75.82% for the mode 8–12 years).

For Burkina Faso, the contribution to the multi-annual scale of the TNA (81.62% for the 20–24-year mode) and the TSA (78.91% for the 18–24-year mode) is greater than that of the AMO (59.34% for the 20–24 mode). For the interannual and quasi-decennial scale the influence of TSA is stronger. It reaches 67.21% for the 8–16-year mode, 73.84% for the 4–8-year mode and 79.98% for the 4–6-year mode.

Mohino et al. [53] show that the climatic drought of the 1970s and 1980s corresponded to a negative multi-decadal oscillation of the Atlantic, favorable to a low rise in the convergence zone over Africa, whereas the 1950s and rainy 1960s, like the recent small rainfall recovery in the years 1990–2000, corresponded to a multi-decennial Atlantic oscillation returning to the positive phase. When studying the general rain trend, we have already highlighted the short periods of rainfall variability in the regions studied. The cycle of climatic drought started from the year 1970. This period extended until 1998 in Senegal but did not exceed the beginning of the 1990s for Senegal (Figure 2). The search for the impact of the surface temperature of the Atlantic Ocean over this period is illustrated in the graphs presented for the two countries (Figure 5). The influence of AMO seems intensified over this period for the 8–16-year mode for Senegal and Burkina Faso (even if for this last country we mentioned a weak global relationship above). For the same mode, we also highlighted a strong link with TNA during the dry period of the latter country (Figure 5). For the other modes this connection sometimes appears to be time-shifted or not clearly established.

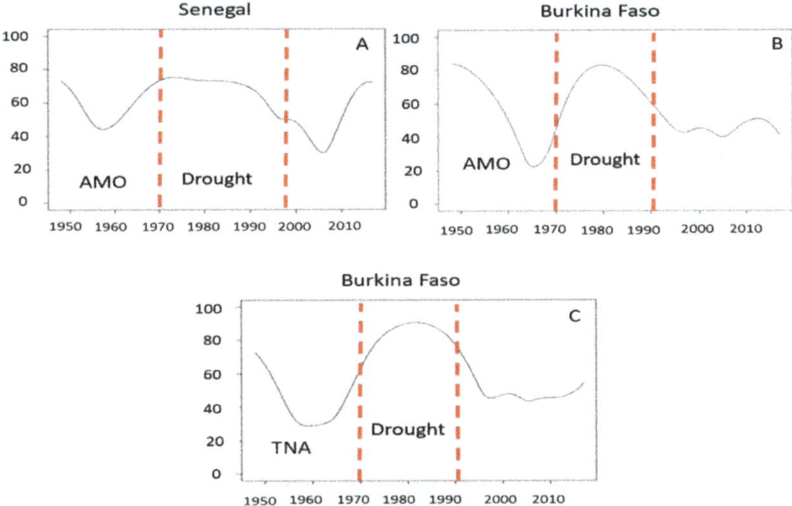

Figure 5. Percentage of consistency between precipitation and the AMO and TNA index for mode 8–16-years.

The connection between the drought period and ocean surface temperatures seems even more evident when one considers the percentage of dry years extracted from the graphic matrix (Figure 2).

Charts of Figure 6 show how the North Atlantic Ocean surface temperature (TNA) and the dry years' index (percentage of stations with a reduced centered index of cumulated precipitations lower than the fourth quintile in the studied area) vary. Variation curves show the reverse connection. Dry year indices are positive in 1970–1999, whereas the Atlantic Ocean surface temperature has negative values in 1971–1994 (the ocean was colder).

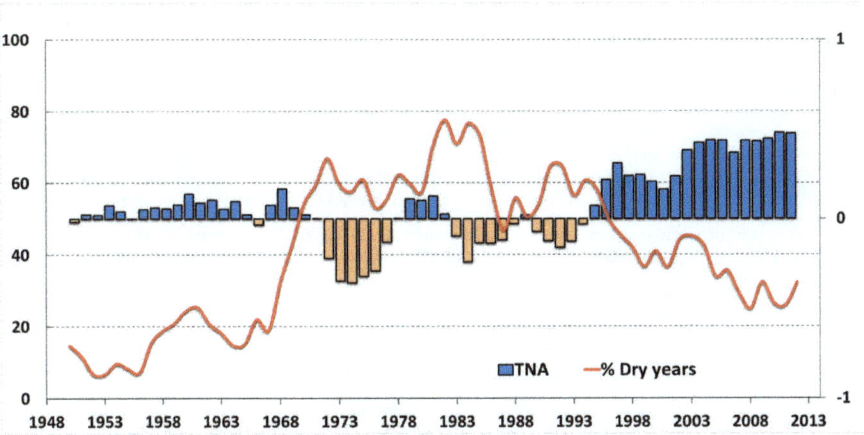

Figure 6. Evolution of dry years (%) and temperature of the Atlantic Ocean (TNA)—moving average over five years and the 1948–2014 interval; data source: https://www.esrl.noaa.gov/psd/data/correlation/tna.data.

In the last period of the series studied, the trend is reversed again, and, as the ocean surface temperature increases, the indices of dry years decrease in favor of rainier periods. Thus, when quasi-decadal oscillations are positive and marked by a higher amplitude (warmer ocean), they correspond to a return of rains after Sahelian droughts and to an intensification of the West African monsoon rain cycle [54]. The graphical matrix (Figure 2) shows this return starting with 1990 in Burkina Faso, 1999-Senegal, and 2003-Mauritania. This explains the reverse connection with rainfall cycles in the studied regions.

5. Conclusions

The analysis of rainfall trend evolution shows that, following a long Sahelian drought, rains returned to this part of West Africa. The observations made in the entire Sahel region point to a major change that occurred in the mid-1990s (brutal alternation of dry and wet years) [54,55], which made some scientists use the term "ecologization" [56,57]. Debates on this issue are still underway within the scientific community, as there is some hesitation in evoking large-scale climate changes in the Sahel region. Therefore, regional contrasts have been clearly established, which is in line with the climate predictions established by the GIEC (2013) [37] for this area as well [48,54]. On a finer scale, the spatial organization described in numerous studies [58] shows a persistence of drought conditions in the extreme west of the Sahel as compared to eastern and central zones (a situation also encountered in the last years in other states such as Niger and Mali) [59,60].

The vulnerability of the area to these changes is, furthermore, acknowledged on a large scale, since much of its economy relies on the precipitation. Some specialists [61–64] have pointed out that the return of rains to the Sahel, even though they do not reach the wet cycle level of the 1950s, is directly related to the current level of greenhouse gases in the atmosphere. At the same time, these researchers minimize the impact of sea surface temperatures on local rainfalls. Another research shows, however, a greater importance of the influence of the ocean in the West African and Sahelian rainfall phenomena [14,65–67].

This connection with the ocean did not appear very strong according to the analysis that we have conducted on the monthly precipitation recorded in Senegal and Burkina Faso. Thus, we have highlighted the strongest coherences on a multi-annual scale with TNA and on an inter-annual scale for TSA in Burkina Faso. In Senegal, on the other hand, the influence of the AMO seems stronger on the pluviometry of this country, for the multi-annual and quasi-decennial modes. Finally, the relationship with the Sahelian drought seems more evident in the quasi-decadal mode (8–16 years). This connection is also very well illustrated in Figure 6.

Author Contributions: Each author has a half of contribution. Both authors have the same degree of contribution in terms of research concepts and methodologies. Z.N. analyzes the statistical-mathematical data strings by the Bertin type matrix method. It also has contributions in the comparative analysis and validation of AMO. TNA. TSA data. O.M. contributed to the article also through concept and research methodology. The editing and translation of this material have been reviewed and completed by O.M. Also, authors contributed to making the changes required by the reviewers. All authors have read and agreed to the published version of the manuscript.

Funding: The payment for the publication was made yesterday, 17.06.2020, by the author of the correspondence. The bank transfer was made, around 9.30 am, from Banca Transilvania to Credit Suisse (Switzerland). INVOICE ID water-817749 was passed on the transfer order.

Conflicts of Interest: The authors declare no conflict of interest.

References

1. Organisation Météorologique Mondiale (OMM). *Déclaration de l'OMM sur L'état du Climat en 2012*; World Meteorological Organization (WMO): Geneva, Switzerland, 2013; Volume 1108, p. 32. Available online: https://library.wmo.int/doc_num.php?explnum_id=7811 (accessed on 14 June 2020).
2. Boé, J.; Terray, L.; Habets, F.; Martin, E. Statistical and dynamical downscaling of the Seine basin climate for hydro-meteorological studies. *Int. J. Clim.* **2007**, *27*, 1643–1655. [CrossRef]
3. Tramblay, Y.; Neppel, L.; Carreau, J.; Sanchez-Gomez, E. Extreme value modelling of daily areal rainfall over Mediterranean catchments in a changing climate. *Hydrol. Process.* **2012**, *26*, 3934–3944. [CrossRef]

4.   Min, S.-K.; Zhang, X.; Zwiers, F.W.; Hegerl, G.C. Human contribution to more-intense precipitation extremes. *Nature* **2011**, *470*, 378–381. [CrossRef]

5.   Westra, S.; Fowler, H.J.; Evans, J.; Alexander, L.; Berg, P.; Johnson, F.; Kendon, E.J.; Lenderink, G.; Roberts, N.M. Future changes to the intensity and frequency of short-duration extreme rainfall. *Rev. Geophys.* **2014**, *52*, 522–555. [CrossRef]

6.   Panthou, G.; Vischel, T.; Lebel, T. Recent trends in the regime of extreme rainfall in the Central Sahel. *Int. J. Clim.* **2014**, *34*, 3998–4006. [CrossRef]

7.   Donat, M.G.; Lowry, A.L.; Alexander, L.; O'Gorman, P.A.; Maher, N. More extreme precipitation in the world's dry and wet regions. *Nat. Clim. Chang.* **2016**, *6*, 508–513. [CrossRef]

8.   Biasutti, M. Rainfall trends in the African Sahel: Characteristics, processes, and causes. *Wiley Interdiscip. Rev. Clim. Chang.* **2019**, *10*, e591. [CrossRef]

9.   Organisation Météorologique Mondiale. *Le Climat Dans le Monde 2001–2010, Une Décennie D'Extrêmes Climatiques, Rapport de Synthèse*; World Meteorological Organization (WMO): Geneva, Switzerland, 2013; Volume 1119, p. 15. Available online: https://library.wmo.int/doc_num.php?explnum_id=7831 (accessed on 14 June 2020).

10. Organisation Météorologique Mondiale. *Déclaration de l'OMM sur L'état du Climat Mondial en 2016*; World Meteorological Organization (WMO): Geneva, Switzerland, 2016; Volume 1189, p. 44. Available online: https://library.wmo.int/doc_num.php?explnum_id=3500 (accessed on 14 June 2020).

11. Blöschl, G.; Hall, J.; Parajka, J.; Perdigão, R.A.P.; Merz, B.; Arheimer, B.; Aronica, G.T.; Bilibashi, A.; Bonacci, O.; Borga, M.; et al. Changing climate shifts timing of European floods. *Science* **2017**, *357*, 588–590. [CrossRef]

12. Westra, S.; Alexander, L.; Zwiers, F.W. Global Increasing Trends in Annual Maximum Daily Precipitation. *J. Clim.* **2013**, *26*, 3904–3918. [CrossRef]

13. Fontaine, B.; Roucou, P.; Camara, M.; Vigaud, N.; Konare, A.; Sanda, S.; Diedhiou, A.; Janicot, S. *Variabilité pluviométrique, changement climatique et régionalisation en région de mousson africaine*; La Météorologie Météo et climat: Paris, France, 2012; pp. 41–48, ISSN 0026-1181.

14. Caniaux, G.; Giordani, H.; Redelsperger, J.-L.; Wade, M.; Bourles, B.; Bourras, D.; De Coëtlogon, G.; Du Penhoat, Y.; Janicot, S.; Key, E.; et al. Les avancées d'AMMA sur les interactions océan-atmosphère. *Météorologie* **2012**, *8*, 17–24. [CrossRef]

15. Lamb, P.J. Large-scale tropical Atlantic surface circulation patterns associated with sub-saharan weather anomalies. *Tellus* **1978**, *30*, 240–251. [CrossRef]

16. Folland, C.K.; Palmer, T.N.; Parker, D.E. Sahel rainfall and worldwide sea temperatures, 1901–1985. *Nature* **1986**, *320*, 602–607. [CrossRef]

17. Folland, C.K.; Parker, D.E.; Kates, F.E. Worldwide marine temperature fluctuations 1856–1981. *Nature* **1984**, *310*, 670–673. [CrossRef]

18. Hastenrath, S. The Relationship of Highly Reflective Clouds to Tropical Climate Anomalies. *J. Clim.* **1990**, *3*, 353–365. [CrossRef]

19. Zhang, R.; Delworth, T.L. Impact of Atlantic multidecadal oscillations on India/Sahel rainfall and Atlantic hurricanes. *Geophys. Res. Lett.* **2006**, *33*, 1–5. [CrossRef]

20. Ting, M.; KushniriD, Y.; Seager, R.; Li, C. Forced and Internal Twentieth-Century SST Trends in the North Atlantic. *J. Clim.* **2009**, *22*, 1469–1481. [CrossRef]

21. Tourre, Y.M.; Paz, S.; Kushnir, Y.; White, W.B. Low-frequency climate variability in the Atlantic basin during the 20th century. *Atmos. Sci. Lett.* **2010**, *11*, 180–185. [CrossRef]

22. Martin, E.R.; Thorncroft, C. Sahel rainfall in multimodel CMIP5 decadal hindcasts. *Geophys. Res. Lett.* **2014**, *41*, 2169–2175. [CrossRef]

23. Martin, E.R.; Thorncroft, C.D. The impact of the AMO on the West African monsoon annual cycle. *Q. J. R. Meteorol. Soc.* **2013**, *140*, 31–46. [CrossRef]

24. Dittus, A.J.; Karoly, D.J.; Donat, M.G.; Lewis, S.C.; Alexander, L. Understanding the role of sea surface temperature-forcing for variability in global temperature and precipitation extremes. *Weather Clim. Extrem.* **2018**, *21*, 1–9. [CrossRef]

25. Nouaceur, Z.; Murarescu, O.; Mură Rescu, O. Rainfall Variability and Trend Analysis of Annual Rainfall in North Africa. *Int. J. Atmos. Sci.* **2016**, *2016*, 1–12. [CrossRef]

26. Nouaceur, Z.; Laignel, B.; Turki, I. Changements climatiques au Maghreb: Vers des conditions plus humides et plus chaudes sur le littoral algérien? *PhysioGéo* **2013**, *7*, 307–323. [CrossRef]

27. Nouaceur, Z. Evaluation des changements climatiques au Maghreb. Etude du cas des regions du quart nord-est algerien. In Proceedings of the 23rd Colloque de l'Association Internationale de Climatologie, 'Risques et Changements Climatiques' 2010, Rennes, France, 1–4 September 2010; pp. 463–468.

28. Amyay, M.; Nouaceur, Z.; Tribak, A.; Okba, K.; Taous, A. Caracterisation des evenements pluviometriques extremes dans le Moyen Atlas marocain et ses marges. In Proceedings of the XXV eme Colloque International de Climatologie 2012, Grenoble, France, 5–8 September 2012; pp. 75–80.

29. Laignel, B.; Nouaceur, Z.; Jemai, H.; Abida, H.; Ellouze, M.; Turki, I. Vers un retour des pluies dans le nord-est tunisien? In Proceedings of the XXVIIe Colloque de l'Association Internationale de Climatologie 2014, Dijon, France, 2–5 July 2014; pp. 727–732.

30. Nouaceur, Z.; Laignel, B.; Turki, I. Changement climatique en Afrique du Nord: vers des conditions plus chaudes et plus humides, dans le Moyen Atlas Marocain et ses marges. In Proceedings of the Actes du XXVII Colloque International de Climatologie 2014, Dijon, France, 2–5 July 2014; pp. 399–405.

31. Nouaceur, Z.; Laignel, B.; Turki, I. Changement climatique au Sahel: des conditions plus chaudes et plus humides en Mauritanie? *Secheresse* **2013**, *24*, 85–95.

32. Nouaceur, Z.; Laignel, B. Caracterisation des evenements pluviometriques extremes sur la rive Sud du bassin mediterraneen: etude du cas du quart nord-est algerien. In Proceedings of the Actes du XXVIII Colloque International de Climatologie 2015, Liege, Belgium, 1–4 July 2015; pp. 573–578.

33. Grossmann, A.; Morlet, J. Decomposition of Hardy Functions into Square Integrable Wavelets of Constant Shape. *SIAM J. Math. Anal.* **1984**, *15*, 723–736. [CrossRef]

34. Labat, D. Recent Advances in Wavelet Analyses: Part 1. A Review of Concepts. *J. Hydrol.* **2005**, *314*, 275–288. [CrossRef]

35. Rossi, A. Analyse spatio-temporelle de la variabilité hydrologique du bassin versant du Mississippi: rôles des fluctuations climatiques et déduction de l'impact des modifications du milieu physique. In *Thèse de Géologie–Hydrologie*; Université de Rouen: Rouen, France, 2010; p. 329. Available online: https://tel.archives-ouvertes.fr/tel-00690189/file/THESE_A._ROSSI_-_MISSISSIPPI.pdf (accessed on 14 June 2020).

36. Zamrane, Z. Recherche d'indices de variabilité climatique dans des séries hydroclmatiques au Maroc: identification, positionnement temporel, tendances et liens avec les fluctuations climatiques: Cas des grands bassins de la Moulouya, du Sebou et du Tensift. In *Sciences de la Terre*; Université Montpellier: Montpellier, France, 2016; p. 197, NNT 2016MONTT181, HAL ID TEL 019690063. Available online: https://tel.archives-ouvertes.fr/tel-01690063/document (accessed on 14 June 2020).

37. Maraun, D. What Can We Learn from Climate Data? Methods for Fluctuation, Time/Scale and Phase Analysis. Ph.D. Thesis, Universität Potsdam, Potsdam, Germany, 2006.

38. Torrence, C.; Compo, G.P. A practical guide to wavelet analysis. *Bull. Am. Meteorol. Soc.* **1998**, *79*, 61–78. [CrossRef]

39. Maraun, D.; Kurths, J. Cross wavelet analysis: significance testing and pitfalls. *Nonlinear Process. Geophys.* **2004**, *11*, 505–514. [CrossRef]

40. Anctil, F.; Coulibaly, P. Wavelet analysis of the interannual variability in southrn Quebec streamflow. *J. Clim.* **2004**, *17*, 163–173. [CrossRef]

41. Labat, D.; Ronchail, J.; Guyot, J.L. Recent advances in wavelet analyses: Part 2—Amazon, Parana, Orinoco and Congo discharges time scale variability. *J. Hydrol.* **2005**, *314*, 289–311. [CrossRef]

42. Okonkwo, C. An Advanced Review of the Relationships between Sahel Precipitation and Climate Indices: A Wavelet Approach. *Int. J. Atmos. Sci.* **2014**, *2014*, 1–11. [CrossRef]

43. Baidu, M.; Amekudzi, L.K.; Aryee, J.; Annor, T. Assessment of Long-Term Spatio-Temporal Rainfall Variability over Ghana using Wavelet Analysis. *Climate* **2017**, *5*, 30. [CrossRef]

44. Dieppois, B.; Durand, A.; Fournier, M.; Massei, N.; Sebag, D.; Hassane, B. Variabilité des précipitations au Sahel central et recherche du forcage climatique par analyse du signal: La station de Maïne-Soroa (SE Niger) entre 1950 et 2005. *Pangea* **2010**, *47/48*, 27–35.

45. Dieppois, B.; Durand, A.; Fournier, M.; Diedhiou, A.; Fontaine, B.; Masssei, N.; Nouaceur, Z.; Sebag, D. Variabilité Basse-Fréquence des précipitations au Sahel et des températures de surface de l'océan Atlantique au cours du dernier siècle. In Proceeding of Actes du XXVe colloque de l'Association Internationale de Climatologie Grenoble, Grenoble, France, 5–8 September 2012; pp. 219–224.

46. Dieppois, B.; Durand, A.; Fournier, M.; Diedhiou, A.; Fontaine, B.; Massei, N.; Nouaceur, Z.; Sebag, D. Low-frequency variability and zonal contrast in Sahel rainfall and Atlantic sea surface temperature teleconnections during the last century. *Theor. Appl. Clim.* **2014**, *121*, 139–155. [CrossRef]

47. Lavaysse, C.; Flamant, C.; Janicot, S.; Parker, D.J.; Lafore, J.-P.; Sultan, B.; Pelon, J. Seasonal evolution of the West African heat low: a climatological perspective. *Clim. Dyn.* **2009**, *33*, 313–330. [CrossRef]

48. Fontaine, B.; Roucou, P.; Gaetani, M.; Marteau, R. Recent changes in precipitation, ITCZ convection and northern tropical circulation over North Africa (1979–2007). *Int. J. Climatol.* **2011**, *31*, 633. [CrossRef]

49. Meteo—France. *La variabilité climatique naturelle dans l'Atlantique Nord, au Cours du 20 ème Siècle, Influencerait la Puissance Destructrice des Cyclones Tropicaux et L'intensité des Sécheresses au Sahel*; Communiqué de Presse; Meteo France: Paris, France, 2010; p. 3. Available online: https://drive.google.com/file/d/1r5-vbwe-sDGRXcxZO2DUb2_X-W2VawUE/view?usp=sharing (accessed on 14 June 2020).

50. Giannini, A.; Saravanan, R.; Chang, P. Oceanic Forcing of Sahel Rainfall on Interannual to Interdecadal Time Scales. *Science* **2003**, *302*, 1027–1030. [CrossRef]

51. Held, I.M.; Delworth, T.L.; Lu, J.; Findell, K.L.; Knutson, T.R. Simulation of Sahel drought in the 20th and 21st centuries. *Proc. Natl. Acad. Sci. USA* **2005**, *102*, 17891–17896. [CrossRef]

52. Biasutti, M.; Giannini, A. Robust Sahel drying in response to late 20th century forcings. *Geophys. Res. Lett.* **2006**, *33*, 1–4. [CrossRef]

53. Mohino, E.; Janicot, S.; Bader, J. Sahel rainfall and decadal to multi-decadal sea surface temperature variability. *Clim. Dyn.* **2010**, *37*, 419–440. [CrossRef]

54. Caminade, C.; Terray, L. Twentieth century Sahel rainfall variability as simulated by the ARPEGE AGCM, and future changes. *Clim. Dyn.* **2009**, *35*, 75–94. [CrossRef]

55. Biasutti, M. Forced Sahel rainfall trends in the CMIP5 archive. *J. Geophys. Res. Atmos.* **2013**, *118*, 1613–1623. [CrossRef]

56. Martin, E.R.; Thorncroft, C.; Booth, B.B.B. The Multidecadal Atlantic SST—Sahel Rainfall Teleconnection in CMIP5 Simulations. *J. Clim.* **2014**, *27*, 784–806. [CrossRef]

57. Skinner, C.B.; Ashfaq, M.; Diffenbaugh, N.S. Influence of Twenty-First-Century Atmospheric and Sea Surface Temperature Forcing on West African Climate. *J. Clim.* **2012**, *25*, 527–542. [CrossRef]

58. Abdou, A. *Variabilité et Changements du Climat au SAHEL: ce que L'observation Nous Apprend sur la Situation Actuelle, Grains de sel*; Inetr réseaux: Ouagadougou, Burkina Faso, 2010; Volume 49, pp. 13–14. Available online: http://www.inter-reseaux.org/IMG/pdf_p13_14_Agrhymet.pdf (accessed on 14 June 2020).

59. Giannini, A.; Biasutti, M.; Verstraete, M. A climate model-based review of drought in the Sahel: Desertification, the re-greening and climate change. *Glob. Planet. Chang.* **2008**, *64*, 119–128. [CrossRef]

60. Hickler, T.; Eklundh, L.; Seaquist, J.W.; Smith, B.; Ardö, J.; Olsson, L.; Sykes, M.T.; Sjöström, M. Precipitation controls Sahel greening trend. *Geophys. Res. Lett.* **2005**, *32*, 4. [CrossRef]

61. Lebel, T.; Cappelaere, B.; Galle, S.; Hanan, N.P.; Kergoat, L.; Levis, S.; Vieux, B.; Descroix, L.; Gosset, M.; Mougin, E.; et al. AMMA-CATCH studies in the Sahelian region of West-Africa: An overview. *J. Hydrol.* **2009**, *375*, 3–13. [CrossRef]

62. Met Office Hadley Center. *Climat Sahélien: Rétrospective et Projections*; Met Office: Devon, UK, 2010; p. 18. Available online: https://www.oecd.org/fr/csao/publications/47093854.pdf (accessed on 14 June 2020).

63. Ozer, P.; Hountondji, Y.; Niang, A.; Karimoune, S.; Laminou, M.O.; Salmon, M. Désertification au Sahel: historique et perspective. In *BSGLg, Bulletin de la Société Géographique de Liège 54*; 2010; pp. 69–84, ISSN 0770-7576, E-ISSN 2507-0711. Available online: https://popups.uliege.be/0770-7576/index.php?id=942&file=1 (accessed on 14 June 2020).

64. Funk, C.C.; Rowland, J.D.; Adoum, A.; Eilerts, G.; White, L. A climate trend analysis of Chad. *U.S. Geol. Surv. Fact Sheet* **2012**, *3070*, 4. [CrossRef]

65. Dong, B.; Sutton, R. Dominant role of greenhouse-gas forcing in the recovery of Sahel rainfall. *Nat. Clim. Chang.* **2015**, *5*, 757–760. [CrossRef]

66. Taylor, C.M.; Belušić, D.; Guichard, F.; Parker, D.J.; Vischel, T.; Bock, O.; Harris, P.; Janicot, S.; Klein, C.; Panthou, G. Frequency of extreme Sahelian storms tripled since 1982 in satellite observations. *Nature* **2017**, *544*, 475–478. [CrossRef]
67. Bernard, F.; Pascal, R.; Moctar, C.; Nicolas, V.; Abdourahamane, K.; Seidou, I.S.; Arona, D.; Serge, J. Variabilité pluviométrique, changement climatique et régionalisation en région de mousson africaine. In *La Météorologie, Série 8, N Special AMMA*; Météo et Climat, Météo-France: Paris, France, 2012; pp. 41–48, ISSN 0026–1181.

Article

North German Lowland Lakes Miss Ecological Water Quality Standards—A Lake Type Specific Analysis

Jacqueline Rücker [1,*], Brigitte Nixdorf [1], Katrin Quiel [2] and Björn Grüneberg [1,3]

[1] Department of Freshwater Conservation, Brandenburg University of Technology Cottbus—Senftenberg, Bad Saarow, Seestraße 45, D-15526 Bad Saarow, Germany; nixdorf@b-tu.de (B.N.); bjoern.grueneberg@landeslabor-bbb.de (B.G.)

[2] Department W14, State Office of Environment (LfU), Seeburger Chaussee 2, D-14476 Potsdam, OT Groß Glienicke, Germany; w14@lfu.brandenburg.de

[3] Berlin-Brandenburg State Laboratory (LLBB), Department IV-2, Rudower Chaussee 39, 12489 Berlin, Germany

* Correspondence: j.ruecker@b-tu.de; Tel.: +49-33631-894-55

Received: 2 October 2019; Accepted: 19 November 2019; Published: 2 December 2019

Abstract: Despite great efforts in point source reductions due to improved wastewater treatment since 1990, more than 70% of the lakes in Germany have not yet achieved the "good ecological status" according to the European Water Framework Directive (WFD). To elicit lake type-specific causes of this failure, we firstly analyzed the ecological status of 183 lakes in NE Germany (Federal State of Brandenburg), as reported to the European Commission in 2015. Secondly, long-term data of two typical lakes (a very shallow polymictic lake with a large and a deep stratified lake with a small catchment area in relation to lake volume) and nutrient load from the common catchment were investigated. About 64%–83% of stratified and even 96% of polymictic shallow lakes in Brandenburg currently fail the WFD aims. Excessive nutrient emissions from agriculture were identified as the main cause of this failure. While stratified deep lakes with small catchments have the best chances of recovery, the deficits in catchment management are amplified downstream in lake chains, so that especially shallow lakes in a large catchment are unlikely to reach good ecological conditions. If the objectives of the WFD are not questioned, agricultural practices and approaches in land use have to be fundamentally improved.

Keywords: water quality; catchment; nutrient load; agriculture; European Water Framework Directive; shallow lake; stratified lake

---

1. Introduction

The pollution and shortage of freshwater resources are worldwide problems. The European Water Framework Directive takes account of the high value of water and the need to protect this precious resource when it starts with the following words: "Water is not a commercial product like any other but, rather, a heritage, which must be protected, defended and treated as such." [1]. The WFD was established by the European Commission (EC) in 2000 to ensure sustainable water management based on River Basin Management Plans (RBMP) and programs of measures. It provides a legislative framework, which commits the member states of the European Union to preventing deterioration of the aquatic environment and achieving good status of all water bodies (groundwater, rivers, lakes, transitional waters, and coastal waters). This includes the biological, hydromorphological, physicochemical, and chemical quality of water bodies. The aim of the WFD is to achieve a "good ecological status" for natural waters. Initially set for 2015 (by the end of the First RBMP), this target is now to be achieved by 2027, but there is growing concern that this is a long way from being achieved in many countries. Continued efforts are, therefore, needed to integrate water policy into other policy areas such as agriculture, urban planning, energy, and climate [2]. Eutrophication is still one of the major environmental problems across Europe. Agricultural or diffuse losses (agriculture plus background) account for more than 60% of the total nitrogen (N) load. For phosphorus (P), point sources tend to be the most significant source in European countries. However, as point source discharges have been reduced markedly during the last 15 years, agriculture has sometimes become the main P source [3]. In contrast to point sources, the quantification of nutrient losses from diffuse sources is a challenging task and several nutrient models have been developed worldwide in an attempt to describe and quantify nutrient transfers from fields to the aquatic environment [4].

For the assessment of lake water quality, the WFD requests the inclusion of different groups of organisms (biological quality elements). Therefore, national assessment methods have been developed during the implementation of the WFD [5], which have been intercalibrated subsequently in geographical intercalibration groups (GIGs) [6,7]. Water quality assessment requires the comparison of the present state of a water body to a lake type-specific reference state, which is defined as a slight deviation from natural conditions due to anthropogenic impact.

In Germany, the typification system by the Federal State Working Group Water (LAWA) [8] is used. Most of the WFD-relevant lakes ≥50 ha (which represents less than 10% of the total number of lakes) are located in the ecoregion of the German lowlands or in the Lowland Central/Baltic (L-CB) according to the intercalibration [7]. Phytoplankton is the biological quality element most widely used for the assessment of lakes. The assessment system for this biological quality element in German lakes (Phyto-See-Index, PSI) considers total biomass, algal classes, and indicator species of phytoplankton. A manual [9] and a Microsoft Access-based tool for calculations [10] are available. The sampling instruction for standing waters was also standardized to ensure better comparability of data [11]. Furthermore, assessment systems were developed for submerged macrophytes and microphytobenthos (PHYLIB) [12] and for elements of the aquatic fauna, such as macrozoobenthos [13] and fish [14].

By the end of the first RBMP in 2015, most German lakes (74%) did not achieve the "good ecological status" [15], which is below the European average [16]. Therefore, the first aim of this study is to identify the causes of missed targets. For this purpose, data of about 200 natural lakes ≥ 50 ha in the Federal State of Brandenburg (North German lowlands), a region rich in surface waters were analyzed. About 3000 lakes and 33,000 km of river course cover 2% of the state area. Intact landscapes and ecosystems, including lakes and their water quality are of significant economic importance for tourism and recreation, but also for transportation, agriculture, and fisheries. In the past century, eutrophication through the excessive input of nutrients from intensive agriculture and insufficient treated municipal wastewater caused a massive deterioration of water quality in most lakes of the area studied. However, the German reunification in 1990 marked a change characterized by the beginning of substantial restoration measures in the catchment areas, especially by improved treatment of wastewater.

The second aim of the study is to analyze the influence of morphometry, land use, catchment size, and hydrology (water residence time) on water quality. To that end, long-term water quality data (1994–2018) of two lakes were analyzed to study their response after load reduction. Both lakes are situated in the same catchment (Scharmützelsee region, Brandenburg). Nutrient emissions from the catchment and resulting loads to the surface waters were estimated by an adapted nutrient export coefficient approach and verified by empirical models. These lakes represent two of the most frequent lake types occurring in North German lowlands: The very shallow polymictic lake with a large catchment area and the deep stratified lake with a small catchment in relation to lake volume. Both lakes are part of a chain of lakes, which is a typical element of this landscape.

Specifically, we want to answer the following scientific questions: How do deep and shallow lakes respond to a reduction in nitrogen and phosphorus loads and in-lake concentrations as limiting nutrients? Do the ecological quality elements macrophytes and phytoplankton respond in a similar manner on the nutrient reduction? How efficient is nutrient recycling in shallow lakes to oppose nutrient loading reductions [17]? Finally, statistical analyses and case studies in combination allow to derive fundamental lake type-specific causes for missing the WFD water quality objectives and commenting of the requirements for the achievement and adequacy of these objectives.

2. Materials and Methods

2.1. Study Sites

2.1.1. Brandenburg Lakes

Brandenburg is a Federal State in the northeastern part of Germany (Figure 1a) with an area of 29,654 km^2 and two and a half million inhabitants. Berlin, the capital of Germany, and Brandenburg constitute the common capital region of Berlin-Brandenburg containing six million inhabitants, with 4.4 million in the affluent suburbs.

Brandenburg is naturally rich in running and standing waters: A network of waterways consisting of 33,000 km of river course and 3000 lakes mainly belonging to the catchment areas of the rivers Elbe and Oder. Surface waters cover 1000 km^2, i.e., 2% of the state area. The distribution of lakes in Brandenburg is shown in Figure 1b. There has been a change in land use and transport demands for construction materials, fuel, and agricultural products over the last few centuries. Small rivers were made navigable for ships and were linked still further by canals from the 17th century onwards. Large-scale agricultural land modifications led to the irrigation or drainage of large areas [18]. The natural wealth of lakes has been further increased by the formation of mining lakes in southeast Brandenburg following large-scale opencast lignite mining. Some of the lakes developed are bigger and deeper than the largest natural lakes.

The supply of water in Brandenburg is limited due to the influence of the continental climate. The annual rainfall (576 mm, long-term mean 1981–2010, Meteorological Observatory of DWD in Lindenberg, 10 km apart from Scharmützelsee region [19]) is the lowest in Germany. Most of the lakes are discharge regulated to prevent water level dropdown at times of low rainfall. Climate scenarios calculated for Brandenburg predict an increase in the average temperature, whereas mean annual precipitation remains unchanged. Due to an increase in evaporation and decreased discharge, this change in the regional climate will be reflected in the water balance [18]. The management of Brandenburg's water through extraction, discharge, and storage can have an even bigger influence on the water balance than climate change.

2.1.2. Catchment Characteristics of the Scharmützelsee Region

The lakes for the long-term study are situated in the Scharmützelsee region in the central part of Brandenburg, about 60 km southeast of Berlin (Figure 1, Table 1 for coordinates). They originate from the Weichsel glaciation during the last stage of the Pleistocene, about 12,000 years ago, and are located at the border of the southern plateau of the Berlin glacial valley. Carbonate-rich glacial till and nutrient-poor sands characterize the surface geology. The lakes were formed by a combination of ice and meltwater erosion. Small rivers and navigable channels connect most of the lakes in the region, which is named after the largest lake, Scharmützelsee. The water level is regulated, and four water gates enable boating to the river Dahme, a tributary entering the river Spree in Berlin. Consequently, the quality of water discharged from the Scharmützelsee region influences downstream lakes and rivers in the catchment areas of the River Spree, Havel, and Elbe flowing to the North Sea. The total size of the catchment studied is 392 km^2.

Land use characteristics are given in Figure 2. The mapping is based on satellite data of the years 2000 and 2006 (CORINE Land Cover data (CLC2006) [20]. Half of the catchment area is covered by forest (53%), and one third is used for agriculture (31%). The portion of surface waters is relatively high (9%), but the area-specific discharge is low (2.5 L km^{-2} s^{-1}). Urban and industrial areas cover 6% of the catchment. The population density, with 58 inhabitants per km^2, is relatively low. The main use of water in the catchment is related to tourism and fisheries. With small regional differences,

94%–100% of the inhabitants are connected to the sewers [21]. The wastewater of 24,319 population equivalents [22] is treated in one large and two small wastewater treatment plants (WWTP). They emit about 3536 m^3 day^{-1} of treated wastewater. Mean concentrations in the effluent of the large WWTP are 0.3 mg TP L^{-1} and 3.5 mg TN L^{-1} (means 2011–2018, based on data of the WWTP operating companies). Drinking water is extracted mainly from groundwater.

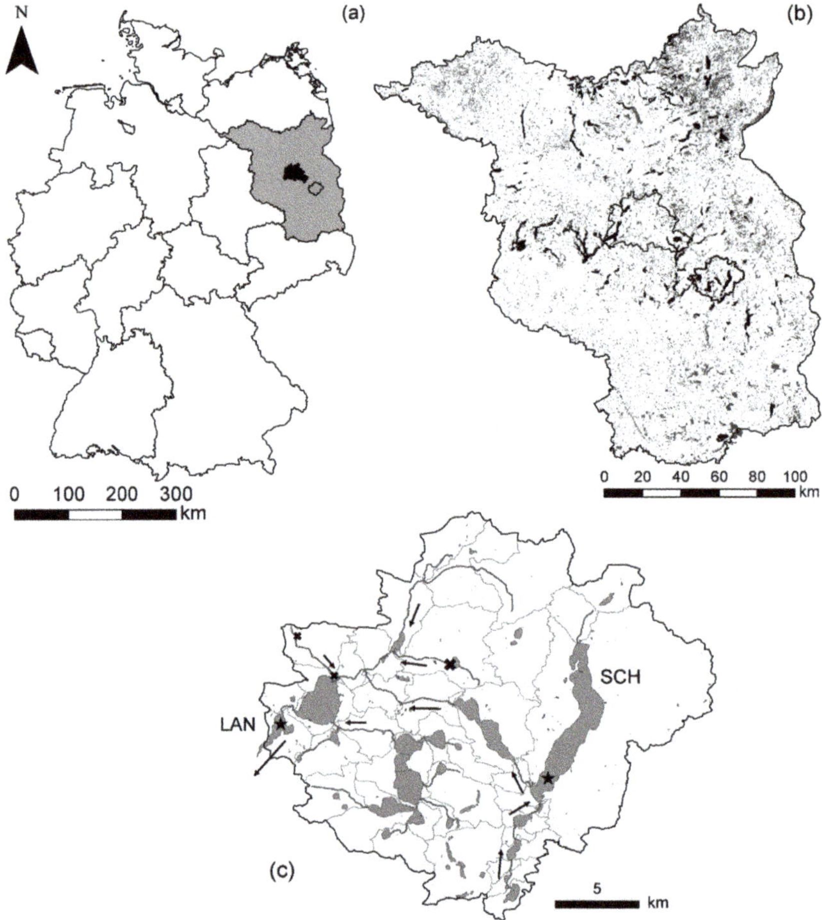

Figure 1. Federal States of Germany (**a**). Brandenburg is highlighted in grey and the capital, Berlin, in black. The catchment Scharmützelsee region is surrounded by a line. Distribution of lakes in Brandenburg (**b**). Lakes to be assessed according to the European Water Framework Directive [1] and have an area ≥50 ha are shown in black. Scharmützelsee region is marked. Catchment of Scharmützelsee region (**c**). Arrows indicate the flow direction, crosses mark the effluents of wastewater treatment plants, and black asterisks mark the sampling locations in Lake Scharmützelsee (SCH) and Lake Langer See (LAN). Source of geographical data: [23–27].

Table 1. Morphological and hydrological characteristics of the lakes studied. Hydrological data are given as mean and range of annual values 1994–2018.

	Scharmützelsee (SCH)	Langer See (LAN)
Geographic coordinates (WGS84)	52°15′ N, 14°03′ E	52°14′ N, 13°47′ E
Catchment area (km^2)	127	392
Lake area (km^2)	12.1	1.3
Lake volume (Mio. m^3)	108.2	2.84
Catchment to volume ratio (m^{-1})	1.2	138
Shore length (km)	30.3	8.4
Maximum depth (m)	29.5	3.8
Mean depth (m)	8.9	2.2
Mean daily discharge (m^3 s^{-1})	0.35 (0.22–0.47)	1.03 (0.70–1.44)
Hydraulic residence time (a)	10.3 (7.4–15.8)	0.09 (0.06–0.13)
Type of mixis	dimictic	polymictic
Lake type according to WFD [8,9]	13 (stratified hard water lake with small catchment)	11.2 (very shallow hard water lake with large catchment)

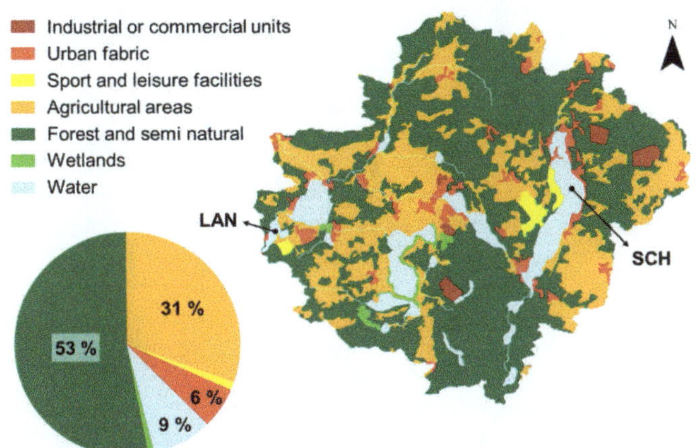

Figure 2. Land use of the catchment of Scharmützelsee region. The lakes in the case study are labeled. The pie chart summarizes the types of land use (industrial and urban areas are grouped together). Map created from CORINE Land Cover data (CLC2006) [20].

2.1.3. Lakes in the Catchment: Deep Lake Scharmützelsee and Shallow Lake Langer See

Lake Scharmützelsee is one of the largest natural lakes in Brandenburg, based on its area (Table 1). The elongated lake is 10 km long and about 1 km wide. The depth of the lake increases from 7 m maximum depth in the polymictic northern basin over a temporarily thermally stratified middle basin with up to 11 m depth to a maximum depth of 29.5 m in the dimictic southern basin (A bathymetric map is given in [28]). The total lake is considered for the assessment of water quality according to the WFD. It belongs to lake type 13, based on the German classification system for WFD: Calcareous, stratified lakes with a small catchment (in relation to lake volume) in the German lowlands ([8], Table 2). Water inflow is predominantly (88%) from groundwater sources [28], although surface water from a chain of four smaller lakes enters the larger lake at its southernmost point (Figure 1c). The lake outflow has been regulated by a weir and an outlet lock since the 18th century. The hydraulic residence time is about ten years (Table 1).

Lake Langer See is a polymictic shallow lake with a maximum depth of 3.8 m and is much smaller than Lake Scharmützelsee (Table 1). A bathymetric map is given in [29]. However, being at the end of the lake chain, it receives the complete load from the Scharmützelsee catchment. The discharge is high in relation to the lake volume, resulting in a mean hydraulic residence time of only 34 (23–47) days (long-term annual mean 1994–2018). With values around 30 days, the lake is at the transition between lake type 11.2 (very shallow, hard water lake with large catchment) and type 12 (riverine lake), according to the German assessment system for phytoplankton ([9], Table 2). Nevertheless, due to a low discharge in the vegetation period (used for assessment), the hydraulic residence time is usually above 30 days (30.5–104.1 days, average 55.1 days, long-term annual mean 1994–2018) and this lake was classified as type 11.2.

Grüneberg and co-authors [28] give details of the nutrient loading history of Lake Scharmützelsee. There was an early limnological survey of this region in the 1930s. In the paper evolving from this study, Wundsch [30] described Lake Scharmützelsee as relatively clear with an oxygen-poor, but not oxygen-free, hypolimnion without the formation of hydrogen sulfide (H_2S) [31]. Thus, a pristine mesotrophic status can be assumed. This situation had changed when Müller [32] studied the lake in 1949 and 1950, as H_2S in the hypolimnion in August and September up to a depth of 12 m was found. The summer Secchi depth decreased from about 3 m in the 1930s [30] to 2.3–1.1 m at the beginning of the 1950s [32]. Thus, Lake Scharmützelsee changed from mesotrophic to eutrophic conditions [28,33,34].

There are no reports on the water quality in Lake Langer See from early times. A natural eutrophic status can be assumed due to its shallowness and the high hydraulic and nutrient load. Almost complete coverage of the lake bottom by submerged macrophyte plants is likely, but not reported. However, the trophic development of all lakes in the region, especially Lake Scharmützelsee and the downstream lakes, must be similar: Eutrophication has increased since the 1960s. Excessive sewage discharge, especially between the 1950s and the 1980s, and the introduction of phosphate-containing detergents in the 1960s resulted in a deterioration of water quality. Another source of nutrient input into lakes was fish food. Lake Storkower See, situated upstream of Lake Langer See, was used for fish farming for over a decade before 1990. This latter year marks political changes in eastern Germany and the beginning of substantial restoration measures in the catchment. The most important element of these management measures was the reduction of point discharge, for example, by the construction of a modern wastewater treatment plant and a central sewage system. The portion of inhabitants connected to the sewers increased from about 25% at the end of the 1980s to almost 100% nowadays. The Department of Freshwater Conservation has studied the response of the lakes to the nutrient load reduction due to restoration measures in the catchment since 1993.

2.2. Methods and Database

2.2.1. Monitoring Program of the Brandenburg Water Authority 2006–2014

About 200 lakes in Brandenburg of a size ≥50 ha are monitored every three to seven years to meet the demands of the WFD. The State Office of the Environment (Landesamt für Umwelt (LfU), formerly the Landesamt für Umwelt, Gesundheit und Verbraucherschutz (LUGV)) of the Ministry of Rural Development, Environment and Agriculture of the Federal State of Brandenburg provided raw data, derived water quality indices and reports of lake monitoring programs for the years 2006 to 2014, as well as management plans and programs of measures for the Elbe and Oder river basins for the period 2015–2021 [35,36].

The assessment of phytoplankton was carried out using the PSI tool [9]. Submerged macrophytes and microphytobenthos were assessed separately or in combination with the PHYLIB tool [12]. The macrophyte assessment was supplemented by the MIB tool (Makrophytenindex Brandenburg [37]), a method developed and adapted to Brandenburg lakes and suitable for lakes smaller than 50 ha.

The Ecological Quality Class (EQC) is a combination of the different indices, whereas its value is determined by the worst single index. Vegetation means of total phosphorus (TP) and total nitrogen (TN) concentrations were calculated from monthly means for the vegetation period from April to October. Depending on the measuring cycle, the data for individual lakes were raised between 2006 and 2014.

Table 2. Classification of calcareous ($Ca^{2+} \geq 15$ mg L^{-1}) German lowland lakes $\geq$50 ha, according to [8] (DE-type), and for the assessment of phytoplankton (PP-type), according to [9]. Classification is based on mixing behavior, the influence of the catchment expressed as volume quotient (VQ, catchment area (km^2) divided by lake volume (10^6 m^3)), mean depth (z_m, lake volume divided by lake area) and hydraulic residence time (RT in days, lake volume divided by mean annual discharche). The European intercalibration lake type (IC type) is given if it exists (Lowland Central/Baltic Geographical Intercalibration Group [38]).

DE Type	PP Type	Mixis	VQ	z_m	RT	IC Type
10	10.1		1.5–15			
	10.2	stratified	>15			L-CB1
13	13		≤1.5			
11	11.1		>1.5	>3 m		
	11.2	polymictic	>1.5	≤3 m		L-CB2
14	14		≤1.5			
12	12	riverine	>1.5		3–30 d	

A total of 183 natural lakes were selected for this study. They include lake types 10 to 14 (hard water lakes in the ecoregion North German lowlands, following [8]. These are the following subtypes based on the phytoplankton assessment method: Deep stratified lakes with stable summer stratification with small (type 13), large (type 10.1) or very large (type 10.2) catchments compared to lake volume. Shallow, polymictic lakes are classified based on their mean depth (z_m) as subtype 11.1 ($z_m \geq 3$ m) or very shallow lakes (subtype 11.2 $z_m < 3$ m). Riverine lakes (type 12) are lakes with short hydraulic residence time (<30 d) and relatively large catchments. Type 14 includes shallow lakes ($z_m \geq 3$ m) with a small catchment (Table 2). All stratified lakes represent lake type L-CB1 of the Central/Baltic Geographic Intercalibration Group. Among polymictic lakes, only type 11.2 overlaps with the Central/Baltic GIG type L-CB2 (Table 2). Lakes types 11.1 and 12 were not included in the European lake types due to hydraulic retention time <1 year, and type 14 due to hydraulic retention time >1 year [38].

2.2.2. Estimation of Nutrient Input from the Catchment in the Scharmützelsee Region

The quantification of nutrient inputs into surface waters and the localization of hotspots of nutrient emissions is the basis for the development of management plans for surface waters in Brandenburg. Consequently, the State Office of the Environment has adapted the export coefficient approach (e.g., [39,40]) to requirements in catchments in the Federal State of Brandenburg [41]. Based on extensive geographical data (e.g., soil types, land use, crop, drainage, and groundwater) and data from the literature (e.g., retention coefficients, emission rates), the annual inputs of phosphorus and nitrogen into surface waters are estimated. These calculations contain runoff as surface runoff or measured discharges, emission rates from point sources (WWTP and urban areas) and emissions from diffuse sources from agricultural or urban areas via groundwater and surface runoff, atmospheric deposition, and a natural groundwater background. In the first step, the nutrient emissions from the sub-catchments are estimated based on land use information. Net nutrient emissions are calculated considering nutrient retention processes in the soil and in the groundwater. In the second step, loads from sub-catchments are summarized along the sections of running waters following the natural flow direction of the water taking the nutrient retention in surface waters into account. The calculations presented are based on land use information and WWTP discharges for the year 2011.

2.2.3. Long-Term Study of Water Quality in Lake Scharmützelsee and Lake Langer See 1993–2018

Lake Scharmützelsee and Lake Langer See have been sampled monthly to biweekly at their deepest point since July 1993 (Figure 1c). Water transparency was measured with a white Secchi disk of 20 cm diameter. Water samples were taken at half-meter intervals with a 2.3 L LIMNOS® sampler and mixed in a vat. Mixed water samples were prepared over the whole water column for polymictic Lake Langer See. For Lake Scharmützelsee, mixed samples were prepared volume-weighted either for the whole water column during periods of complete mixing (usually November to April) or separately for the upper mixed part (epilimnion) during thermal stratification periods (usually May to October). Aliquots of the mixed samples were analyzed to determine the concentrations of TP, TN, and chlorophyll a according to German standard methods [42]. Phytoplankton composition and biovolume were estimated using an aliquot fixed with Lugol's solution studied under an inverse microscope, according to [43,44]. Biovolume was converted to biomass assuming a specific density of 1 mg mm^{-3}.

The German Trophic Index, according to LAWA, was calculated by using the Microsoft Access tool [45]. This index summarizes trophic parameters (vegetation means of chlorophyll a, Secchi depth and TP, and TP during spring overturn) to seven classes ranging from oligotrophic to hypertrophic. The assessment of water quality was carried out with the PSI tool for phytoplankton [10] and with the PHYLIB tool for submerged macrophytes [12]. Macrophyte mapping was carried out once in 2011 by Lanaplan, Nettetal [46], based on the PHYLIB method.

Different empirical models were applied to analyze lake-catchment interactions, which describe the relationship between in-lake nutrient concentration, nutrient input, and hydraulic conditions. The mean annual inflow concentrations of TP and TN were calculated from the in-lake nutrient concentrations and the mean hydraulic residence time. For Lake Scharmützelsee, the models of Vollenweider (1976) [47], OECD (1982, combined data set) [48], Nürnberg (1984) [49] and Sas (1989) [50], the latter with the empirical net sedimentation coefficient according to Brett and Benjamin (2008) [51] were applied. Details on the models are given in [28]. The TP load for Lake Langer See was estimated with the models of Vollenweider [47] as well as the combined data model and the shallow lakes model of OECD [48]. The TN load calculations were carried out with the OECD [48] model for the combined dataset. Mean hydraulic residence time was calculated from daily discharge data measured at the outlet lock of Lake Scharmützelsee, kindly supplied by the Water and Shipping Authority, Berlin. The resulting area-specific discharge of this sampling station and a second station in the catchment measured by LfU were used to estimate the discharge for Lake Langer See. Critical loads of TP and TN were calculated, as described above, using the lake type-specific target values as input data, i.e., TP and TN concentrations at which lakes achieved the "good ecological status" according to the PSI derived by Dolman et al. (2016) [52]. The TP-target value for type 13 was multiplied by 1.5 to consider the mean difference between annual values (used by the empirical models) and vegetation mean (used by [52]) for Lake Scharmützelsee. The relation between annual and vegetation mean for TN in Lake Scharmützelsee, and TP and TN in Lake Langer See, was about 1.

3. Results

3.1. Ecological Status of Lakes in Brandenburg, According to WFD in 2014

The ecological status of 183 Brandenburg lakes, according to WFD, as reported to the European Commission in 2014, is analyzed here. The lakes studied represent lake types typical for the North German lowlands, respectively, central Europe (Central/Baltic GIG), with 83 (45%) stratified and 100 (55%) polymictic lakes (Figure 3). Stratified lakes are most frequent with a volume quotient between 1.5 and 15 m^2 lake area per m^3 lake volume (type 10.2, 54 lakes) and very shallow polymictic lakes (type 11.2, 45 lakes). Type 14 (polymictic lakes with a small catchment) comprised only three lakes and will not be discussed further.

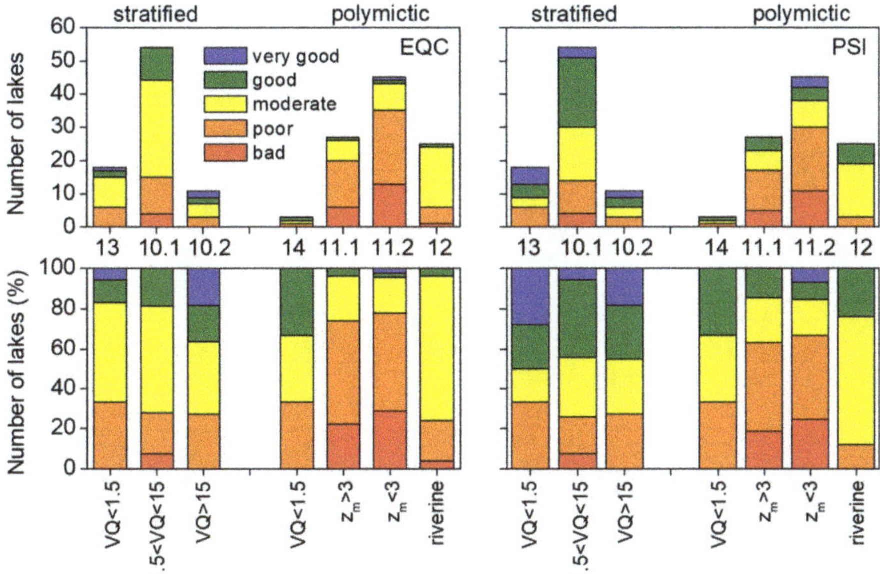

Figure 3. Assessment of 183 natural lakes in Brandenburg ≥50 ha according to the Ecological Quality Class (EQC, **left**) and Phyto-See-Index (PSI, **right**) as reported to the EU in 2014. The EQC summarizes indicator values of different assessment tools for diverse biocomponents (phytoplankton, benthic diatoms, submerged macrophytes, and microphytobenthos). Lakes are classified in sub-types according to PSI [9]. Numbers of lake types and distinctive features are given. VQ—volume quotient (catchment area divided by lake volume (m^2/m^3)), z_m—mean depth (m).

According to the assessment of phytoplankton (PSI, Figure 3), 44%–50% of stratified lakes have reached the "good" or "very good" ecological state. This portion was much smaller for the polymictic lakes (type 11 and 12), with only 15%–24% in "good" ecological state based on the PSI. The high proportion of 63%–67% of lakes in a "poor" and "bad" state for lake types 11.1 and 11.2 was striking. The proportion of "poor" and "bad" lakes was the smallest in riverine lakes (12% in type 12). The ecological quality (EQC, Figure 3) as a more comprehensive evaluation, including other biological quality elements, is even worse than the PSI (only phytoplankton). In total, only a minority of 12% of all lakes achieved the EQC of "good" and "very good". This portion was higher among stratified lakes (17%–36%). However, polymictic shallow lakes (types 11.1, 11.2 and 12) are especially at risk: Only 4% of 97 lakes have reached the "good ecological status" (EQC).

TP and TN concentrations of the lakes were analyzed for the year of assessment to investigate the causes for the failure to reach the water quality goals. This was compared to lake type-specific target values for TP and TN to achieve the "good ecological status," according to PSI (vertical lines in Figure 4) derived by Dolman et al. (2016) [52]. The higher the TP concentrations, the more likely is an ecological status worse than "good". Whereas the TP concentrations in lakes of "good" and "very good" ecological status are mostly relatively close to the lake type-specific target values, the TN concentrations tend to be more above the TN target values, especially in lakes of type 12, and also in the other polymictic lakes (type 11.1 and 11.2). Only some stratified lakes (and very few polymictic lakes) reach the TN target values. Generally, the TN concentrations are often far above the TN target values for all lake types, independent of the present ecological state.

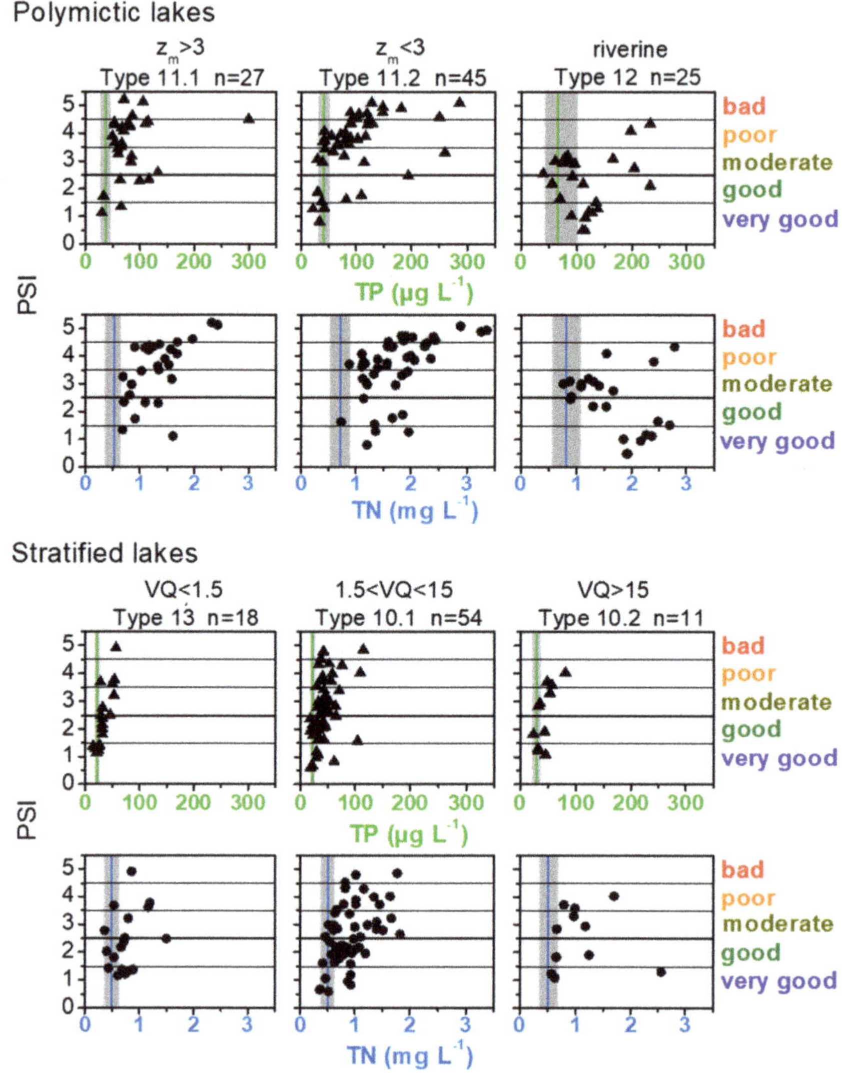

Figure 4. The PSI and concentrations of total phosphorus (TP, triangles) and total nitrogen (TN, circles) of the vegetation period (April–October) for 180 lakes in Brandenburg (without lake type 14). Numbers of lake types and distinctive features are given. VQ—volume quotient (catchment area divided by lake volume (m^{-1})), z_m—mean depth (m). Vertical lines indicate lake type-specific TP and TN target values for achieving good ecological status according to Dolman et al. (2016) [52] (range of error—grey area). The bold horizontal line is the boundary between "good" and "moderate ecological status". (Figure modified from [36]).

3.2. Nutrient Inputs from the Catchment—Case Study Scharmützelsee Region

Net P and N emissions for sub-catchments in the Scharmützelsee region were analyzed based on the land use information for the catchment without considering nutrient retention in the surface water bodies (Figure 5). It turns out that N emissions from sources related to agricultural usage (surface runoff or subsurface transport, Figure 5c) are much higher than from sources related to urban areas (Figure 5d). The N emissions were especially high (>10 kg ha^{-1} yr^{-1}) in the western part of the Scharmützelsee region, where intensively used arable land is concentrated (Figure 2). In contrast to this, the P emissions from urban areas (Figure 5b) partly exceed the emissions from agricultural areas (Figure 5a). The main urban point source of P is the biggest wastewater treatment plant treating the wastewater of 22,202 population equivalents [22]. The sub-catchment where the WWTP is situated stands out with an annual P emission of 0.18 kg ha^{-1} yr^{-1}.

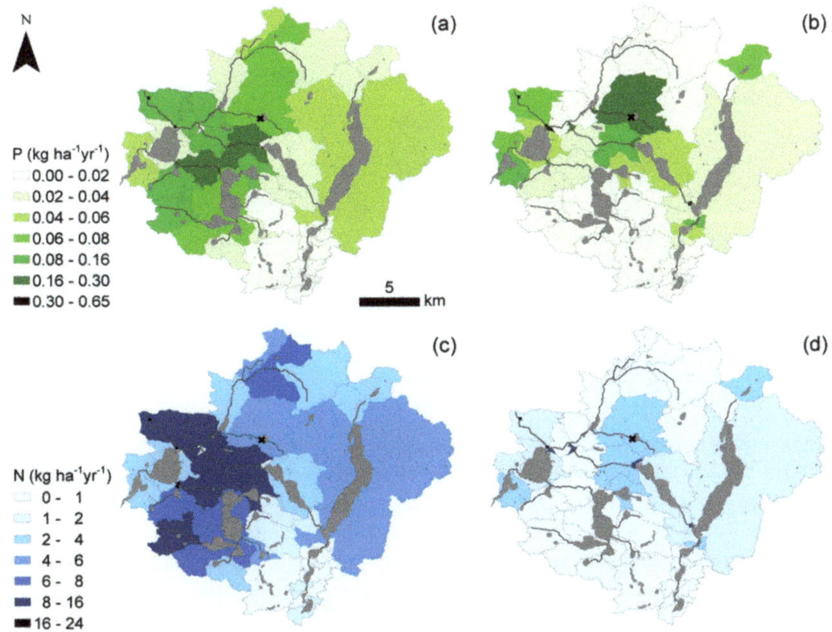

Figure 5. Phosphorus (**a,b**) and nitrogen net emissions (**c,d**) in kg ha^{-1} yr^{-1} from agricultural (**a,c**) and urban areas (**b,d**) for the sub-catchments in the Scharmützelsee region calculated without nutrient retention in the surface water bodies, according to the method of LUGV [41].

Only a part of the nutrients emitted enters the lakes downstream due to the different retention rates for N and P in the surface water bodies considered in the method of LUGV [41]. The resulting nutrient loads are presented in Figure 6. The portion for N input from agricultural sources is relatively high at 55% for Lake Scharmützelsee and 62% for Lake Langer See. By contrast, about half of the P emissions originate from atmospheric deposition and natural background (groundwater), and the portion of urban sources (about 20%) is somewhat higher than for agricultural P sources (about 15%). The total nutrient input calculated from land use is much lower for upstream Lake Scharmützelsee (0.41 t P yr^{-1}, 5.73 t N yr^{-1}) than for Lake Langer See at the end of the catchment (1.76 t P yr^{-1}, 20.96 t N yr^{-1}). Applying these estimates, the empirical models predict in-lake TP concentrations of about 9 µg L^{-1} and 35 µg L^{-1} for Lake Scharmützelsee and Lake Langer See, respectively (mean values of four or three models).

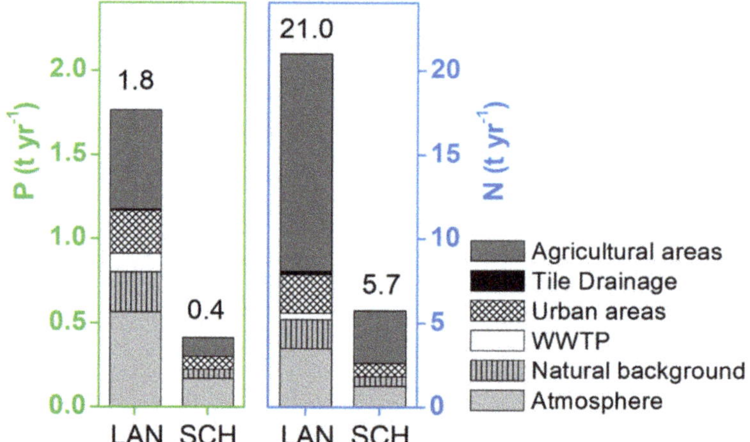

Figure 6. Phosphorus (**left**) and nitrogen (**right**) input to Lake Scharmützelsee (SCH) and Lake Langer See (LAN) from different sources, calculated according to the method of [41]. Nutrient retention in the soil and in water bodies is considered. WWTP—wastewater treatment plant.

3.3. Long-Term Development of Lake Water Quality after Reduction of Nutrient Input from the Catchment—Case Studies Lake Scharmützelsee and Lake Langer See

The analysis of long-term data revealed decreasing TP and TN concentrations in both lakes (Figure 7). This corresponds to the declining nutrient loads derived from the empirical models (Table 3). Two periods of lake development could be clearly distinguished: 1994–2003 with higher in-lake nutrient concentrations and 2004–2018 with lower concentrations. Nutrient input from the catchment must have changed because the hydrologic conditions were similar during these periods. Grüneberg et al. [28] described these phenomena as the "transient phase" and the "recovery phase" for Lake Scharmützelsee, but they also apply to Lake Langer See.

However, the response to nutrient load reduction from the catchment was different for each lake. Whereas nutrient concentrations in Lake Scharmützelsee decreased continuously, phytoplankton biomass remained relatively high until 2002 (Figure 7). A drastic shift in phytoplankton biomass was observed in 2003, and the trophic status changed from highly eutrophic to mesotrophic. Since then, in-lake nutrient concentrations have remained at an almost constant level of TP: 24 µg L^{-1} and TN: 577 µg L^{-1} (mean 2004–2018, Table 4). That is close to the lake type-specific target values for the "good ecological status" (PSI) [52] of TP: 21 (18–25) µg L^{-1} and TN: 480 (350–620) µg L^{-1}, respectively. The ecological status based on PSI improved from "poor" to "moderate". The lake even reached "good" ecological quality in 2005 and 2006, and touched it in 2014 and 2016 but remains more or less in the "moderate" status. During this short period of two years, the portion of cyanobacteria of the total phytoplankton biomass decreased to 17% or 24%, respectively, compared to 35% to 70% in the highly eutrophic phase before 2003. The shift in phytoplankton composition during the summer was the most noticeable feature of the response to nutrient reduction in Lake Scharmützelsee. Mean seasonal courses of main phytoplankton groups illustrate the change in seasonal patterns for both trophic periods (Figure 8). The dominance of cyanobacteria over the whole summer in eutrophic years was replaced by a clear water phase usually lasting from May to July. The seasonal pattern and biomass of spring phytoplankton bloom remained almost unchanged.

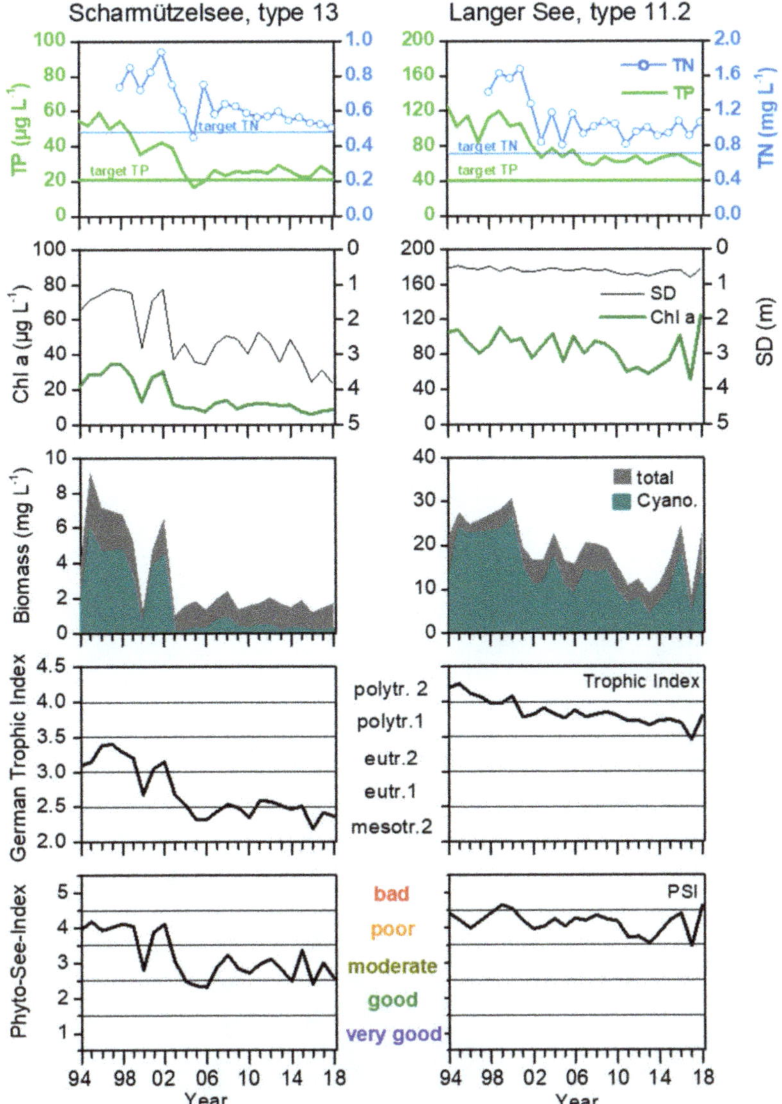

Figure 7. Long-term trophic development of Lake Scharmützelsee (**left**) and Lake Langer See (**right**) from 1994 to 2018. From the top to bottom: Vegetation mean (April–October) of concentration of TP and TN. Vertical lines indicate lake type-specific target values to reach good ecological quality, according to Dolman et al. (2016) [52], for both nutrients, chlorophyll a (Chl a) and Secchi depth (SD), biomass of total phytoplankton and cyanobacteria, German Trophic Index [45] and ecological quality as Phyto-See-Index [9].

Table 3. Phosphorus (P) and nitrogen (N) load for Lake Scharmützelsee and Lake Langer See for the periods 1994–2003 and 2004–2018 calculated by empirical models from mean annual discharge and mean annual concentrations of P and N in the upper mixed part of the water column as input data. Mean and standard deviation of two to four empirical models is given for the P load. The N load was derived from only one model [48]. The calculation of critical P and N loads is based on lake type-specific target values (vegetation mean) of Dolman et al. (2016) [52] (in italics). Due to the difference between annual mean and vegetation mean, the P-target value for Lake Scharmützelsee was multiplied by a factor of 1.5.

	Lake Scharmützelsee (Type 13)			Lake Langer See (Type 11.2)		
	1994–2003	2004–2018	*Target Value* Critical Load	1994–2003	2004–2018	*Target Value* Critical Load
	Input data, mean (minimum–maximum)					
Mean annual discharge (10^6 m^3 yr^{-1})	11.7 (8.6–14.7)	10.5 (6.9–13.9)		33.5 (23.5–45.3)	31.7 (22.2–43.9)	
Total P (µg L^{-1})	59.6 (48.6–78.6)	38.4 (33.9–44.4)	*21 (18–25)*	99.9 (71.4–124.6)	67.7 (55.3–76.4)	*41 (32–51)*
Total N (mg L^{-1})	0.87 (0.82–0.94)	0.65 (0.55–0.74)	*0.48 (0.35–0.62)*	1.27 (0.81–1.49)	1.06 (0.84–1.17)	*0.71 (0.52–0.88)*
	Nutrient load, mean (±standard deviation)					
P load (t yr^{-1})	3.3 (±1.0)	1.9 (±0.4)	1.6 (±0.3)	6.5 (±2.8)	3.9 (±1.3)	2.5 (±0.7)
Area-specific P load (g m^{-2} yr^{-1})	0.27 (±0.08)	0.16 (±0.03)	0.14 (±0.02)	4.6 (±2.0)	2.8 (±0.9)	1.7 (±0.5)
N load (t yr^{-1})	32.8 (±5.3)	20.8 (±3.9)	14.8 (±1.8)	50.6 (±14.1)	36.2 (±8.9)	24.1 (±4.8)
Area-specific N load (g m^{-2} yr^{-1})	2.7 (±0.4)	1.7 (±0.3)	1.2 (±0.2)	35.8 (±10.0)	25.7 (±6.3)	17.0 (±3.4)

Table 4. Limnological characteristics of the lakes studied and results of assessment of biological components according to the WFD. The range of vegetation means (April–October) of total phosphorus, total nitrogen, chlorophyll a concentrations, Secchi depth, and phytoplankton biovolume, and assessment values are given for the years 2011–2018, assessment of submerged macrophytes for 2011.

	Scharmützelsee	Langer See
Lake type acc. to WFD [8,9]	13 (stratified hard water lake with small catchment)	11.2 (very shallow hard water lake with large catchment)
Total phosphorus (μg L^{-1})	17–29	58–77
Total nitrogen (mg L^{-1})	0.45–0.75	0.81–1.18
Secchi depth (m)	2.4–3.8	0.5–0.8
Chlorophyll a (μg L^{-1})	6–14	51–125
Phytoplankton biovolume (mm^3 L^{-1})	1.2–2.4	9.1–24.2
German Trophic Index and trophic status acc. to LAWA [45]	2.2–2.6 (mesotrophic – eutrophic)	3.5–3.9 (polytrophic 1)
Assessment of phytoplankton (PSI) [9]	2.3–3.4 (moderate)	3.5–4.6 (poor)
Assessment of submerged macrophytes (PHYLIB) [12]	3.0 (moderate)	4.9 (bad)

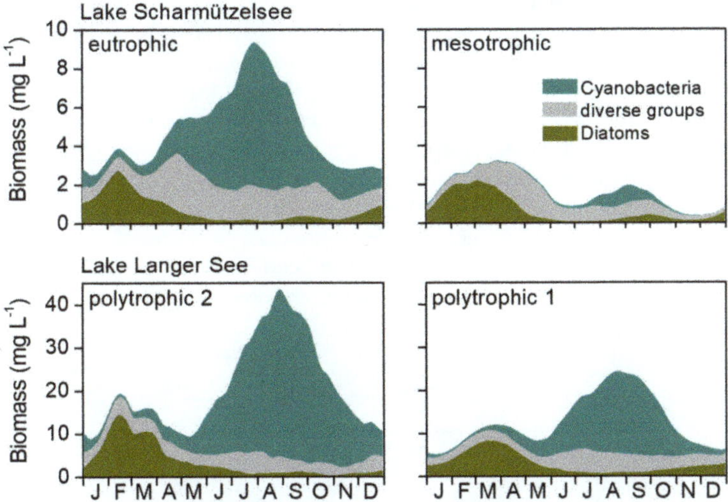

Figure 8. Mean seasonal courses of biomass of different taxonomic groups of phytoplankton. Data are averaged for eutrophic years (1993–2002, except 2000) and mesotrophic years (2003–2018) for Lake Scharmützelsee (**top**) and for years when Lake Langer See (**bottom**) was either in the polytrophic 2 status (1993–2000) or polytrophic 1 status (2001–2018), according to the German Trophic-Index [45].

Whereas summer phytoplankton before 2003 was characterized by cyanobacteria with fine filaments (Oscillatoriales: *Pseudanabana* spp., *Planktolyngbya* spp., *Limnothrix* spp., Nostocales: *Aphanizomenon gracile, Cylindrospermopsis raciborskii*) [33], a shift to a more diverse phytoplankton consisting of dinoflagellates, green algae, diatoms and cyanobacteria (now essentially Chroococcales and Nostocales) was observed in the mesotrophic years. However, cyanobacteria still occurred in a substantial portion of total phytoplankton biomass in the late summers of recent years, and local people claimed to observe a temporal and local occurrence of surface scums formed by N-fixing species of the order Nostocales (mainly *Dolichospermum flos-aquae* (formerly *Anabaena flos-aquae*) and *Aphanizomenon flos-aquae*). These species are responsible for the lake not achieving the "good ecological status" based on PSI despite the obvious improvements (Table 4).

Another consequence of lower phytoplankton biomass in Lake Scharmützelsee is the improvement of underwater light conditions for submerged macrophyte vegetation. Hilt et al. (2010) [34] described the recolonization of the lake by submerged macrophytes after eutrophic years with very low submerged vegetation. The charophyte *Nitellopsis obtusa* reappeared in 2004 and presently forms wide submerged meadows mainly at a depth of 2–4 m [46]. The number of submerged plant species increased from 6 in 1994 to 26 in 2011. In the same time, the lake bottom area covered by submerged macrophytes increased from 7% to 24%. However, the assessment of submerged macrophytes by the PHYLIB tool [12] also revealed only a "moderate ecological status" (Table 4). The failure here is caused by the absence of indicator species for the "good ecological status". Moreover, species such as *Nitellopsis obtusa* give only a "good" assessment when they settle in depths below 4 m.

In the shallow Lake Langer See, nutrient concentrations also decreased in a similar way to Lake Scharmützelsee from 1994 onwards and remain on an almost constant level at TP: 65 µg L^{-1} and TN: 1 mg L^{-1} since 2004. In contrast to Lake Scharmützelsee, these concentrations are well above the lake type-specific target values [52] of TP: 41 (32–51) µg L^{-1} and TN: 710 (520–880) µg L^{-1}, respectively (Figure 7). Trophic status improved by one class from polytrophic 2 to polytrophic 1, according to the German Trophic index classification system [45]. This refers to the hypertrophic status according to the OECD (1982) [48]. The biomass of phytoplankton and cyanobacteria showed a decreasing trend until 2013 (Figure 7). This change is also illustrated by a comparison of mean seasonal courses of the biomass of main phytoplankton groups in Figure 8. Here, the phytoplankton biomass was averaged for the years 1993–2000 (polytrophic 2) and 2001–2018, when the lake reached the lower trophic status (polytrophic 1). Both the biomass of summer and spring phytoplankton bloom decreased in Lake Langer See. Cyanobacteria growth still starts in May, but the biomass maximum in late summer is lower. In the years 2014–2018, phytoplankton biomass and chlorophyll concentrations reached again the level of the 1990s (except 2017, Figure 7). This is presumably due to the sharp reduction in discharge resulting from unusually long periods of drought. The hydraulic residence time in the vegetation time increased to more than 90 days in the vegetation period of the years 2015, 2016, and 2018. The correlation of the chlorophyll concentrations (vegetation mean) with the hydraulic residence time (May–October) showed a clearly positive trend for the years 2004–2018 in Lake Langer See (r^2 = 0.6, Figure 9). Contrastingly, Lake Scharmützelsee showed a decreasing trend.

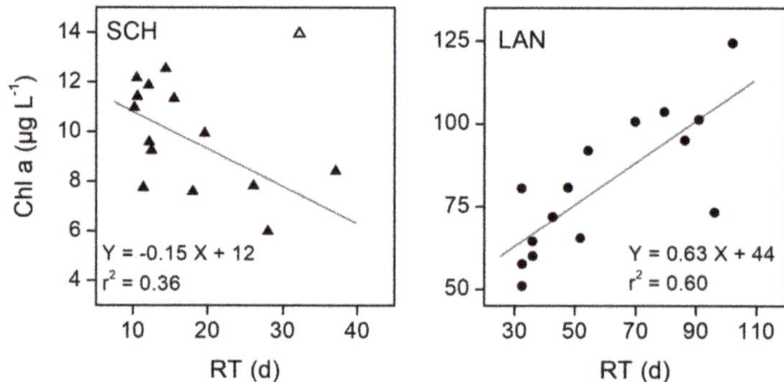

Figure 9. Vegetation mean of chlorophyll a concentration in dependence of hydraulic residence time (May–October) for Lake Scharmützelsee (**left**) and Lake Langer See (**right**) for the years 2004–2018. One value was excluded from the correlation analysis (open triangle).

Among cyanobacteria, a drastic decline in *Planktothrix agardhii* after 2001 was striking. Since this is a microcystin-producing species, this shift was accompanied by a decrease in the concentrations of that cyanotoxin. Lake Langer See contributed a large portion of toxin data to several studies [52–54], although not explicitly mentioned in these papers. The biovolume of other filamentous cyanobacteria, such as *Pseudanabana* spp., *Planktolyngbya* spp., *Limnothrix* spp. and *Aphanizomenon gracile*, remained almost unchanged after 2001 [55]. Phytoplankton assessment based on PSI [9] improved slightly and is now between the "poor" and "moderate" ecological status. Submerged macrophytes seem to have expanded in the last few years, however, their biomass is still too low and species indicating eutrophic ("poor") conditions dominate. That is why the assessment according to PHYLIB [12] yields a "bad" ecological quality (Table 4).

4. Discussion

4.1. Lakes Miss Water Quality Goals due to High Nutrient Loads—Ecological Status of Lakes in Brandenburg in 2014

Despite of great efforts in catchment restoration and thus a general decrease in nutrient emissions into lakes since 1990, the ecological quality of the majority of lakes is still inappropriate. Only 12% of the 183 natural lakes studied in Brandenburg reached the "good ecological status" (EQC), according to the WFD in 2014, which is below the average of 26% for all natural German lakes ≥50 ha [15]. This difference is due to a high number of shallow lakes in the North German lowlands, which comprises the lake types with the largest water quality deficits (Figures 3 and 4) [56].

The main reason for the failure to meet the WFD water quality goals is still too high nutrient loading. Especially the N inputs due to intensive agriculture especially in lowland (peaty) areas with artificially intensified drainage become more evident. The extent of excess in-lake TN concentrations is obvious (Figure 4). This seems to be contradictory to the results of a statistical analysis of nutrient concentrations in lakes of the ecoregion German lowlands: Dolman and co-authors [52] found that the phytoplankton of about half of the lakes is limited by N during the vegetation period. These situations are characteristic for shallow lakes during summer when denitrification reduces nitrate concentration while P release from the sediment increases P availability [57]. The major portion of TN is bound in phytoplankton biomass or in dissolved organic substances. Furthermore, the N demand of primary producers will be covered by internal nutrient input due to intensive recycling of freshly settled organic material in the water and especially in the upper sediment layer, detectable as a high release of ammonia from the sediment [58]. Therefore, phytoplankton growth can be apparently N-limited even in lakes with high TN concentrations if P supply is sufficiently high. Consequently, management measures to reduce nutrient inputs (P and N) are recommended [59]. Considering the fast N recycling and the high N emissions from catchments resulting in a high deviation between target values and concentrations (Figure 4), phytoplankton control by P seems a more viable management option [57]. Management is successful if positive feedback is induced: Lower trophic status with less phytoplankton biomass results in a lower intensity of lake internal remineralization of organic matter and an increasing P sorption in sediments [60]. However, Germany belongs to the minority of countries (38%) that have not yet set target values for nitrogen, although the use of only one nutrient for lake assessment is increasingly questioned [61].

If other parameters (e.g., silica concentration or underwater light) are limiting, lakes can reach a good assessment for phytoplankton despite their P or N concentrations exceeding the target values (see also [61]). Some shallow lakes and especially riverine lakes in our study have been assessed as "good" or better, albeit TN and TP concentrations that were above the lake type-specific target values (Figure 4). Phytoplankton growth in these lake types is known to be controlled by light or dilution due to high discharge rather than by nutrient concentrations. This was also reported by Dolman and co-authors [52] analyzing a dataset of 369 German lowland lakes. Low concentrations of silica can control the biomass of diatoms and chrysophytes in spring and can, therefore, favor the development of cyanobacteria [62]. In addition to these "bottom up" control mechanisms, phytoplankton biomass and composition will also be influenced by "top down" control due to zooplankton grazing [63] or the filtration activity of benthic macroinvertebrates such as mussels. The interaction between submerged macrophytes and phytoplankton by different mechanisms (increased sedimentation, reduced resuspension, nutrient uptake, excretion of allelochemicals, providing shelter for phytoplankton grazers) is an additional factor regulating phytoplankton dynamics, especially in shallow lakes [64], but was also shown for deep Lake Scharmützelsee by Hilt et al. (2010) [34]. However, the recovery of submerged macrophytes after their eutrophication-related loss and subsequent restoration measures in shallow lakes and their catchments is a long-lasting process whose mechanisms are not yet fully understood [65]. In general, the way in which food web interactions, including macrozoobenthos, fish, and water birds, and interspecies competition between plants influence phytoplankton dynamics is not quantified for most lakes.

However, it is even more difficult to quantify the impact of climate change on water quality. Climate change is known to exacerbate the eutrophication process in lakes through several direct and indirect processes [66,67]. This is especially appropriate for lakes in the study region, regarding the climatic conditions with low precipitation and negative net hydrological budget (chapter 2.1.1.). Increasing water temperatures, decreasing water levels and summer discharges, cyanobacteria, and macrophyte mass developments are expected [68]. Consequently, with global warming, lakes become more sensitive to nutrient loading which underlines the requirement for strict catchment management.

4.2. Effects of Lake Morphometry and Catchment Size—Case Study Scharmützelsee Region

The Scharmützelsee region is an example of the successful restoration of a catchment in Brandenburg and the entire eastern part of Germany after political changes in 1990. The reduction of nutrient emissions, primarily by eliminating point sources, caused a significant improvement of water quality. Nutrient input from point sources is no longer relevant for Lake Scharmützelsee, although Grüneberg and co-authors [28] assumed a delayed P input via groundwater contaminated by septic systems. However, point sources still contribute 6% of TP input and 3% of TN input for Lake Langer See, caused mainly by the effluent of the main regional wastewater treatment plant.

The relative importance of nutrient transport pathways and the absolute quantity of nutrient loads differ for the two lakes due to their different geographical locations and different land use of the upstream catchment. Baseline nutrient imports gain in relevance with decreasing size of catchment, flushing rate, and decreasing trophic level. This was shown for Lake Scharmützelsee receiving 54% of P input via groundwater and atmospheric deposition (Figure 6). This was only 45% for Lake Langer See. Nutrient input from agriculture is more important for Lake Langer See, especially for N (63% for Lake Langer See, 55% for Lake Scharmützelsee). Lake Langer See, the last lake of the catchment, receives about twice as much TP and TN compared to Lake Scharmützelsee in the central part of the region (Table 4). However, the nutrient loads estimated from land use (Figure 6) were only half of the estimates from empirical models (Table 4) for Lake Langer See. The deviation between the different approaches was even a factor 4 (TN) to 5 (TP) for Lake Scharmützelsee. Based on the TP load estimates from land use, the empirical models predict in-lake TP concentrations that would refer to the oligotrophic status for Lake Scharmützelsee (9 μg L^{-1}) or to the mesotrophic status for Lake Langer See (35 μg L^{-1}) according to OECD (1982) [48]. This corresponds to neither the current nor the historical status of Lake Scharmützelsee nor the potential natural status in dependence of lake morphometry which is the mesotrophic status [69]. However, the method of LUGV (2015) [41] was developed for the whole of Brandenburg to identify hotspots of nutrient input into surface waters. It reflects the proportions of different nutrient entry pathways quite well, depending on the present land use. Differences for small sub-catchments are possible if assumptions in this country-wide method do not meet local conditions. Exemplarily, the mean groundwater concentration of 58 μg L^{-1} TP used is much smaller compared to the 100 μg L^{-1} used by Grüneberg et al. (2011) [28]. Since Lake Scharmützelsee is mainly (88%) fed by groundwater, a difference in background concentrations could explain the huge difference between the applied methods and models. The results from the empirical models prove that the total nutrient loads from the LfU study are underestimated, while the relative contribution of the different nutrient emission pathways in the sub-catchments is plausible. This confirms the results of different studies which found that including local know-how, soft information, and more detailed field investigations improves modeling of diffuse nutrient pollution and water quality [70,71].

Whereas the elimination of point sources caused a rather abrupt nutrient load reduction for Lake Scharmützelsee [28], this was probably a more gradual process for Lake Langer See at the end of the catchment. Lake Scharmützelsee responded to the reduction of nutrient loading with a distinct change in trophic status and biota after about 13 years. This time delay is in the range of 10 to 15 years as reported by Jeppesen et al. (2005) [72] for a variety of lakes independent of hydraulic residence time. It was caused by internal load, by delayed seepage from septic systems [28] and due to the long hydraulic residence time. Contrastingly, Lake Langer See is immediately influenced by changes of inflowing water quality due to large catchment size relative to lake volume or area (Table 1) and short hydraulic residence time (Tables 1 and 3). This lake acts more like a plankton reactor. It could be shown that phytoplankton biomass is strongly influenced by the hydraulic residence time, the longer the residence time, the more phytoplankton can grow (Figure 9). Due to its shallowness, frequent mixing of the water column and, consequently, sufficient underwater light supply, nutrients are converted into phytoplankton biomass very efficiently, indicated by a TP to chlorophyll a ratio around 1 [17]. This lake depends on the water quality of all upstream lakes. In the course of lake chains, seasonal effects (summer nutrient release) can be amplified [73].

Despite the improvement of water quality classes by one step, both lakes are not in the "good ecological status" according to WFD. In-lake concentrations are still too high and biomass and species composition do not match the lake type-specific reference status. The latter is the mesotrophic status for deep and thermally stratified Lake Scharmützelsee with a relatively small catchment size and the eutrophic status for the very shallow, polymictic Lake Langer See with a relatively large catchment and low hydraulic residence time. According to the empirical models, a further reduction of TP load by 20% would be necessary to meet the target value to achieve the good ecological status [52] in Lake Scharmützelsee. However, the present TP load for Lake Langer See should be reduced by almost 45%. Since the contribution of point sources is only small, nutrient inputs from diffusive sources, especially for N from agricultural areas has to be reduced. However, some studies confirm, that nutrient loads should be close to natural background concentrations to meet WFD water quality goals. This means that land use would have to be drastically changed, which would be unrealistic from a socio-economic point of view [2,71].

5. Conclusions

Following a period of intense anthropogenic eutrophication of surface waters in the last century, great efforts have been undertaken in catchment restoration over the last few decades, mainly by improving sewage collection and wastewater treatment. Despite significant improvements, only a minority of 12% of lakes in Brandenburg (North German lowlands) meet the ecological quality requirements of the European Water Framework Directive by the end of the first River Basin Management Plan of the EC in 2015.

In-lake nutrient concentrations are still too high. The TN concentration especially deviates significantly from lake type-specific target values, a clear indication of elevated diffusive imports from the catchments. Thus, catchment size in relation to lake area or volume becomes a major predictor for the probability of lakes reaching at least a "good" ecological quality. With the present level of diffuse nutrient emissions, many lakes fail the objectives of the WFD. This applies especially to shallow lakes with a largely agriculturally used catchment. The deficits in catchment management resulting in elevated nutrient emissions are amplified downstream lake chains, with downstream lakes depending on delayed recovery and amplified seasonal effects (summer nutrient release) of all upstream lakes. These lakes may fail to meet water quality goals in the long-term. Consequently, either quality goals or land use practices require revision, as it was recently discussed by Carvalho and co-authors [2]. This is of large practical relevance as deficits in nutrient import reduction may increasingly be compensated by expensive substitute lake internal (technical) restoration measures to meet the objectives of the WFD. Often, these measures erroneously combat an "internal nutrient load" which is normally not significant if coupled with a sufficiently low import from the catchment [74].

Strategies to reduce nutrient emissions are generally known and measures to reduce especially N inputs in hotspots and sensitive areas have been published recently by the German Advisory Council on the Environment (Sachverständigenrat für Umweltfragen, SRU) [75,76]: Existing legal instruments for local protected-area management should be used to reduce agricultural fertilizer use. Buffer zones should be established around nature conserves or sensitive areas. Intensive agriculture in lowland peaty areas is questionable. A reduction of N emissions from agriculture should be accompanied by a reformation of the common agricultural policy leading to an increase in agri-environmental payments and an intensification of environmental requirements for agricultural subsidies (maintaining permanent grassland, setting far-reaching requirements for ecological focus areas, and crop diversification). A further step is to reform and strictly enforce the fertilizer regulations, and the SRU recommends that the legal requirements related to good agricultural practice must be specified.

Despite the frequent occurrence of N limitation, especially in shallow lakes [52], there are several arguments for directing the management to reduce P availability: Many lakes seem to be closer to P targets than to N targets, and catchment-wide reduction of DIN emission is difficult owing to its chemical mobility. However, N-limited lakes should benefit from reduced N availability, and there are indications that high N availability has adverse effects on reed and submerged macrophytes [57]. These secondary effects are likely to contribute to the difference between phytoplankton only and EQC assessments found in our study. In practice, most catchment measures have multiple effects on both N and P.

Author Contributions: Conceptualization, J.R. and B.N.; methodology, J.R., K.Q., B.G.; investigation, J.R., K.Q., B.G.; data curation, J.R., K.Q.; writing—original draft preparation, J.R.; writing—review and editing, B.N., B.G., K.Q.; visualization, J.R.; project administration, B.N.; funding acquisition, B.N., J.R., B.G.

Funding: This work was supported by the German Ministry for Education and Research (BMBF) as part of the projects NITROLIMIT I and II (grant numbers 033L041A and 0033W015AN) and the preliminary phase of the project AgriTip (01LC1717A). The long-term study of the two lakes was realized within the BMBF projects NITROLIMIT I and II, NOSTOTOX (0330792 B) and the German-Israeli Water cooperation (02WT0985), CYLIN funded by KWB Berlin (with the financial support of Veolia Waters, BWB) and the DFG project AZ NI347/5-1, funded by the German Research Foundation.

Acknowledgments: We are grateful to the State Office of Environment (LfU) of the Ministry of Rural Development, Environment and Agriculture of the Federal State of Brandenburg, namely to Antje Barsch, Jens Päzolt and Jörg Schönfelder, for supplying data and comprehensive material on lake assessment and management, land use and nutrient emissions. The Water and Shipping Authority of Berlin and the LfU kindly provided discharge data. We thank the staff of the Research Station of the Department of Freshwater Conservation in Bad Saarow for their reliable work in the long-term study and Claudia Wiedner for the coordination of NITROLIMIT and earlier research projects. We gratefully acknowledge the work of Annette Tworeck (LBH Hoehn, Freiburg) who did most of the phytoplankton analyses for the two lakes, furthermore Ute Hahmann (LBH), Ute Mischke and Paul Zippel. We are grateful to Klaus van de Weyer (Lanaplan, Nettetal) and colleagues for the macrophyte mapping and Philip Saunders for proofreading.

Conflicts of Interest: The authors declare no conflict of interest. The funders had no role in the design of the study; in the collection, analyses, or interpretation of data; in the writing of the manuscript, or in the decision to publish the results.

References

1. European Union (EU). *Directive 2000/60/EC of the European Parliament and of the Council of 23 October 2000 Establishing a Framework for Community Action in the Field of Water Policy*; Official Journal L 327; EU, Publications Office of the European Union: Luxembourg, 2000; Available online: http://eur-lex.europa.eu/resource.html?uri=cellar:5c835afb-2ec6-4577-bdf8-756d3d694eeb.0004.02/DOC_1&format=PDF (accessed on 14 February 2018).

2. Carvalho, L.; Mackay, E.B.; Cardoso, A.C.; Baattrup-Pedersen, A.; Birk, S.; Blackstock, K.L.; Borics, G.; Borja, A.; Feld, C.K.; Ferreira, M.T.; et al. Protecting and restoring Europe's waters: An analysis of the future development needs of the Water Framework Directive. *Sci. Total Environ.* **2019**, *658*, 1228–1238. [CrossRef] [PubMed]

3. European Environment Agency (EEA). *Source Apportionment of Nitrogen and Phosphorus Inputs into the Aquatic Environment*; EEA Report No. 7/2005; EEA, Office for Official Publications of the European Communities: Copenhagen, Denmark, 2005; Available online: https://www.eea.europa.eu/publications/eea_report_2005_7 (accessed on 27 November 2019).

4. Kronvang, B.; Borgvang, S.A.; Barkved, L.J. Towards European harmonised procedures for quantification of nutrient losses from diffuse sources—The EUROHARP project. *J. Environ. Monit.* **2009**, *11*, 503–505. [CrossRef] [PubMed]

5. Birk, S.; Bonne, W.; Borja, A.; Brucet, S.; Courrat, A.; Poikane, S.; Solimini, A.; Van De Bund, W.; Zampoukas, N.; Hering, D. Three hundred ways to assess Europe's surface waters: An almost complete overview of biological methods to implement the Water Framework Directive. *Ecol. Indic.* **2012**, *18*, 31–41. [CrossRef]

6. Solheim, A.L.; Rekolainen, S.; Moe, S.J.; Carvalho, L.; Phillips, G.; Ptacnik, R.; Penning, W.E.; Toth, L.G.; O'Toole, C.; Schartau, A.-K.L.; et al. Ecological threshold responses in European lakes and their applicability for the Water Framework Directive (WFD) implementation: Synthesis of lakes results from the REBECCA project. *Aquat. Ecol.* **2008**, *42*, 317–334. [CrossRef]

7. Poikane, S.; van den Berg, M.; Hellsten, S.; de Hoyos, C.; Ortiz-Casas, J.; Pall, K.; Portielje, R.; Phillips, G.; Solheim, A.L.; Tierney, D.; et al. Lake ecological assessment systems and intercalibration for the European Water Framework Directive: Aims, achievements and further challenges. *Procedia Environ. Sci.* **2011**, *9*, 153–168. [CrossRef]

8. Länderarbeitsgemeinschaft Wasser (LAWA-AO). In *RaKon Monitoring Teil B, Arbeitspapier I—Gewässertypen und Referenzbedingungen*; Stand 02.02.2016; LAWA-AO, MELUND: Kiel, Germany, 2016; Available online: http://www.gewaesser-bewertung.de/files/rakon_b-arbeitspapier-i_stand_20160202.pdf (accessed on 14 February 2017).

9. Mischke, U.; Riedmüller, U.; Hoehn, E.; Nixdorf, B. Method Description of the Assessment of Lakes and Reservoirs with Phytoplankton and the Phyto-See-Index in Germany. In *User Handbook. Excerpt of Original Version December*; Electronic Publication: Berlin, Germany, 2016; Available online: http://www.gewaesser-bewertung.de/files/english_handbook_german_lake_assessment_method_description_psi_dec2016-1.pdf (accessed on 27 November 2019).

10. Mischke, U.; Böhmer, J.; Riedmüller, U.; Deneke, R.; Maier, G. Auswertungsprogramm PhytoSee Version 7.0 zur Berechnung des Phyto-See-Index (PSI) für die Ökologische Bewertung von Natürlichen, Künstlichen und Erheblich Veränderten Seen in Deutschland Gemäß EG-Wasserrahmenrichtlinie; Erweitert um Das PhytoLoss Modul 1.2 zur Einbindung von Zooplanktonbefunden. Stand 15.12.2017. 2017. Available online: http://www.gewaesser-bewertung.de/files/phytosee-tool_7.0_dez2017.zip (accessed on 14 February 2018).

11. Nixdorf, B.; Hoehn, E.; Riedmüller, U.; Mischke, U.; Schönfelder, I. Probenahme und Analyse des Phytoplanktons in Seen und Flüssen zur ökologischen Bewertung gemäß der EU-WRRL. In *Handbuch Angewandte Limnologie—Methodische Grundlagen*; III-4.3.1; Ergänzungslieferung 4/10, Wiley-VCH Verlag GmbH & Co. KgaA: Weinheim, Germany, 2010; pp. 1–24.

12. Schaumburg, J.; Schranz, C.; Stelzer, D.; Vogel, A. *Verfahrensanleitung für die ökologische Bewertung von Seen zur Umsetzung der EG-Wasserrahmenrichtlinie: Makrophyten und Phytobenthos. PHYLIB. Stand Februar 2014*; Bayerisches Landesamt für Umwelt. Im Auftrag der LAWA Projekt-Nr. O 10.10: Augsburg/Wielenbach, Germany, 2015; Available online: http://www.gewaesser-bewertung.de/files/verfahrensanleitung_seen2015.pdf (accessed on 14 February 2018).

13. Böhmer, J. *Methodisches Handbuch zur WRRL-Bewertung von Seen Mittels Makrozoobethos Gemäß AESHNA—Handbuch zur Untersuchung und Bewertung von Stehgewässern auf der Basis des Makrozoobenthos vor dem Hintergrund der EG-Wasserrahmenrichtlinie*; Bioforum GmbH: Kirchheim/Teck, Germany, 2017; Available online: http://www.gewaesser-bewertung.de/files/AESHNA_Anleitung_Endfassung_5-5-17.pdf (accessed on 27 November 2019).

14. Ritterbusch, D.; Brämick, U. Verfahrensvorschlag zur Bewertung des ökologischen Zustandes von Seen anhand der Fische (1). *Schriften Instituts Binnenfischerei eV* **2015**, *41*, 69.

15. Bundesministerium für Umwelt und Bau/Umweltbundesamt (BMUB/UBA). *Water Framework Directive—The Status of German Waters 2015*; BMU/UBA: Bonn/Dessau, Germany, 2016; Available online: http://www.gewaesser-bewertung.de/files/wrrl_englische_version_dez_2016.pdf (accessed on 14 February 2017).

16. European Environment Agency (EEA). *European Waters—Assessment of Status and Pressures 2018*; EEA Report No. 7/2018; European Environment Agency, Publications Office of the European Union: Luxembourg, 2018; Available online: https://www.eea.europa.eu/publications/state-ofwater (accessed on 27 November 2019). [CrossRef]

17. Nixdorf, B.; Deneke, R. Why 'very shallow' lakes are more successful opposing reduced nutrient loads. *Hydrobiologia* **1997**, *342*, 269–284. [CrossRef]

18. Ministry of Rural Development, Environment and Agriculture of the Federal State of Brandenburg (MLUL). Landschaftswasserhaushalt. MLUL, Potsdam, Germany. Available online: https://lfu.brandenburg.de/cms/detail.php/bb1.c.327798.de (accessed on 28 November 2019).

19. Deutscher Wetterdienst (DWD). Niederschlag: Vieljährige Mittelwerte 1981–2010. Available online: https://www.dwd.de/DE/leistungen/klimadatendeutschland/mittelwerte/nieder_8110_akt_html.html?view=nasPublication&nn=16102 (assessed on 30 September 2019).

20. Umweltbundesamt (UBA, DLR-DFD). *CORINE Land Cover (CLC2006)*; Umweltbundesamt, DLR-DFD: Weßling, Germany, 2009.

21. Ministry of Rural Development, Environment and Agriculture of the Federal State of Brandenburg (MLUL). *Lagebericht Abwasser 2015*; MLUL: Potsdam, Germany, 2015; Available online: http://www.mlul.brandenburg.de/cms/media.php/lbm1.a.3310.de/kawb2015_teil3.pdf (accessed on 2 July 2016).

22. Ministry of Rural Development, Environment and Agriculture of the Federal State of Brandenburg (MLUL). *Kommunale Kläranlagen und Einleitstellen im Land Brandenburg [kommka.shp; kommeinleit.shp]*; Stand 31.12.2011; MLUL: Potsdam, Germany, 2017; Available online: http://www.metaver.de/search/dls/dataset/365B64CD-55CA-4C65-8F48-8B93B9C06E40/4266D9C1-D20B-4922-B8B2-6BA116243CCE (accessed on 14 February 2018).

23. Bundesamt für Kartographie und Geodäsie (BKG); BKG, Frankfurt/Main, Germany. Available online: http://sg.geodatenzentrum.de/web_download/vg/vg1000-ew_3112/utm32s/shape/vg1000-ew_3112.utm32s.shape.kompakt.zip (accessed on 23 November 2015).

24. Ministry of Rural Development, Environment and Agriculture of the Federal State of Brandenburg (MLUL). Surface catchment areas. MLUL: Potsdam, Germany. Available online: http://www.mugv.brandenburg.de/lua/gis/ezg25.zip (accessed on 9 September 2014).

25. Ministry of Rural Development, Environment and Agriculture of the Federal State of Brandenburg (MLUL). Water network. MLUL: Potsdam, Germany. Available online: http://www.mugv.brandenburg.de/lua/gis/gewnet25.zip (accessed on 9 September 2014).

26. Ministry of Rural Development, Environment and Agriculture of the Federal State of Brandenburg (MLUL). Surface water bodies. MLUL: Potsdam, Germany. Available online: http://www.mugv.brandenburg.de/lua/gis/seen25.zip (accessed on 9 September 2014).

27. Ministry of Rural Development, Environment and Agriculture of the Federal State of Brandenburg (MLUL). Municipal wastewater treatment plants: Potsdam, Germany. Available online: http://www.mlul.brandenburg.de/lua/gis/kommka.zip (accessed on 9 September 2014).

28. Grüneberg, B.; Rücker, J.; Nixdorf, B.; Behrendt, H. Dilemma of non-steady state in lakes—Development and predictability of in-lake P concentration in dimictic Lake Scharmützelsee (Germany) after abrupt load reduction. *Int. Rev. Hydrobiol.* **2011**, *96*, 599–621. [CrossRef]

29. Kleeberg, A.; Freidank, A.; Jöhnk, K. Effects of ice cover on sediment resuspension and phosphorus entrainment in shallow lakes: Combining in situ experiments and wind-wave modeling. *Limnol. Oceanogr.* **2013**, *58*, 1819–1833. [CrossRef]

30. Wundsch, H.H. Beiträge zur Fischereibiologie märkischer Seen. VI. Die Entwicklung eines besonderen Seentypus (H_2S-Oscillatorien-Seen) im Fluß-Seen-Gebiet der Spree und Havel und seine Bedeutung für die fischereibiologischen Bedingungen in dieser Region. *Zeitschrift Fischerei* **1940**, *38*, 443–658.

31. Kleeberg, A. Re-assessment of Wundsch's (1940) 'H_2S-Oscillatoria-Lake' type using the eutrophic Lake Scharmützel (Brandenburg, NE Germany) as an example. *Hydrobiologia* **2003**, *501*, 1–5. [CrossRef]

32. Müller, H. Die produktionsbiologischen Verhältnisse märkischer Seen in der Umgebung Storkows. *Zeitschrift Fischerei Hilfswissenschaften* **1952**, *1*, 95–160.

33. Nixdorf, B.; Mischke, U.; Rücker, J. Phytoplankton assemblages and steady state in deep and shallow eutrophic lakes—An approach to differentiate the habitat properties of Oscillatoriales. *Hydrobiologia* **2003**, *502*, 111–121. [CrossRef]

34. Hilt, S.; Henschke, I.; Rücker, J.; Nixdorf, B. Can submerged macrophytes influence turbidity and trophic state in deep lakes? Suggestions from a case study. *J. Environ. Qual.* **2010**, *39*, 725–733. [CrossRef] [PubMed]

35. Ministerium für Ländliche Entwicklung und Landwirtschaft des Landes Brandenburg (MLUL). *Umsetzung der Wasserrahmenrichtlinie: Beiträge des Landes Brandenburg zu den Bewirtschaftungsplänen und Maßnahmenprogrammen der Flussgebietseinheiten Elbe und Oder für den Zeitraum 2016 bis 2021*; MLUL: Potsdam, Germany, 2016; Available online: https://lfu.brandenburg.de/cms/media.php/lbm1.a.3310.de/wrrl_2016_gesamt.pdf (accessed on 27 November 2019).

36. Rücker, J.; Barsch, A.; Nixdorf, B. „Besser, aber noch nicht gut"—Ökologischer Zustand der Seen in Brandenburg 2014. *WasserWirtschaft* **2015**, *12*, 41–47. [CrossRef]

37. Päzolt, J. Der Makrophytenindex Brandenburg—Ein Index zur Bewertung von Seen mit Makrophyten. *Nat. Landsch. Brandenbg.* **2007**, *16*, 116–121.

38. European Commission, Joint Research Centre (EC, JRC). *Water Framework Directive Intercalibration Technical Report*; Part 2: Lakes; EC, JRC, Office for Official Publications of the European Communities: Luxembourg, 2009; Available online: https://publications.jrc.ec.europa.eu/repository/bitstream/JRC51340/3009_08-vol-lakes-cover.pdf (accessed on 27 November 2019).

39. Johnes, P.J. Evaluation and management of the impact of land use change on the nitrogen and phosphorus load delivered to surface waters: The export coefficient modelling approach. *J. Hydrol.* **1996**, *183*, 323–349. [CrossRef]

40. Behrendt, H.; Huber, P.; Kornmilch, M.; Opitz, D.; Schmoll, O.; Scholz, G.; Uebe, R. Nutrient Emissions into River Basins of Germany. UBA-Texte, 23/00, 2000. Available online: https://www.umweltbundesamt.de/sites/default/files/medien/publikation/long/1837.pdf (accessed on 14 February 2018).

41. Landesamt für Umwelt, Gesundheit und Verbraucherschutz des Landes Brandenburg (LUGV). *Methodik der Nährstoffbilanzierung in Brandenburg als Grundlage für die Ausweisung von Maßnahmen zur Nährstoffreduzierung für den BWPL 2014*; Heft Nr. 144; Fachbeiträge des LUGV: Potsdam OT Groß-Glienicke, Germany, 2015; Available online: http://www.lfu.brandenburg.de/media_fast/4055/fb_144.pdf (accessed on 14 February 2018).

42. DEV (Deutsche Einheitsverfahren zur Wasser-, Abwasser- und Schlammuntersuchung). *D11, E5, D9, C9, E1, H7*; WILEY-VCH Verlag GmbH & Co. KGaA, Beuth Verlag GmbH: Berlin, Germany, 1976–1998.

43. Utermöhl, H. Zur Vervollkommnung der quantitativen Phytoplankton-Methodik. *Mitt. Int. Ver. Theor. Angew. Limnol.* **1958**, *9*, 1–38. [CrossRef]

44. DEV (Deutsche Einheitsverfahren zur Wasser-, Abwasser- und Schlammuntersuchung). *DIN EN 16695:2015-12: Water Quality—Guidance on the Estimation of Phytoplankton Biovolume*, German version EN 16695:2015; WILEY-VCH Verlag GmbH & Co. KGaA, Beuth Verlag GmbH: Berlin, Germany, 2015. [CrossRef]

45. Länderarbeitsgemeinschaft Wasser (LAWA). *Trophieklassifikation von Seen. Richtlinie zur Ermittlung des Trophie-Index nach LAWA für natürliche Seen, Baggerseen, Talsperren und Speicherseen. Empfehlungen Oberirdische Gewässer*; Access-Tool Version, 1.1; LAWA—Bund/Länder Arbeitsgemeinschaft Wasser, Ed.; Kulturbuchverlag: Berlin, Germany, 2014; ISBN 978-3-88961-345-52014. Available online: http://www.gewaesserfragen.de/pdfs/Auswerte-Tool_Trophie-Index_Seen_nach_LAWA_Version_1.1_Apr2015.zip (accessed on 14 February 2017).

46. NITROLIMIT. *NITROLIMIT—Stickstofflimitation in Binnengewässern: Ist Stickstoffreduktion ökologisch sinnvoll und wirtschaftlich vertretbar? Abschlussbericht des BMBF-Verbundprojekts NITROLIMIT I, Mai 2014*; BTU Cottbus–Senftenberg, Lehrstuhl Gewässerschutz: Bad Saarow, Germany, 2014; p. 208. Available online: http://www-docs.tu-cottbus.de/gewaesserschutz/public/downloads/NITROLIMIT_Endbericht_Mai2014.pdf (accessed on 5 July 2016).

47. Vollenweider, R.A. Advances in defining critical loading levels for phosphorus in lake eutrophication. *Memorie Istituto Italiano Idrobiologia* **1976**, *33*, 53–83.

48. Organization for Economic Cooperation and Development (OECD). *Eutrophication of Waters—Monitoring, Assessment and Control*; OECD: Paris, France, 1982.

49. Nürnberg, G.K. The prediction of internal phosphorus load in lakes with anoxic hypolimnia. *Limnol. Oceanogr.* **1984**, *29*, 111–124. [CrossRef]

50. Sas, H. *Lake Restoration by Reduction of Nutrient Loading. Expectations, Experiences, Extrapolations*; Academia Richarz: Sankt Augustin, Germany, 1989; p. 497.
51. Brett, M.T.; Benjamin, M.M. A review and reassessment of lake phosphorus retention and the nutrient loading concept. *Freshw. Biol.* **2008**, *53*, 194–211. [CrossRef]
52. Dolman, A.M.; Mischke, U.; Wiedner, C. Lake-type-specific seasonal patterns of nutrient limitation in German lakes, with target nitrogen and phosphorus concentrations for good ecological status. *Freshw. Biol.* **2016**, *61*, 444–456. [CrossRef]
53. Fastner, J.; Rücker, J.; Stüken, A.; Preußel, K.; Nixdorf, B.; Chorus, I.; Köhler, A.; Wiedner, C. Occurrence of the cyanobacterial toxin cylindrospermopsin in Northeast Germany. *Environ. Toxicol.* **2007**, *22*, 26–32. [CrossRef] [PubMed]
54. Dolman, A.M.; Rücker, J.; Pick, F.R.; Fastner, J.; Rohrlack, T.; Mischke, U.; Wiedner, C. Cyanobacteria and cyanotoxins: The influence of nitrogen versus phosphorus. *PLoS ONE* **2012**, *7*, e38757. [CrossRef] [PubMed]
55. NOSTOTOX. Development of Toxic Nostocales (Cyanobacteria) in the Course of Declining Trophic State and Global Warming. In *Final Report of the Joint Research Project NOSTOTOX*; IGB: Berlin, Germany, 2011; Available online: https://www-docs.b-tu.de/fg-gewaesserschutz/public/projekte/Nostotox_final_2011_web.pdf (accessed on 14 July 2016).
56. German Environment Agency. *Waters in Germany: Status and Assessment*; German Environment Agency: Dessau-Roßlau, Germany, 2017; ISSN 2363-832X. Available online: http://www.gewaesser-bewertung.de/files/171018_uba_gewasserdtl_engl_bf.pdf (accessed on 14 February 2018).
57. Moss, B.; Jeppesen, E.; Søndergaard, M.; Lauridsen, T.L.; Liu, Z. Nitrogen, macrophytes, shallow lakes and nutrient limitation: Resolution of a current controversy? *Hydrobiologia* **2013**, *710*, 3–21. [CrossRef]
58. Nixdorf, B.; Rücker, J.; Grüneberg, B. Importance of internal nitrogen recycling on water quality and cyanobacterial biomass in a shallow polymictic lake (Lake Langer See, Germany). In Proceedings of the 3rd International Conference Water Resources and Wetlands, Tulcea, Romania, 8–10 September 2014; Gâştescu, P., Bretcan, P., Eds.; Asociatia Romana de Limnogeografie: Bukarest, Romania, 2016; pp. 38–46, ISSN 2285-7923. Available online: https://www.limnology.ro/wrw2016/proceedings/4_Brigitte_Nixdorf.pdf (accessed on 14 February 2018).
59. Positionspapier des Projekts NITROLIMIT—Stickstofflimitation in Binnengewässern—Ist Stickstoffreduktion Ökologisch Sinnvoll und Wirtschaftlich Vertretbar? Wiedner, C., Schlief, J., Eds.; NITROLIMIT: Bad Saarow, Germany, 2016; Available online: https://opus4.kobv.de/opus4-btu/frontdoor/index/index/docId/4019 (accessed on 27 November 2019).
60. Rothe, M.; Kleeberg, A.; Grüneberg, B.; Friese, K.; Pérez-Mayo, M.; Hupfer, M. Sedimentary Sulphur: Iron Ratio Indicates Vivianite Occurrence: A Study from Two Contrasting Freshwater Systems. *PLoS ONE* **2015**, *10*, e0143737. [CrossRef]
61. Poikane, S.; Kelly, M.G.; Herrero, F.S.; Pitt, J.-A.; Jarvie, H.P.; Claussen, U.; Leujak, W.; Lyche Solheim, A.; Teixeira, H.; Phillips, G. Nutrient criteria for surface waters under the European Water Framework Directive: Current state-of-the-art, challenges and future outlook. *Sci. Total Environ.* **2019**, *695*, 133888. [CrossRef]
62. Launhardt, A.; Rücker, J.; Nixdorf, B. Control of seasonal phytoplankton dynamics in Lake Scharmützelsee (northeast Germany) by nutrient competition during winter. *Verh. Internat. Verein. Limnol.* **2006**, *29*, 1675–1678. [CrossRef]
63. Jeppesen, E.; Søndergaard, M.; Meerhoff, M.; Lauridsen, T.L.; Jensen, J.P. Shallow lake restoration by nutrient loading reduction-some recent findings and challenges ahead. *Hydrobiologia* **2007**, *584*, 239–252. [CrossRef]
64. Scheffer, M.; Hosper, S.H.; Meijer, M.-L.; Moss, B.; Jeppesen, E. Alternative equilibria in shallow lakes. *Trends Ecol. Evol.* **1993**, *8*, 275–279. [CrossRef]
65. Hilt, S.; Nuñez, A.; Marta, M.; Bakker, E.S.; Blindow, I.; Davidson, T.A.; Gillefalk, M.; Hansson, L.-A.; Janse, J.H.; Janssen, A.B.G.; et al. Response of Submerged Macrophyte Communities to External and Internal Restoration Measures in North Temperate Shallow Lakes. *Front. Plant Sci.* **2018**, *9*, 194. [CrossRef] [PubMed]
66. Dokulil, M.T.; Teubner, K. Eutrophication and Climate Change: Present Situation and Future Scenarios. In *Eutrophication: Causes, Consequences and Control*; Ansari, A., Singh Gill, S., Lanza, G., Rast, W., Eds.; Springer: Dordrecht, The Netherlands, 2010. [CrossRef]
67. Jeppesen, E.; Søndergaard, M.; Liu, Z. Lake Restoration and Management in a Climate Change Perspective: An Introduction. *Water* **2017**, *9*, 122. [CrossRef]

68. Nixdorf, B.; Rücker, J.; Deneke, R.; Grüneberg, B. Gewässer im Klimastress? Eutrophierungsgefahr in Seen am Beispiel der Scharmützelseeregion. *Forum Forschung* **2009**, *22*, 99–106.

69. Länder Arbeitsgemeinschaft Wasser (LAWA). *Gewässerbewertung—Stehende Gewässer—Vorläufige Richtlinie für eine Erstbewertung von Natürlich Entstandenen Seen Nach Trophischen Kriterien 1998*; Kulturbuch: Berlin, Germany, 1999; ISBN 3-88961-225-3. Available online: http://www.lawa.de/documents/Gewaesserbewertung_stehende_Gewaesser_2_4ed_copy_589.pdf (accessed on 14 February 2017).

70. Arheimer, B.; Andersson, L.; Alkan-Olsson, J.; Jonsson, A. Using catchment models to establish measure plans according to the Water Framework Directive. *Water Sci. Technol.* **2007**, *56*, 21–28. [CrossRef]

71. Volk, M.; Liersch, S.; Schmidt, G. Towards the implementation of the European Water Framework Directive? Lessons learned from water quality simulations in an agricultural watershed. *Land Use Policy* **2009**, *26*, 580–588. [CrossRef]

72. Jeppesen, E.; Søndergaard, M.; Jensen, J.P.; Havens, K.E.; Anneville, O.; Carvalho, L.; Coveney, M.F.; Deneke, R.; Dokulil, M.T.; Foy, B.; et al. Lake responses to reduced nutrient loading—An analysis of contemporary long term data from 35 case studies. *Freshw. Biol.* **2005**, *50*, 1747–1770. [CrossRef]

73. Lindim, C.; Becker, A.; Grüneberg, B.; Fischer, H. Modelling the effects of nutrient loads reduction and testing the N and P control paradigm in a German shallow lake. *Ecol. Eng.* **2015**, *82*, 415–427. [CrossRef]

74. Hupfer, M.; Reitzel, K.; Grüneberg, B. Methods for measuring internal loading. In *Internal Phosphorus Loading in Lakes: Causes, Case Studies, and Management*; Steinman, A.D., Spears, B.M., Eds.; J. Ross Publishing: Plantation, FL, USA, 2019.

75. German Advisory Council on the Environment (SRU). *NITROGEN: Strategies for Resolving an Urgent Environmental Problem. Summary. January 2015*; SRU: Berlin, Germany, 2015; Available online: http://www.umweltrat.de/SharedDocs/Downloads/EN/02_Special_Reports/2012_2016/2015_01_Nitrogen_Strategies_summary.pdf?__blob=publicationFile (accessed on 27 November 2019).

76. German Advisory Council on the Environment (SRU). *Sondergutachten Stickstoff: Lösungsstrategien für ein Drängendes Umweltproblem*; SRU: Berlin, Germany, 2015; Available online: http://dip21.bundestag.de/dip21/btd/18/040/1804040.pdf (accessed on 20 July 2016).

 *water*

Article

Long-Term Water Quality Changes as a Result of a Sustainable Restoration—A Case Study of Dimictic Lake Durowskie

Renata Dondajewska [1,*], Katarzyna Kowalczewska-Madura [1], Ryszard Gołdyn [1], Anna Kozak [1], Beata Messyasz [2] and Sławek Cerbin [2]

[1] Department of Water Protection, Faculty of Biology, Adam Mickiewicz University, Umultowska 89, 61-614 Poznań, Poland; madura@amu.edu.pl (K.K.-M.); rgold@amu.edu.pl (R.G.); akozak@amu.edu.pl (A.K.)

[2] Department of Hydrobiology, Faculty of Biology, Adam Mickiewicz University, Umultowska 89, 61-614 Poznań, Poland; messyasz@amu.edu.pl (B.M.); cerbins@amu.edu.pl (S.C.)

*   Correspondence: gawronek@amu.edu.pl; Tel.: +48-61-829-58-80

Received: 18 February 2019; Accepted: 22 March 2019; Published: 25 March 2019

Abstract: Nature-based solutions in lake restoration enable gradual ecosystem reconstruction without drastic and expensive intervention. Sustainable lake restoration involves limited external interference strong enough to initiate and maintain positive changes in the ecosystem. It was introduced in Lake Durowskie, an urban, flow-through lake situated in Western Poland, using hypolimnetic aeration, phosphorus precipitation with small doses of chemicals and biomanipulation in 2009, and is continued until today. Oxygen conditions in the lake hypolimnion after initial deterioration were gradually improved, and finally a shortening of the duration and range of oxygen deficits was observed. Nitrogen transformations were induced in the hypolimnion by water aeration as well, reducing ammonium N (30% during 2013–2017 in comparison to 2008) and increasing nitrates (90% in 2013–2017 in comparison to 2008). Phosphorus content was diminished (19% during 2015–2017 in relation to 2008 for SRP) due to effective iron-binding and a smaller amount of fresh organic matter being decomposed. Its reduction was related to lower phytoplankton biomass, expressed in a decrease of chlorophyll-a concentrations (55% reduction during 2013–2017 in comparison to 2008) and an increase in water transparency (two-fold during 2013–2017 in relation to 2008) throughout the nine years of treatment. A long-term restoration program, based on non-aggressive, multiple in-lake techniques was applied and, despite the lack of a reduction in total external loading, was able to suppress progressive eutrophication.

Keywords: biomanipulation; chlorophyll-a; hypolimnion aeration; nutrients; phosphorus inactivation

---

1. Introduction

It is estimated that almost a half of European lakes remain below the good ecological state (GES), required by the Water Framework Directive [1] as necessary to achieve in the coming years [2]. It is therefore essential for them to take protective and/or restoration measures to fulfill this obligation. However, the costs of restoration are too high to be implemented on a large scale, thus only a small number of lakes are currently under recovery treatment [3]. A new approach to lake restoration is urgently needed, using internal mechanisms of ecosystems to support the return to good ecological status. Traditional methods of lake restoration, consisting of a one-off, very intense interference of the ecosystem, bring only a short-term water quality improvement [4,5]. Meanwhile, nature-based solutions enable gradual ecosystem reconstruction without drastic and expensive human intervention. This term has been adopted by International Union for Conservation of

Nature, to name all actions undertaken to "protect, sustainable manage, and restore natural or modified ecosystems (…) simultaneously providing human well-being and biodiversity benefits". It has a multidimensional potential, covering different human activities, including restoration of degraded ecosystems. As "nature-based", the restoration shall be site-specific, applied at landscape scale and maintain biodiversity (www.iucn.org). Such ecological approach is represented by sustainable lake restoration, which involves limited external interference, but strong enough to initiate and then maintain positive changes in the ecosystem [6]. Water quality improvement leads to increase of biodiversity (ecological benefits) and to social benefits as well, by means of increasing safety for local communities using lake water for drinking/recreation/fish production.

Several methods are usually used at the same time to limit the effectiveness of feedbacks coming from the lake ecosystem to maintain its current degraded state [6]. Some nature-based solutions are used in Polish lakes, contributing to sustainable restoration, such as phosphorus inactivation using small precisely calculated doses of native chemical compounds, e.g., magnesium chloride and/or iron sulfate, deep water oxygenation using wind aerators, biomanipulation with predatory fish fry stocking, enrichment of water overlying the sediments with nitrates from the tributaries, increasing the redox potential of the sediment–water interphase and activating the denitrification process [7–11].

It is taken for granted in lake restoration manuals that the maximum limitation of external nutrient loading is crucial prior to the commencement of restoration [12,13]. In many cases, however, radical and rapid elimination of the external loads is not possible, especially in the case of non-point sources related to agriculture. Even a change in the way of farming or the use of biogeochemical barriers requires years for the system to be developed and the nutrient reserves in the soil to be reduced. Waiting for a decrease of loading from the catchment results in the persistence of hypereutrophy in a lake, hence the application of restoration measures after this time would be much more difficult than at the beginning, when the ecosystem resilience "remembers" the state before degradation [14,15]. Application of several years of sustainable restoration at the beginning of deterioration can easily check ecosystem degradation and maintain good lake water quality.

Lake Durowskie is an example of a gradual improvement in the water quality of inflowing riverine waters. The city of Gołańcz, situated in the upper part of the lake catchment, was equipped with a modern wastewater treatment plant; nevertheless, nutrients deposited in the bottom sediments of lakes in the course of the Struga Gołaniecka are still exerting a strong impact on this river as well as on the water quality of Lake Durowskie. The aim of the local authorities was to recover and maintain a good ecological status of the lake, despite the continuous flow of excessive external loads of nutrients, primarily to preserve the ecosystem services of this lake. The goal of the studies conducted during 2008–2017 was to assess whether sustainable, inexpensive restoration can stop the progressive degradation and restore good water quality, enabling safe recreation for the local community. Additionally, long-term studies were able to determine the longevity of restoration results, usually unattainable for traditional restoration methods.

2. Materials and Methods

2.1. Study Site

Lake Durowskie is an urban lake situated in the Wielkopolska Region (Western Poland) in the city of Wągrowiec (17°12′1″ E, 52°49′6″ N). This dimictic, flow-through, postglacial ribbon-type lake is located in the course of River Struga Gołaniecka, which is a tributary of the River Wełna (River Odra basin). This watercourse, with total length of 28 km, supplies Lake Durowskie with nutrients from the catchment, including a cascade of hypereutrophic lakes situated above (i.e., Lakes Kobyleckie, Bukowieckie, Grylewskie and Laskownickie). The lake surface area is 143 ha, maximum depth 14.6 m, mean depth 4.6 m and volume 11,322,900 m^3. Maximum length is 4340 m, whilst the length of the shoreline is 10,515 m [9]. The lake is surrounded by forests in the northern part, and by urban areas in the south, thus recreational pressure (swimming, sailing, and angling) is severe. A promenade connecting the recreational centers runs along the whole eastern and southern shore line. There are numerous piers for fishing and jetties along the beaches and marinas. The total catchment surface of Durowskie Lake is 236.1 km^2 (covered mainly by rural areas). The adjacent catchment area reaches 1581 ha, 58.3% being occupied by farmlands, 33.5% by forests and 8.2% by urban areas [16].

Progressive eutrophication of Durowskie Lake was observed at the turn of 20th and 21st century, with summer water blooms dominated by cyanobacteria, low transparency, oxygen depletion and the presence of hydrogen sulfide in deeper layers of the water column. Therefore, sustainable restoration was begun in 2009 using three methods: (i) hypolimnetic water aeration with the use of two wind-driven aerators; (ii) phosphorus inactivation in the water column using low doses of iron sulfate (PIX type coagulant) and magnesium chloride; and (iii) biomanipulation, based on pike and pikeperch fry stocking to increase the contribution of predatory fish in the lake ichthyofauna. Aerators were installed in two parts of the lake, one in the deepest part adjacent to the city, the second one at a 12 m depth in the part surrounded by forest [6] (Figure 1). The operation of aerators was based on oxygenation of the hypolimnetic water above the surface of the lake and its re-delivery to the bottom after oxygenation [17]. The small doses, i.e., 4–15 kg of iron sulfate per ha, used for restoration did not coagulate the suspended solids, but inactivated orthophosphates in the water column. The treatment was repeated 3–5 times during the vegetation season, eliminating P from the water column. Pike stocking was conducted irregularly, with a greater quantity of fry in 2011, i.e., 100,000 specimens. The highest amount of pikeperch fry was introduced in 2010, namely 114,000 specimens [6].

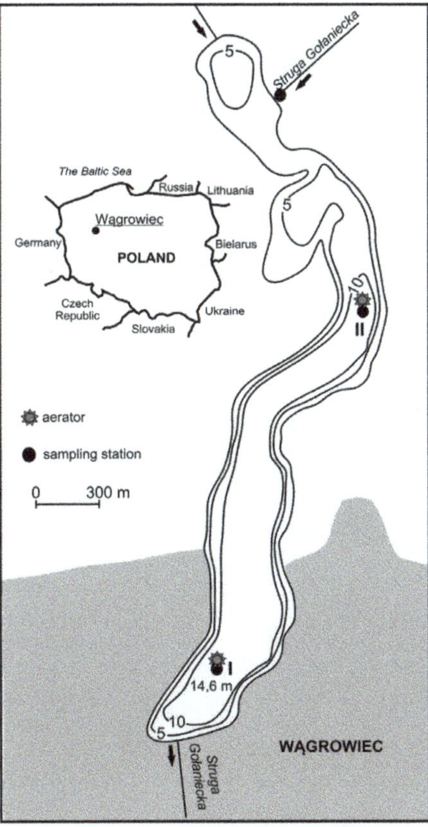

Figure 1. The location of sampling stations and aerators on Lake Durowskie and sampling station on River Struga Gołaniecka.

2.2. Methods

Sampling was conducted monthly from April 2008 to December 2017. Water samples prior to the restoration (2008) were taken at Station I only, while during restoration (2009–2017) at two stations and from the inflow of the Struga Gołaniecka. Both lake stations were located near the aerators: Station I at the deepest place (14.6 m) in the southern part of the lake, while Station II in the northern part at a depth of 12.0 m. Samples for the physicochemical and biological analyses were taken from epi-, meta- and hypolimnion, i.e., at Station I from a depth of 1 m, 7 m and 12 m, while at Station II from a depth of 1 m, 5 m and 8 m. Field measurements (water temperature, oxygen concentration, water saturation with oxygen, conductivity and pH) were conducted using a YSI 556 MPS-meter. Water transparency was measured with a Secchi disc. The concentrations of ammonium nitrogen ($N-NH_4$), nitrate nitrogen ($N-NO_3$), nitrite nitrogen ($N-NO_2$), organic nitrogen (Norg), orthophosphates (SRP) and total phosphorus (TP) were analyzed with the spectrophotometric method according to Polish standards [18]. Concentration of chlorophyll-a was analyzed after filtration through Whatman GF/C filters using the spectrophotometric method with acetone extraction [19], and suspended solids using the weighting method.

Basic statistical calculations were made using STATISTICA version 10.0 software. The non-parametric Kruskall–Wallis test (K-W test) was used to determine the significance of value changes throughout the analyzed period. To relate the environmental variables in riverine and lake waters, a redundancy analysis was carried out using CANOCO 4.5 software.

3. Results

3.1. The Quality of Durowskie Lake Waters

3.1.1. Basic Water Characteristics

Durowskie Lake is a dimictic lake, thermally stratified in summer. Typical water mixing in the water column was observed in spring and autumn. The epilimnion usually reached a depth of 3 m, while the metalimnion down to 7–8 m at Station I and 6 m at Station II. The hypolimnion covered the deepest layer of water, reaching the bottom (Table 1). Mean temperature of the epilimnetic waters usually exceeded 20 °C and even 23 °C in 2012 and 2016. In the metalimnion, mean temperature varied from 11.6 to 16.4 °C, while in the hypolimnion from 5.9 to 8.5 °C, wherein warmer waters were noted at shallower Station II. The functioning of aerators did not affect the stability of thermal stratification, however, an increase in the temperature of hypolimnetic waters was observed. At Station I (depth 12 m) median water temperature was 6.0–6.1 °C in 2009–2011, and 6.5–7.2 °C in 2014–2017, whilst at Station II these values were 6.4–6.8 °C and 6.8–7.9 °C, respectively. This tendency was statistically significant at Station I (K-W test, $p < 0.01$) for years 2009–2017. Water temperatures noted in 2008 were not taken into consideration due to the lack of data from the entire year (only for the period April–November).

Table 1. Range of individual thermal layers of water in subsequent years of research (m) and mean water temperature in the period June–August (°C) at two stations compared to the average air temperature in the same period (°C).

Range of Water Layers (m)		2008	2009	2010	2011	2012	2013	2014	2015	2016	2017
Station I	epi	0–4	0–4	0–3	0–3	0–2	0–3	0–3	0–3	0–3	0–3
	temp.	19.4	20.3	20.9	20.8	22.5	21.3	20.9	20.0	23.1	20.9
	meta	5–8	5–8	4–6	4–7	3–6	4–7	4–8	4–8	4–7	4–7
	temp.	11.6	13.0	13.9	13.6	16.4	12.5	13.1	14.0	14.1	12.9
	hypo	9–14	9–14	7–14	8–14	7–14	8–14	9–14	9–14	8–14	8–14
	temp.	8.1	6.3	6.8	6.6	8.2	5.9	6.9	7.5	6.9	7.7
Station II	epi	na	0–4	0–2	0–3	0–2	0–2	0–3	0–3	0–3	0–4
	temp.		20.9	20.7	21.2	23.4	21.7	20.5	19.6	23.1	20.2
	meta	na	5–7	3–6	4–6	3–6	3–6	4–6	4–6	4–6	5–6
	temp.		13.4	14.6	15.3	16.3	15.4	14.1	15.3	14.3	12.2
	hypo	na	8–12	7–12	7–12	7–12	7–12	7–12	7–12	7–12	7–12
	temp.		7.0	7.6	7.7	8.4	7.1	8.1	8.5	7.5	8.5
Mean air temperature [1]		18.9	18.3	19.2	18.7	18.6	19.2	18.9	19.6	18.9	18.7

na, not applicable; [1] data from Poznań-Ławica Airport (weatheronline.com).

The oxygen concentration in the water column at both stations varied within wide limits from zero to ca. 20 mgO$_2$ L^{-1}. The highest content in each year was usually noted in early spring, with the maximum in March 2012 (Figure 2). The range and duration of oxygen depletion (less than 3 mgO$_2$ L^{-1}) in deep water layers changed from in time. The anoxic zone near the bottom was permanent in 2008, even in spring and autumn, indicating very short and incomplete water mixing. In 2009, deficits were already noted at the depth of 3 m in summer; however, in subsequent years, they were stated in a narrower layer of the water column, and in 2017 were noted only at 7–8 m depth. The duration of oxygen depletion shortened, from April to October at the beginning of the studies, while from June to October in 2017. The concentration of oxygen did not change linearly in subsequent years but fluctuated; however, it showed a tendency to decrease in the epilimnion and increase in meta- and hypolimnion (Figure 3). An increase in the metalimnion, especially at Station I, was confirmed by K-W test ($p < 0.05$) for layers 6–9 m at Station I and for 7 m at Station II. Nevertheless, the increase in the hypolimnion was not statistically significant.

The pH of Lake Durowskie waters varied within wide limits, 6.51–9.07 at Station I and 6.33–8.98 at Station II, with highest values in the epilimnion and lowest near the bottom (Figure 4). Median values fluctuated in all water layers in subsequent years slightly decreasing in time, however, only at the depth of 10 m at Station II was the decreasing tendency statistically significant (K-W test with $p < 0.01$).

A wide range of water conductivity was also noted in Lake Durowskie, with 322–969 µS cm^{-1} at Station I and 310–914 µS cm^{-1} at Station II. The highest annual variability was noted during 2010–2012 (Figure 5), and in 2008—prior to the restoration. The years 2009 and 2013–2014 were characterized by conductivity varying from ca. 600 to 750 µS cm^{-1}, while during 2015–2016 a wider range starting from 350 µS cm^{-1} was noted again. In all water layers representing the epilimnion, metalimnion and hypolimnion at both stations, there was a statistically significant increase in conductivity over time (K-W test with $p < 0.05$).

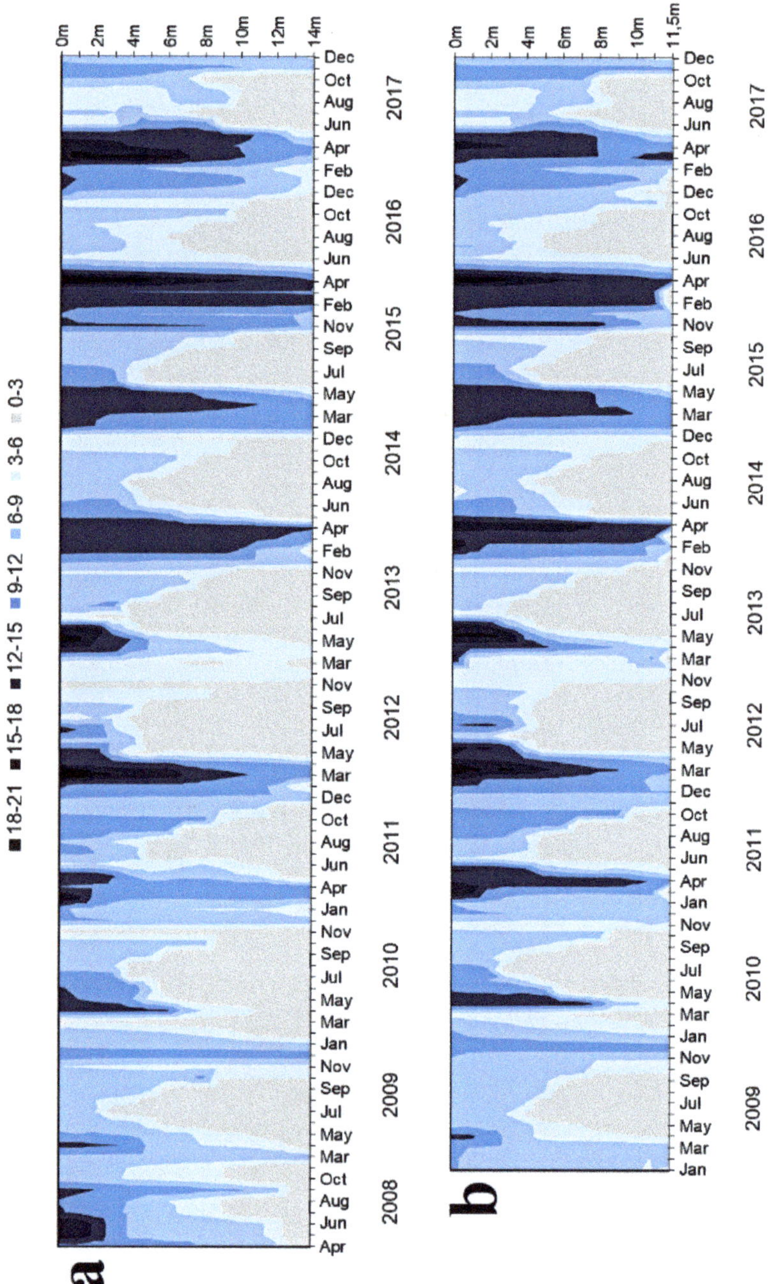

Figure 2. Changes of oxygen concentration (in mgO$_2$ L^{-1}) in Lake Durowskie waters in 2008–2017 at: Station I (**a**); and Station II (**b**).

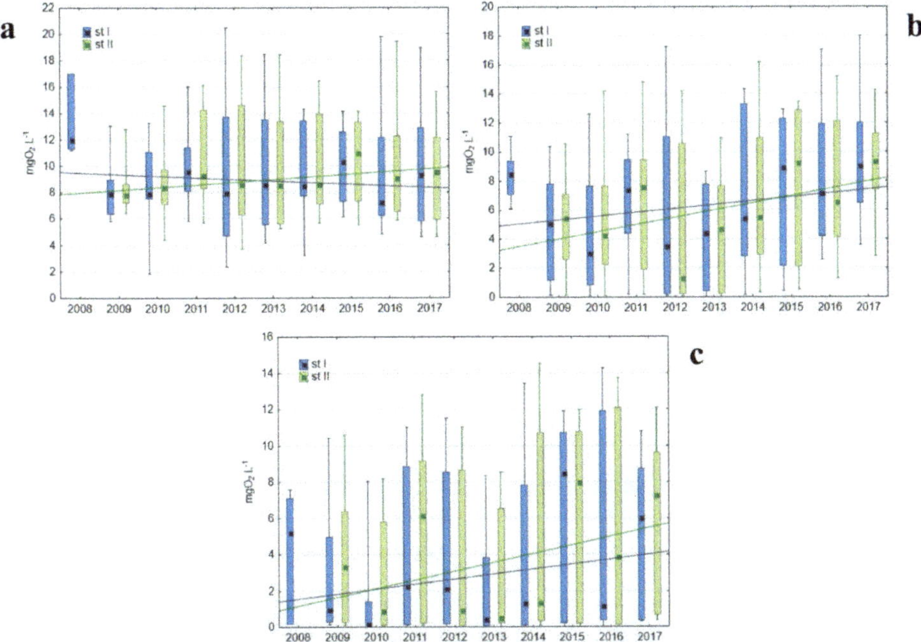

Figure 3. Box plot with min-max concentrations of oxygen in selected water layers at both stations representing: epilimnion (**a**); metalimnion (**b**); and hypolimnion (**c**). The lines indicate trendlines for each station.

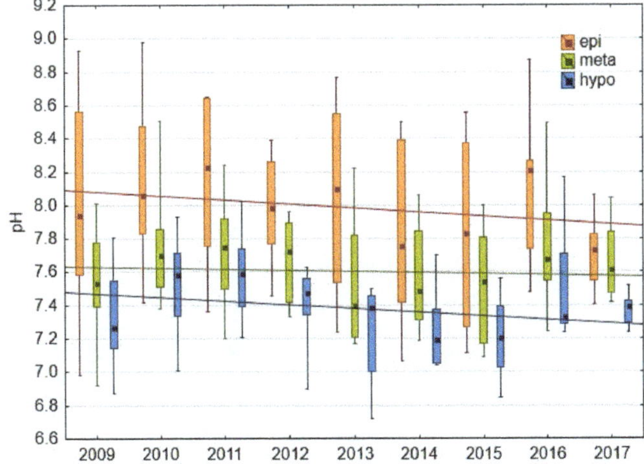

Figure 4. Box plot with min-max of pH in selected water layers representing epilimnion, metalimnion and hypolimnion at both stations. The lines indicate trendlines for each layer.

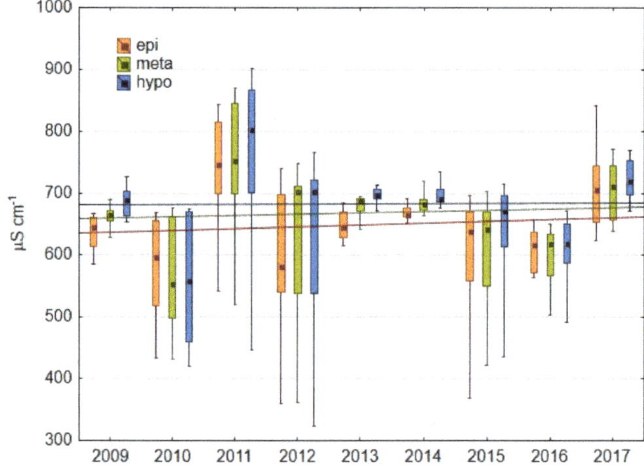

Figure 5. Box plot with min-max of conductivity in selected water layers representing epilimnion, metalimnion and hypolimnion at both stations. The lines indicate trendlines for each layer.

3.1.2. Water Transparency and Chlorophyll-a Content

Water transparency increased gradually throughout the research period, which was confirmed statistically at Station I (K-W test, $p < 0.05$). It was only 1.6 m maximally in 2008, with a median of 0.9 m (Figure 6), while during 2010–2011 it reached 3.8 m, with a median reaching 2.5 m. It was fluctuated in the following years, however, the upward trend continued, especially at Station I (Figure 6). The highest value was recorded in winter 2014, amounting to 5.4 m. The lowest values of about 0.5 m were found in the first half of the research period; later, the minimum values were about 1 m. They were associated with spring water blooms, while in summer they were around 1.5 m. Overall, Secchi disc values increased over twofold during 2014–2017 in comparison to 2008.

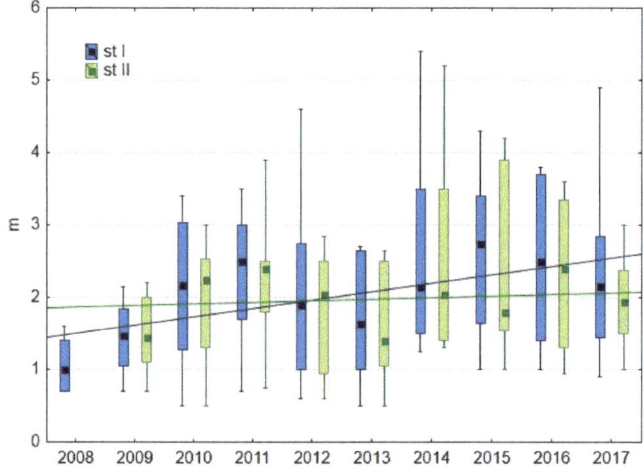

Figure 6. Box plot with min-max of Secchi depth in Lake Durowskie at both stations. The lines indicate trendlines for each station.

A decrease of chlorophyll-a concentrations was noted in Lake Durowskie in the following years, especially in the epilimnion (Figure 7). This tendency was statistically significant at a depth of 1 m at Station I as well as at a depth of 7 m, and in the surface waters at Station II (K-W test, $p < 0.01$). The concentration decreased exponentially in the first years of restoration. In subsequent years, it fluctuated, however, the median remained around 10 mg m^{-3}. Maximum values in the second half of the research period were below 35 mg m^{-3} during spring water bloom, while the minimum in the autumn and winter reached 5 mg m^{-3}. A decrease in chlorophyll-a concentration was 55% during 2013–2017 in relation to 2008 at Station I.

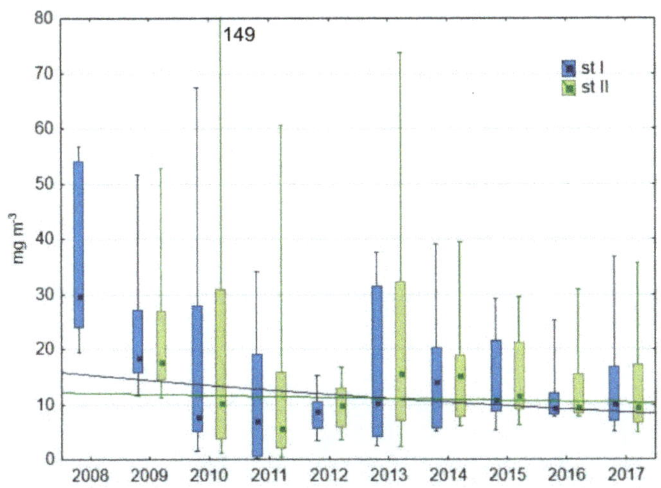

Figure 7. Box plot with min-max of chlorophyll-a in epilimnion at both stations I. The lines indicate exponential trendlines for each station.

3.1.3. Nitrogen Concentrations

Among nitrogen compounds, mineral ones, mainly ammonium N and nitrates were dominant. The mean contribution of mineral forms in total N was 60.5%, while the highest was 72.4% in 2011. It is characteristic that, as a result of the beginning of restoration, concentration of ammonium nitrogen increased very markedly, especially in the hypolimnion (Figure 8a,b). It reached maximum values at the end of summer thermal stratification, especially at deeper Station I, maximally up to 8.49 mgN-NH$_4$ L^{-1} (2012). During restoration a statistically significant decrease of ammonium N content in subsequent years was observed in all water layers at both stations (K-W test, $p < 0.01$). Median reduction in the epilimnion was from over 1 mgN L^{-1} in 2009 to less than 0.4 mgN L^{-1} in 2017, while in the hypolimnion—from over 1.9 to less than 0.6 mgN L^{-1}. It was thus on average about 30% lower during 2013–2017 in comparison to 2008. Nitrate concentration, on the other hand, fluctuated during the research period. During 2008–2010 and 2014–2016, it was less than 2 mgN-NO$_3$ L^{-1}, with a median usually below 1 mgN-NO$_3$ L^{-1}. It increased dramatically in 2011, reaching maximally 6.6 and 8.0 mgN-NO$_3$ L^{-1} at Stations I and II, respectively, thus the median exceeded 4 mgN-NO$_3$ L^{-1}, especially in epi- and metalimnion (Figure 8c,d). A second period of nitrate content increase was noted in 2017 with medians ca. 2 mgN-NO$_3$ L^{-1}. As a result, the average nitrate concentration during 2013–2017 was ca. 90% higher than in 2008.

Organic nitrogen also fluctuated throughout the study period. A slight decreasing trend was observed at both stations (K-W test, $p < 0.01$), especially between 2008 and 2016. During 2008–2011, medians of organic N content varied from 1.46 to 2.49 mgN L^{-1}, diminishing during 2012–2016 to 0.65–1.66 mgN L^{-1} (Figure 9a,b). The year 2017 was characterized by another increase, clearly marked in maximum concentrations reaching 5.86 and 4.55 mgN L^{-1} at Stations I and II, respectively.

Due to the reduction of ammonium N and organic N concentrations during the analyzed period, total N content diminished as well (K-W test, $p < 0.01$; however, two crucial peaks were noted in 2011 and 2017 (Figure 9c,d). Maximum total N reached ca. 10 mgN L^{-1} and median around 7 mgN L^{-1} in 2011. A gradual decrease was observed during 2013–2016, when median values dropped from 3.7–4.5 mgN L^{-1} to 1.9–3.0 mgN L^{-1}.

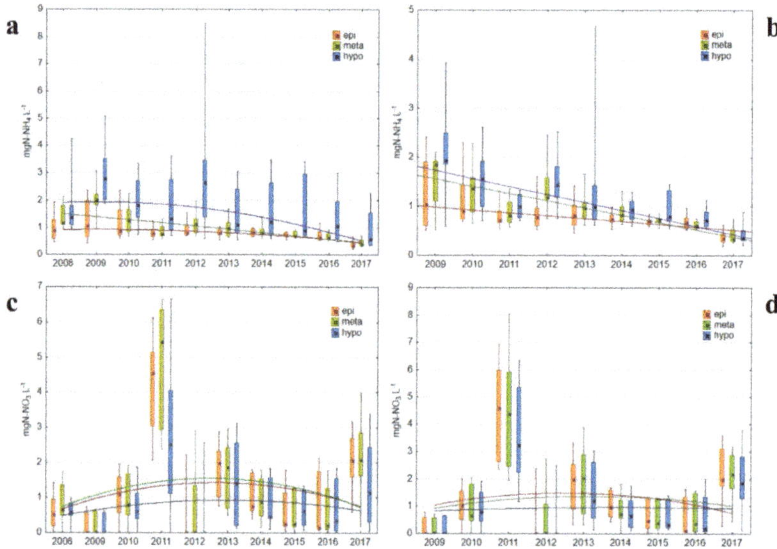

Figure 8. Box plot with min-max of mineral nitrogen compounds in selected water layers: ammonium N at Station I (**a**) and Station II (**b**); and nitrate nitrogen at Station I (**c**) and Station II (**d**). The lines indicate binomial trendlines for water layers in (**a,c,d**), and linear trend in (**b**).

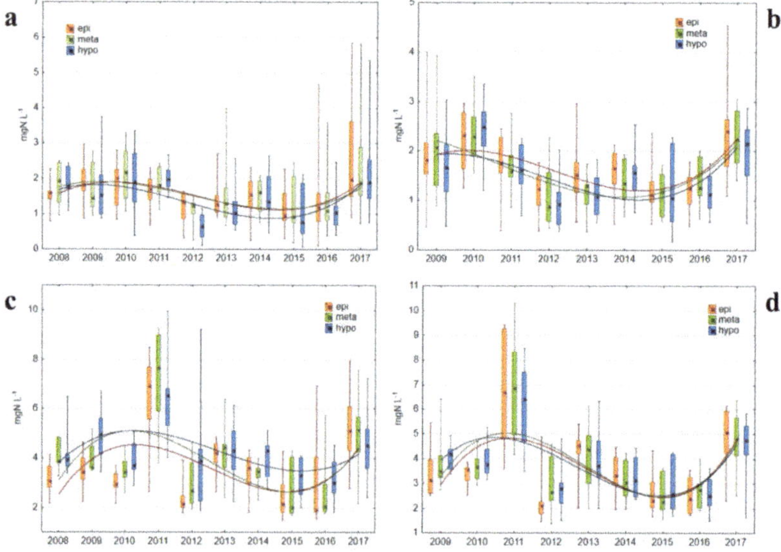

Figure 9. Box plot with min-max of nitrogen compounds in selected water layers: organic N at Station I (**a**) and Station II (**b**); and total N at Station I (**c**) and Station II (**d**). The lines indicate polynomial trendlines for water layers.

3.1.4. Phosphorus Concentrations

SRP concentrations increased significantly in the first few years of restoration, especially in the hypolimnion at deeper Station I (Figure 10a). The maximal value of 0.06 mgP L^{-1} in 2008 rose to 0.23 mgP L^{-1} in 2012, while in the following years it decreased systematically to 0.08 mgP L^{-1}. Median changed from 0.002 mgP L^{-1} in 2008 through 0.088 mgP L^{-1} in 2012 and finally decreased to 0.002 mgP L^{-1} again in 2017. Temporal changes at Station II were similar (Figure 10b); however, during 2011–2014 medians were higher in the epilimnion in comparison to the meta- and hypolimnion. Nevertheless, the SRP concentrations increased from maximally 0.04 mgP L^{-1} in 2009 to 0.08 mgP L^{-1} in 2012, and dropped again to less than 0.03 mgP L^{-1} in 2017. This decrease of SRP content was statistically significant at both stations in all water layers (K-W test, $p < 0.01$). In comparison to 2008, SRP concentration reduction during 2015–2017 was 19% in water column, and even higher in epilimnion (65%).

Similar to SRP, TP concentrations also increased significantly during the first few years of restoration. This was particularly evident in the hypolimnion at Station I, where the median increased from 0.05 mg L^{-1} in 2008 to 0.12 mg L^{-1} in 2012. It decreased again to 0.06 mg L^{-1} in subsequent years. A similar trend of changes was also observed in the epilimnion and metalimnion, although the range of changes was much smaller (Figure 10c). At Station II, concentrations in all three thermal layers were similar, and changes in the entire study period showed fluctuations. In the first years of restoration, concentrations increased, then decreased, but in the last two years again slightly increased (Figure 10d). Interestingly, the maximum concentrations at both stations at the end of the study period were higher than at the beginning. Nevertheless, the median values were lower at the end of the study than at the beginning.

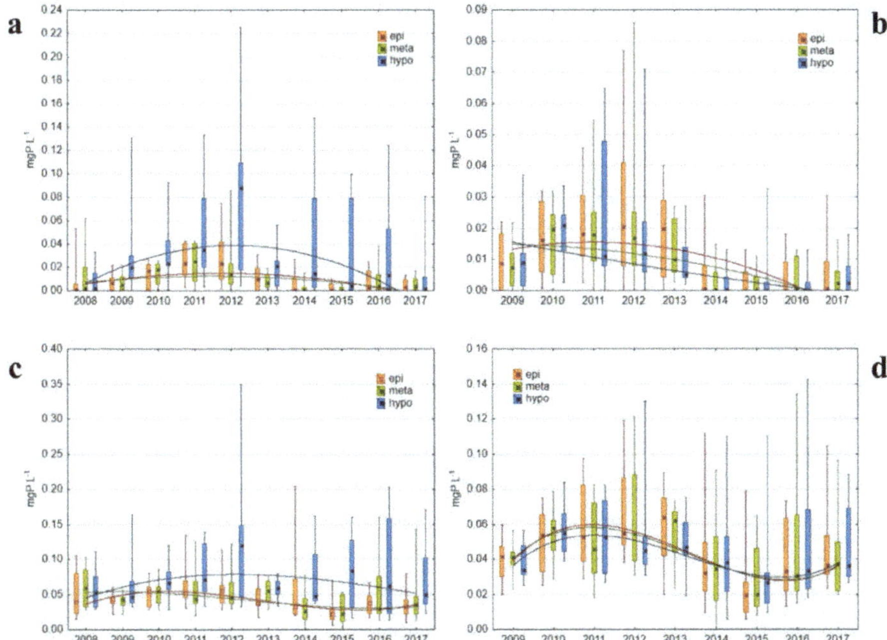

Figure 10. Box plot with min-max of phosphorus compounds in selected water layers: SRP at Station I (**a**) and Station II (**b**); and total P at Station I (**c**) and Station II (**d**). The lines indicate binomial trendlines for water layers at a and b, and polynomial trendlines for water layers at c and d.

3.2. The Quality of River Struga Gołaniecka Waters

Spring and summer were characterized by higher water temperature (maximum over 25 °C) and pH (over 8), lower oxygen concentration (usually less than 10 mgO$_2$ L^{-1}) and conductivity (less than 600 µS cm^{-1}). During winter the temperature fell to ca. 1 °C and oxygen content reached maximum values (usually over 15 mgO$_2$ L^{-1}, Table 2).

A decreasing tendency throughout the research period was noted for chlorophyll-a concentration (K-W test, $p < 0.05$). Maximum values during 2009–2010 were over 80 mg m^{-3}, decreasing to ca. 50 mg m^{-3} during 2011–2014, and finally to less than 30 mg m^{-3} during 2016–2017.

Ammonium N concentration increased in winter each year, reaching ca. 2 mgN-NH$_4$ L^{-1} (maximally 3.03 mgN-NH$_4$ L^{-1} in 2013), while in the vegetation season it diminished to 0.5–1.0 mgN-NH$_4$ L^{-1}. A slight decrease was noted during the analyzed period, although it was not statistically significant. Nitrate N fluctuated, showing two distinct maxima in 2011 and 2017 and a smaller third in 2013, exciding 10, 8 and 5 mgN-NO$_3$ L^{-1}, respectively. Its concentrations were low in other years, less than 1.7 mgN-NO$_3$ L^{-1}. These peaks influenced TN concentrations. Medians were usually 2.6–3.7 mgN L^{-1}, but sometimes exceeded 5 mgN L^{-1} (2011, 2013, and 2017), together with maximum annual values (Table 2). A slight, statistically significant (K-W test, $p < 0.05$) decrease of organic N was also noted.

SRP and TP concentration fluctuated in the River Struga Gołaniecka over time. First, it increased during 2010–2013, and decreased in the following years. This tendency was statistically significant (K-W test, $p < 0.01$) in the case of SRP, while for TP an increase was noted (K-W test, $p < 0.05$), especially during 2015–2017. Maximum SPR concentrations reached almost 0.070 mgP L^{-1} in 2012, while TP—0.200 mgP L^{-1} in 2017.

Table 2. The variability of medians and maximum values of water quality characteristics in the River Struga Gołaniecka flowing into Lake Durowskie.

Parameter	2009	2010	2011	2012	2013	2014	2015	2016	2017
temperature	13.3	9.8	7.8	10.8	15.4	15.2	10.0	12.8	11.0
(°C)	22.1	27.1	22.2	25.6	25.7	25.5	20.9	23.9	21.4
oxygen	8.2	9.7	10.7	9.5	10.5	9.7	11.0	7.6	9.7
(mgO$_2$ L^{-1})	17.7	16.1	17.5	15.9	16.2	16.8	13.9	18.3	15.7
pH	8.3	8.4	8.5	8.0	8.3	8.1	7.7	8.2	7.8
	9.2	9.0	8.6	8.8	8.9	8.6	8.5	8.9	8.4
conductivity	685	587	716	566	700	710	647	598	820
(µS cm^{-1})	791	744	903	792	796	746	715	687	859
chlorophyll-a	26.5	41.9	8.1	25.3	18.6	26.5	28.9	19.5	19.6
(mg m^{-3})	88.9	51.8	54.7	38.0	47.8	45.2	38.5	30.3	28.6
ammonium N	0.99	1.24	0.88	1.03	1.02	0.91	0.86	0.97	0.54
(mgN-NH$_4$ L^{-1})	2.59	2.76	2.26	3.03	1.79	2.07	2.37	1.93	1.22
nitrates	0	0.77	2.82	0	2.44	1.22	0.34	0.19	2.94
(mgN-NO$_3$ L^{-1})	0.88	3.04	10.51	1.00	5.57	2.72	1.26	1.62	8.08
organic N	1.6	2.0	1.8	1.4	1.1	1.7	1.2	1.3	2.0
(mgN L^{-1})	3.9	3.5	3.2	1.8	3.6	2.2	2.3	1.7	3.7
TN	3.7	3.7	5.7	2.5	5.7	4.4	2.6	2.7	5.6
(mgN L^{-1})	5.2	4.1	11.8	4.1	7.2	5.0	4.6	4.3	11.1
SRP	0.011	0.025	0.023	0.022	0.019	0.003	0.004	0.006	0.007
(mgP L^{-1})	0.025	0.058	0.056	0.069	0.029	0.020	0.011	0.039	0.022
TP	0.050	0.061	0.046	0.049	0.046	0.027	0.030	0.028	0.048
(mgP L^{-1})	0.072	0.122	0.080	0.121	0.101	0.067	0.175	0.125	0.200

4. Discussion

Freshwater ecosystems are characterized by both top-down (TD) and bottom-up (BU) regulation, as nutrients play an important limiting role for primary production [20] and higher trophic levels may also influence the lower ones through trophic cascade [21]. A combination of both these pervasive ways of ecosystem productivity control occurs in every lake and might be applied in water quality management [22]. Sustainable lake restoration links the reduction of nutrient concentrations with the chemical method, supported by water aeration (BU), and biomanipulation, aiming to change fish fauna, influencing the plankton community (TD). Combining the "classic" in-lake restoration methods, i.e., biological and physicochemical, into a consistent restoration program increases the chance of success; nevertheless, it is still quite uncommon. Single cases, mainly from European lakes [23], have proved that such an approach results in water quality improvement, both in shallow and in deep lakes. The Lake Durowskie case study follows this approach and is an example of long-term lake restoration, applying multiple methods on a non-aggressive level, harmless to biota.

4.1. Oxygen Conditions As a Result of Hypolimnetic Aeration

Hypolimnetic anoxia is a common phenomenon in summer-stratified lakes, however, in the course of eutrophication, it lengthens in time and extends in water volume, resulting in a number of processes of further lake water degradation, e.g., increased release of phosphorus from sediments [24], as well as ammonia N [25] and compounds associated with taste, odor and color, e.g., hydrogen sulfide, iron and manganese [26]. The depletion of dissolved oxygen in bottom waters also affects fish fauna and zooplankton by the reduction of cold-water and a dark habitat [27]. Water aeration or oxygenation is used to combat the adverse effects of eutrophication, although several studies have questioned its efficiency [28–30], indicating that there is no increase in dissolved oxygen content in the hypolimnion. This effect results from a very high oxygen demand in deep water layers and sediment–water interphase due to a multiannual organic matter sedimentation. A similar outcome was observed in Lake Durowskie. After initial deterioration, oxygen conditions in the lake hypolimnion gradually improved. The deep oxygen deficit occurred less and less during summer, and its range from the bottom included an increasingly smaller part of the water column. A similar shortening of the duration and range of oxygen deficits has been observed as an effect of hypolimnetic aeration in other lakes [31,32]. Initially, oxygen supplied by a pulverizing aerator to the sediment–water interphase caused the intensification of organic matter mineralization. The oxygen demand for microorganisms was greater than the oxygen supply capacity of the aerator. Therefore, in the initial period, the oxygen deficit even deepened. However, the reduced phytoplankton biomass in subsequent years delivered to the bottom less and less amounts of fresh organic matter, further decreasing oxygen demand [29].

It is of paramount importance that the type of aerator used in Lake Durowskie fulfills the requirements of sustainable restoration. A pulverizing aerator is characterized by high specific capacity, aeration intensity adjusted to changes in wind direction and force, but most of all—the use of a renewable energy source [17]. Its application in lake restoration together with other methods results in water quality improvement of i.a. Swarzędzkie Lake [33,34] and contributed to changes in the TN:TP ratio, indicating phosphorus deposition in sediments in other lakes [30]. Nevertheless, hypolimnetic aeration seems to be a supportive, rather than the main restoration method, helping to alleviate eutrophication symptoms, thus its combination with other methods is crucial.

4.2. Nitrogen Transformations in Relation to Oxygen and Temperature

The presence of low concentrations of oxygen above the bottom even during summer—ranging 0–2 mg O_2 L^{-1}—favored the nitrification process, leading to the reduction of ammonium nitrogen concentration and increasing the concentration of nitrates. Ammonia N release in the anoxic hypolimnion of eutrophic and hypertrophic lakes exceeds 15 $mgNm^{-2}d^{-1}$, and its accumulation in deep water layers results from the lack of nitrification and decreased assimilation [25]. Low oxygen concentrations due to water aeration allow oxidizing ammonium N to nitrates, which in turn provide substrate for denitrification [31]. An intense denitrification process in the initial period of restoration was indicated by the total depletion of nitrates in the hypolimnion of the lake. Respiratory denitrification is considered as an important nitrogen removal pathway [35], which can be accelerated by artificial aeration [31]. The process of fermentative dissimilatory nitrate reduction to ammonium (DNRA) might well be responsible for ammonium N content decrease, although it is believed that it is favored in nitrate-limited environments rich in labile carbon, reductive conditions and higher temperature [31,35,36], thus it might contribute to the diminishment of ammonium N only at the beginning of restoration process. Low concentration of oxygen in the water overlying the sediment, however, allowed the annamox process to take place, i.e., the reduction of ammonium nitrogen to N_2 [37]. The presence and growing importance of this process in subsequent years of the restoration was demonstrated by the decreasing concentration of ammonium nitrogen in the hypolimnion. However, little is known about annamox in freshwaters, but its crucial role in deep, large, oligotrophic lakes is suspected [35]. As the oxygen content increased, the role of DNRA and denitrification in nitrogen removal decreased in favor of nitrification, which was confirmed by growing concentrations of nitrates in lake water, including the hypolimnion. Both ammonium N and nitrates correlated with oxygen content in the hypolimnion (r = -0.52, $p < 0.01$ and r = 0.357, $p < 0.01$, respectively).

Nitrogen cycling in lake waters is microbe-mediated, thus strongly related to water temperature. Observed transformations, therefore, corresponded not only to changes in oxygen concentrations but to temperature increase in the hypolimnion as well. The change of hypolimnetic temperature was slight (ca. 1 °C in the case of medians); nevertheless it definitely influenced the microorganisms and their activity in the sediment–water interphase. Nitrogen transformations were accelerated, as the correlation between ammonium N/nitrates and temperature was noted (r = 0.347, $p < 0.01$ and r = -0.360, $p < 0.01$, respectively). Such changes are often observed during hypolimnetic aeration/oxygenation [29,31,38] and may stimulate the decomposition of organic matter, resulting in further increased oxygen demand. Long-term treatment, including water aeration and other measures aiming at the reduction of fresh organic matter deposition in the sediments are required to suppress the potential effects of elevated temperature.

As the concentration of individual nitrogen forms depends on many microbiological processes in sediments and in water, which are affected by temperature and oxygen content and additionally depend on the load carried to the lake with inflow waters, their concentration in the lake varies with time depending on these various factors. Therefore, they are subject to linear but not polynomial changes, however with a tendency to decrease fluctuations over time as the ecosystem stabilizes.

4.3. Phosphorus–Oxygen–Chlorophyll-a Interactions

The changes of SRP concentrations at the deeper Station I remained under the influence of internal P loading, especially in the case of the hypolimnion. The release of phosphorus from bottom sediments was much higher at the beginning of lake restoration (2010–2013) as a result of more intense and longer oxygen depletion in sediments during the vegetation season in comparison to the following years (2014–2016), when low oxygen concentrations were present [9]. The reduction of iron compound as well as other metals, and the release of phosphates adsorbed on them are a consequence of the lack of oxygen in the sediment–water interphase [39]. An additional factor stimulating internal P loading during the first years of restoration was the aforementioned large amount of fresh organic matter of planktonic origin deposited in sediments prior to the treatment, as well as phytoplankton inflowing with the River Struga Gołaniecka, also sedimented and decomposed by sediment bacteria. The latter was indicated by high chlorophyll-a concentrations in riverine waters, especially in 2009–2010, diminishing gradually as the water flowed through the lake. Hypolimnetic aeration resulted in oxygenation-induced stimulation of mineralization [29]. A decrease of chlorophyll-a content during the restoration indicated that fresh organic matter deposition was reduced as well, due to the less available orthophosphates in the lake waters, inactivated by chemicals, and the decrease of external SRP loading. Consequently, internal

P loading depending on aerobic microbial decomposition of organic matter decreased, although the oxygen concentration increased. This was confirmed by the decrease of P content in the waters overlying the sediments in subsequent years [9]. Such an additional feedback, related to less internal P loading impact on the phytoplankton biomass, is probably one of the most important mechanisms regulating water quality improvement, especially in the case of water transparency.

The reduction of SRP concentrations at both stations was a result of its inactivation with the use of small doses of iron sulfate and magnesium chloride, aiming at only SRP precipitation, not entire suspended solids. The treatment was repeated 3–5 times during the vegetation season, eliminating excessive SRP available for phytoplankton. This kind of treatment mitigates the potentially negative impact of chemical compounds on biota as well as lessening the costs of restoration [33]. Apart from iron coagulant, magnesium chloride was also added to bind both orthophosphates and ammonium N in insoluble ammonium- magnesium monophosphate, so-called struvite ($MgNH_4PO_4 \cdot 6H_2O$), deposited in sediments [11]. A simultaneous diminishment in SRP and ammonium N content in all water layers and at both stations indicated the success of the restoration treatments. Further P accumulation in sediments could be controlled by nitrates—as their content in the hypolimnion increased—due to: (i) maintaining a high redox potential of sediments; (ii) decreasing the sulfate reduction and hydrogen sulfide formation; and (iii) allowing iron to bind phosphorus effectively as iron phosphate or via adsorption to ferric oxide hydroxides [39]. A positive influence of nitrate treatment has been stated, e.g., in Lake Uzarzewskie, in which cold nitrate-rich waters are directed into hypolimnion [7].

Hypolimnetic aeration also affected phosphorus concentrations in the hypolimnion [28] by increasing both oxygen and temperature in deep waters. Both SRP and TP concentrations in the lake hypolimnion were negatively correlated with oxygen content (r = −0.39 and −0.32, respectively, $p < 0.01$), a common relation in freshwater, resulting from P retention in sediments during aerobic conditions on oxidized iron compounds. The high sensitivity of iron to redox conditions indicates the need of water aeration/oxygenation during P precipitation with this inactivating agent [40], and many applications of different Fe compounds have confirmed that lowering P concentrations without additional aeration or destratification is usually unsuccessful [41–43]. SRP and TP correlation with the temperature of the hypolimnion was positive (r = 0.279 and 0.298, respectively, $p < 0.01$), indicating a possible increase of P content in deep water layers due to its aeration. This phenomenon is true as it is not only oxygen that controls P release, and this paradigm—valid for many years—has been questioned [44,45]. Temperature accelerates many alternative mechanisms of P release, e.g., decomposition of organic matter or accumulation/release of bacterial P, therefore its increase affects P concentrations. As previously mentioned, restoration techniques combined in "sustainable restoration" aim at the reduction of organic matter deposition and thus the suppression of increased temperature impact. This is of paramount importance in the light of climate changes, believed to be manifested in freshwater temperature boost, influencing both internal P loading and phytoplankton (especially cyanobacteria) proliferation [46–48]. Adverse changes induced by global warming are crucial mainly for shallow lakes and shallow parts of deeper lakes as well, thus of lesser importance in Lake Durowskie with its steeply sloped lake basin and limited range of epilimnion. Nevertheless, future studies should focus on the shallower northern part of the lake fed by the River Struga Gołaniecka and one more small stream, with a depth less than 5 m, as recent findings suggest that such parts of freshwater reservoirs are considered to be much more important in internal P loading and deterioration of water quality [49]. The northern zone of the lake together with the River Struga Gołaniecka might be responsible for the less well defined results of lake restoration, including fluctuations in TP content at Station II.

As the influence of the River Struga Gołaniecka on the concentration of phosphorus at Station II is much more pronounced than at Station I, the changes of phosphorus over time can overlap with the variable release from the bottom sediments, causing clear polynomial fluctuations.

4.4. The Influence of River Struga Gołaniecka on Lake Waters

Nitrate concentrations were strongly correlated with their content in the inflowing River Struga Gołaniecka (r = 0.917 for Station I and r = 0.926 for Station II, *p* < 0.05). The high variability of nitrate amounts in both lake and riverine waters resulted from year-to-year changes in precipitation. Increased rainfalls caused greater flushing of nitrates from the soils of the agricultural river catchment [50]. Nevertheless, an annual shift in time in relation to precipitation was observed, i.e., nitrate content corresponded to the rainfall from the previous year (r = 0.805, *p* < 0.05) as a result of the retention effect of lakes situated in a cascade along the course of the River Struga Gołaniecka (Figure 11). Very high nitrate concentrations occurring periodically in the river (up to 10.5 mgN L^{-1}) distinctly influenced the lake content of TN (r = 0.934, *p* < 0.01).

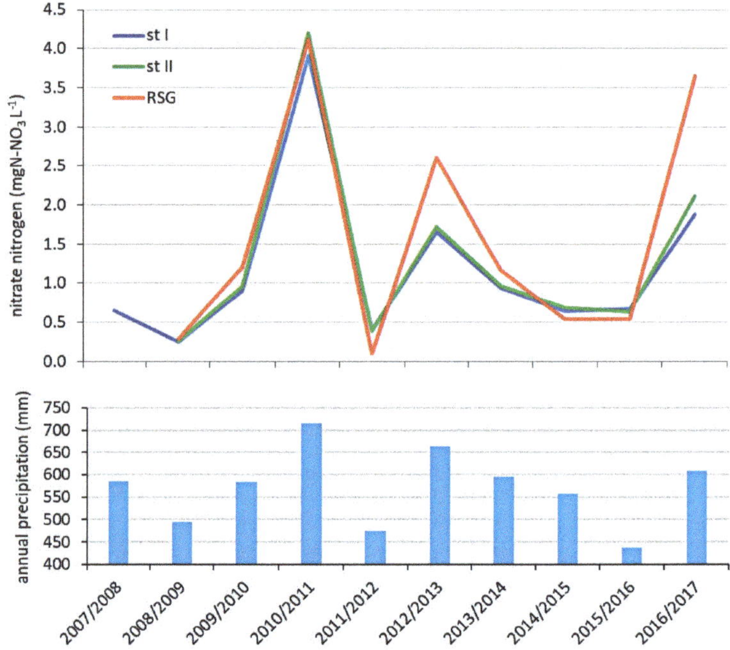

Figure 11. The dependence of nitrate concentrations at Station I, II and River Struga Gołaniecka (RSG) on the sum of annual precipitation from the previous year (first year on horizontal axis).

SRP at Station II varied in particular years in correlation with concentrations of SRP in the inflowing River Struga Gołaniecka (r = 0.975, *p* < 0.01), which were, on the other hand, weakly correlated with precipitation (r = 0.380) (Figure 12). Medians at Station II were always lower than in the river due to multiple physical, chemical and biological processes, e.g., phosphate sorption on suspended solids, sedimentation, uptake by phytoplankton and further deposition with organic matter to sediments [51,52]. At neither Station I nor Station II were TP concentrations statistically correlated with phosphorus concentrations in inflowing riverine waters (r = 0.388 at Station I and 0.075 at Station II), although a distinct relation between TP and precipitation was observed at Station II (r = 0.647, *p* < 0.05), indicating the role of external sources of phosphorus in the river catchment, including strongly eutrophicated lakes situated in the river course.

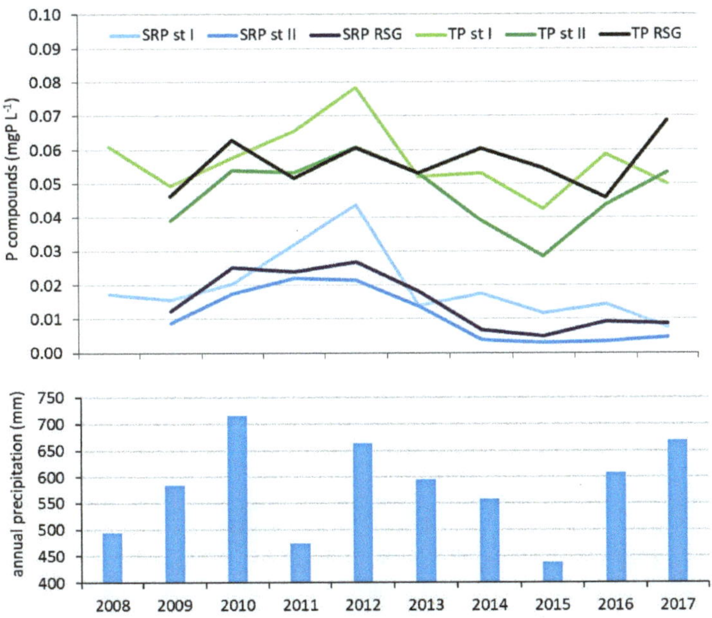

Figure 12. The dependence of SRP and TP concentrations in lake water at both stations on the content in River Struga Gołaniecka (RSG) and annual precipitation.

Canonical RDA analysis indicated that only a few water quality features in the lake were statistically dependent on the water quality of the inflowing River Struga Gołaniecka (Figure 13). Apart from the aforementioned nitrates, these included temperature, conductivity and oxygen concentration at Station II and SRP at Station I. The relationship in the case of temperature resulted in fact not from the impact of riverine water temperature, but rather from the influence of air temperature on both river and lake waters. This correlation was strong, despite the summer thermal stratification, thus values noted in cold months as well as in the epilimnetic waters in summer (similar to river temperature) had a decisive impact. In the case of oxygen content, the lack of any relation between inflowing waters and Station I derived from the deficits noted in the lake hypolimnion, while a much stronger correlation was found for the shallower Station II, hence the impact of anaerobic zone was less pronounced. These dependencies are not direct, as in the case of temperature, but they arise from atmospheric influence (oxygen diffusion into water) on the one hand, and phytoplankton impact (oxygen production in photosynthesis) on the other hand. The latter factor also contributed to the separate oxygen content at Station I as a lower amount of phytoplankton biomass in the southern part of the lake was responsible for less oxygen in the water, in comparison to the Struga Gołaniecka and Station II. Conductivity at Station II was clearly dependent on ion content in the inflowing river due to the short distance between those two points. During the water flow through the lake to Station I some of the ions were assimilated by biota and underwent many physical and chemical processes, thus water conductivity decreased. Long-term fluctuations of this water quality variable resulted from water discharge related to precipitation. Similar to the case of nitrates, the strength of this relationship increased after taking into account the annual shift of rainfall data (r = 0.542 at Station I and r = 0.653 at Station II).

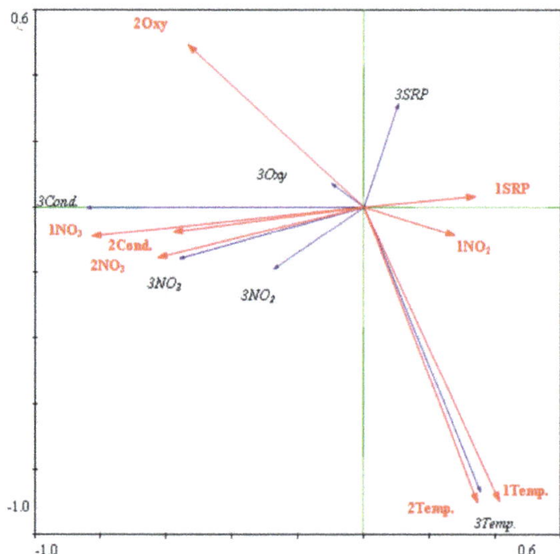

Figure 13. RDA biplot illustrating the influence of the River Struga Gołaniecka (Nos. 3) on mean values of water quality variables in the depth profile of the lake at Station I (Nos. 1) and Station II (Nos. 2).

4.5. Supportive Role of Biomanipulation in Water Quality Improvement

Physicochemical methods of lake restoration were supported in Lake Durowskie by the biological method, i.e., biomanipulation, mainly based on pike fry stocking, aiming at the increase of food pressure on zooplanktivorous and benthivorous fish hatch. As a result of this biomanipulation predation on large zooplankton, fish-induced resuspension and nutrient translocation from sediments to water via feeding and extraction should be lower [22]. Increasing Secchi depth as a consequence of chlorophyll-a reduction indicated that intense changes of fish fauna had occurred as well as a top-down effect as it was expected. This result is strengthened by stocking repeated annually. Changes in phytoplankton structure were noted as early as 2010, as cyanobacteria domination was eliminated in favor of diatoms, dinophytes and chrysophytes. Reconstruction of this community composition favored the larger species, not grazed by crustaceans, whose number increased due to the biomanipulation measures [16]. The effect of biomanipulation could be deepened by hypolimnetic oxygenation, as already noted in stratified lakes where improved oxygen conditions allow some fish fauna (e.g., perch) to forage over a larger sediment surface, thus passing through the benthic phase before becoming piscivores was easier [5].

5. Conclusions

Combining physicochemical and biological methods of lake restoration in a sustainable, cost-effective manner in an urban, dimictic, flow-through lake resulted in water quality improvement. Both nitrogen and phosphorus concentrations decreased over the course of nine years of treatment aimed at bottom-up (nutrient reduction) and top-down (biomanipulation) control of primary production. The changes in water chemistry influenced biota, especially planktonic communities, which was manifested in increased Secchi depth and lower chlorophyll-a concentrations. Diminished fresh organic matter deposition influenced mineral nutrients in deep water layers. Nitrogen transformations were also induced in the hypolimnion by water aeration, decreasing ammonium N and increasing nitrates as the oxygen conditions were slightly improved. At the same time, the duration and range of oxygen depletion zones were reduced. The long-term restoration program, based on non-aggressive, multiple in-lake techniques applied despite the lack of total external loading reduction, allowed the progressive eutrophication to be suppressed. However, it is essential that the program is continued throughout the following years in the light of predicted climate changes as well as potential variability in the amount of future external loading. Together with the in-lake restoration, monitoring actions shall be conducted, involving some remote systems as well, to map the chlorophyll-a and water transparency, allowing more rapid reaction to water quality deterioration by means of phosphorus precipitation.

Author Contributions: Conceptualization, R.D. and R.G.; Methodology and investigation, R.D., K.K.M., B.M., S.C., and R.G.; Writing—original draft preparation, R.D.; Writing—review and editing, R.G.; and Formal analysis, A.K.

Funding: This research received financial support from the Wągrowiec Municipality.

Acknowledgments: The authors would like to thank to Robert Kippen for the proofreading.

Conflicts of Interest: The authors declare no conflict of interest.

References

1. EU Directive 2000/60/EC of the European parliament and of the council of 23 October 2000 establishing a framework for community action in the field of water policy. *Off. J. Eur. Communities* **2000**, *L327*, 1–72.
2. European Environment Agency. *European Waters. Assessment of Status and Pressures 2018*; Report no 7/2018; European Environment Agency: Luxembourg, 2018. [CrossRef]
3. Verdonschot, P.F.M.; Spears, B.M.; Feld, C.K.; Brucet, S.; Keizler-Vlek, H.; Borja, A.; Elliott, M.; Kernan, M.; Johnson, R.K. A comparative review of recovery processes in rivers, lakes, estuarine and coastal waters. *Hydrobiologia* **2013**, *704*, 453–474. [CrossRef]
4. Annadotter, H.; Cronberg, G.; Aagren, R.; Lundstedt, B.; Nilsson, P.-A.; Ströbeck, S. Multiple techniques for lake restoration. *Hydrobiologia* **1999**, *395/396*, 77–85. [CrossRef]
5. Jeppesen, E.; Sammalkorpi, I. Lakes. In *Handbook of Ecological Restoration, Vol. 2, Restoration in Practice*; Perrow, M.R., Davy, A.J., Eds.; Cambridge University Press: Cambridge, UK, 2002; pp. 297–324.
6. Gołdyn, R.; Podsiadłowski, S.; Dondajewska, R.; Kozak, A. The sustainable restoration of lakes—towards the challenges of the Water Framework Directive. *Ecohydrol. Hydrobiol.* **2014**, *14*, 68–74. [CrossRef]
7. Dondajewska, R.; Kozak, A.; Kowalczewska-Madura, K.; Budzyńska, A.; Gołdyn, R.; Podsiadłowski, S.; Tomkowiak, A. The response of a shallow hypertrophic lake to innovative restoration measures—Uzarzewskie lake case study. *Ecol. Eng.* **2018**, *121*, 72–82. [CrossRef]
8. Kowalczewska-Madura, K.; Dondajewska, R.; Gołdyn, R.; Podsiadłowski, S. The influence of restoration measures on phosphorus internal loading from the sediments of a hypereutrophic lake. *Environ. Sci. Pollut. Res.* **2017**, *24*, 14417–14429. [CrossRef]
9. Kowalczewska-Madura, K.; Dondajewska, R.; Gołdyn, R.; Kozak, A.; Messyasz, B. Internal phosphorus loading from the bottom sediments of the dimictic lake during its sustainable restoration. *Water Air Soil Pollut.* **2018**, *229*, 280. [CrossRef]
10. Kowalczewska-Madura, K.; Dondajewska, R.; Gołdyn, R. Internal phosphorus loading in eutrophic lakes in Western Poland. In *The Handbook of Environmental Chemistry. Polish River Basins and Lakes—Part I: Hydrology and Hydrochemistry*; Korzeniewska, E., Harnisz, M., Eds.; Springer: Berlin/Heidelberg, Germany, 2019; in press.
11. Dondajewska, R.; Gołdyn, R.; Kowalczewska-Madura, K.; Kozak, A.; Romanowicz-Brzozowska, W.; Rosińska, J.; Budzyńska, A.; Podsiadłowski, S. Hypertrophic lakes and the results of their restoration in Western Poland. In *The Handbook of Environmental Chemistry. Polish River Basins and Lakes—Part II: Biological Status of Water Management*; Korzeniewska, E., Harnisz, M., Eds.; Springer: Berlin/Heidelberg, Germany, 2019; in press.
12. Jeppesen, E.; Søndergaard, M.; Liu, Z. Lake restoration. In *Routledge Handbook of Ecological and Environmental Restoration*; Allison, S.K., Murphy, S.D., Eds.; Routlege: London, UK; New York, NY, USA, 2017; pp. 226–242.
13. O'Sullivan, P.E.; Reynolds, C.S. *Lakes Handbook: Lake Restoration and Rehabilitation*; Blackwell Publishing: Malden, MA, USA, 2005; Volume 2.
14. Carpenter, S.R.; Cottingham, K.L. Resilience and restoration of lakes. *Ecol. Soc.* **1997**, *1*, 2. [CrossRef]
15. Gunderson, L.H. Ecological resilience—In theory and application. *Annu. Rev. Ecol. Syst.* **2000**, *31*, 425–439. [CrossRef]
16. Gołdyn, R.; Messyasz, B.; Domek, P.; Windhorst, W.; Hugenschmidt, C.; Nicoara, M.; Plavan, G. The response of Lake Durowskie ecosystem to restoration measures. *Carpath. J. Earth Environ.* **2013**, *8*, 43–48.

17. Podsiadłowski, S.; Osuch, E.; Przybył, J.; Osuch, A.; Buchwald, T. Pulverizing aerator in the process of lake restoration. *Ecol. Eng.* **2018**, *121*, 99–103. [CrossRef]

18. Elbanowska, H.; Zerbe, J.; Siepak, J. *Physico-Chemical Water Analyses*; AMU Press: Poznań, Poland, 1999. (In Polish)

19. Wetzel, R.G.; Likens, G.E. *Limnological Analyses*, 2nd ed.; Springer-Verlag Inc.: New York, NY, USA, 1991.

20. McQueen, D.J.; Post, J.R.; Mills, E.L. Trophic relationships in freshwater pelagic ecosystems. *Can. J. Fish. Aquat. Sci.* **1986**, *43*, 1571–1581. [CrossRef]

21. Carpenter, S.R.; Kitchell, J.F.; Hodgson, J.R. Cascading trophic interactions and lake productivity. *Bioscience* **1985**, *35*, 634–639. [CrossRef]

22. Taylor, J.M.; Vanni, M.J.; Flecker, A.S. Top-down and bottom-up interactions in freshwater ecosystems: Emerging complexities. In *Trophic Ecology: Bottom-Up and Top-Down Interactions Across Aquatic and Terrestrial Systems*; Hanley, T.C., La Pierre, K.J., Eds.; Cambridge University Press: Cambridge, UK, 2015; pp. 55–85. [CrossRef]

23. Jeppesen, E.; Søndergaard, M.; Lauridsen, T.L.; Davidson, T.A.; Liu, Z.; Mazzeo, N.; Trochine, C.; Özkan, K.; Jensen, H.S.; Trolle, D.; et al. Biomanipulation as a restoration tool to combat eutrophication: Recent advances and future challenges. *Adv. Ecol. Res.* **2012**, *47*, 411–488. [CrossRef]

24. Søndergaard, M.; Jensen, J.P.; Jeppesen, E. Role of sediments and internal loading of phosphorus in shallow lakes. *Hydrobiologia* **2003**, *506–509*, 135–145. [CrossRef]

25. Beutel, M.W. Inhibition of ammonia release from anoxic profundal sediments in lakes using hypolimnetic oxygenation. *Ecol. Eng.* **2006**, *28*, 271–279. [CrossRef]

26. Singleton, V.L.; Little, J.C. Designing hypolimnetic aeration and oxygenation systems—A review. *Environ. Sci. Technol.* **2006**, *40*, 7512–7520. [CrossRef]

27. Ekau, W.; Auel, H.; Pörtner, H.-O.; Gilbert, D. Impacts of hypoxia on the structure and processes in pelagic communities (zooplankton, macro-invertebrates and fish). *Biogeosciences* **2010**, *7*, 1669–1699. [CrossRef]

28. Gächter, R.; Müller, B. Why the phosphorus retention of lakes does not necessarily depend on the oxygen supply to their sediment surface. *Limnol. Oceanogr.* **2003**, *48*, 929–933. [CrossRef]

29. Liboriussen, L.; Søndergaard, M.; Jeppesen, E.; Thorsgaard, I.; Grünfeld, S.; Jakobsen, T.S.; Hansen, K. Effects of hypolimnetic oxygenation on water quality: Results from five Danish lakes. *Hydrobiologia* **2009**, *625*, 157–172. [CrossRef]

30. Siwek, H.; Włodarczyk, M.; Czerniawski, R. Trophic state and oxygen conditions of waters aerated with pulverising aerator: The results from seven lakes in Poland. *Water* **2019**, *10*, 219. [CrossRef]

31. Holmroos, H.; Horppila, J.; Laakso, S.; Niemistö, J.; Hietanen, S. Aeration-induced changes in temperature and nitrogen dynamics in a dimictic lake. *J. Environ. Qual.* **2016**, *45*, 1359–1366. [CrossRef]

32. Grochowska, J.; Gawrońska, H. Restoration effectiveness of a degraded lake using multi-annual artificial aeration. *Pol. J. Environ. Stud.* **2004**, *13*, 671–691.

33. Rosińska, J.; Kozak, A.; Dondajewska, R.; Gołdyn, R. Cyanobacterial blooms before and during the restoration process of a shallow urban lake. *J. Environ. Manag.* **2017**, *198*, 340–347. [CrossRef] [PubMed]

34. Rosińska, J.; Kozak, A.; Dondajewska, R.; Kowalczewska-Madura, K.; Gołdyn, R. Water quality response to sustainable restoration measures—Case study of urban Swarzędzkie Lake. *Ecol. Indic.* **2018**, *84*, 437–449. [CrossRef]

35. Burgin, A.J.; Hamilton, S.K. Have we overemphasized the role of denitrification in aquatic ecosystems? A review of nitrate removal pathways? *Front. Ecol. Environ.* **2007**, *5*, 89–96. [CrossRef]

36. Nizzoli, D.; Carraro, E.; Nigro, V.; Viaroli, P. Effect of organic enrichment and thermal regime on denitrification and dissimilatory nitrate reduction to ammonium (DNRA) in hypolimnetic sediments of two lowland lakes. *Water Res.* **2010**, *44*, 2715–2724. [CrossRef] [PubMed]

37. Op den Camp, H.J.M.; Jetten, M.S.M.; Strous, M. Annamox. In *Biology of the Nitrogen Cycle*; Bothe, H., Ferguson, S.J., Newton, W.E., Eds.; Elsevier: Amsterdam, The Netherlands, 2007; pp. 245–262.

38. Prepas, E.E.; Burke, J.M. Effects of hypolimnetic oxygenation on water quality in Amiska Lake, Alberta, a deep stratified lake with high internal phosphorus loading rates. *Can. J. Fish. Aquat. Sci.* **1997**, *54*, 2111–2120. [CrossRef]

39. Søndergaard, M.; Wolter, K.D.; Ripl, E. Chemical treatment water and sediments with special references to lakes. In *Handbook of Ecological Restoration. Vol. 1. Principles of Restoration*; Perrow, M.R., Davy, A.J., Eds.; Cambridge University Press: Cambridge, UK, 2002.

40. Cooke, G.D.; Welch, E.B.; Martin, A.B.; Fulmer, D.G.; Hyde, J.B.; Schrieve, G.D. Effectiveness of Al, Ca and Fe salts to control of internal loading in shallow and deep lakes. *Hydrobiologia* **1993**, *253*, 323–335. [CrossRef]

41. Boers, P.; Van der Does, J.; Quaak, M.; Van der Vlugt, J. Phosphorus fixation with iron (III) chloride: A new method to combat internal phosphorus loading in shallow lakes? *Arch. Hydrobiol.* **1994**, *129*, 339–351.

42. Jaeger, D. Effects of hypolimnetic water aeration and iron-phosphate precipitation on the trophic level of Lake Krupunder. *Hydrobiologia* **1994**, *275–276*, 433–444. [CrossRef]

43. Deppe, T.; Benndorf, J. Phosphorus reduction in shallow hyper-eutrophic reservoir by in-lake dosage of ferrous iron. *Water Res.* **2002**, *36*, 4525–4534. [CrossRef]

44. Hupfer, M.; Lewandowski, J. Oxygen controls the phosphorus release from lake sediments—A long-lasting paradigm in limnology. *Int. Rev. Hydrobiol.* **2008**, *93*, 415–432. [CrossRef]

45. Nygrén, N.A.; Tapio, P.; Horppila, J. Will the oxygen-phosphorus paradigm persist?—Expert views of the future of management and restoration of eutrophic lakes. *Environ. Manag.* **2007**, *60*, 947–960. [CrossRef] [PubMed]

46. Pearl, H.W.; Hall, N.S.; Calandrino, E.S. Controlling harmful cyanobacterial blooms in world experiencing anthropogenic and climate-induced change. *Sci. Total Environ.* **2011**, *409*, 1739–1745. [CrossRef] [PubMed]

47. Napiórkowska-Krzebietke, A.; Dunalska, J.A.; Zębek, E. Taxa-specific eco-sensitivity in relation to phytoplankton bloom stability and ecologically relevant lake state. *Acta Oecol.* **2017**, *81*, 10–21. [CrossRef]

48. Mantzouki, E.; Lürling, M.; Fastner, J.; de Senerpont Domis, L.; Wilk-Woźniak, E.; Koreivienė, J.; Seelen, L.; Teurlincx, S.; Verstijnen, Y.; Krztoń, W.; et al. Temperature effects explain continental scale distribution of cyanobacterial toxins. *Toxins* **2018**, *10*, 156. [CrossRef] [PubMed]

49. Tammeorg, O.; Möls, T.; Niemistö, J.; Holmroos, H.; Horppila, J. The actual role of oxygen deficit in the linkage of the water quality and benthic phosphorus release: Potential implications for lake restoration. *Sci. Total Environ.* **2017**, *599–600*, 732–738. [CrossRef]

50. Saunders, D.L.; Kalff, J. Nitrogen retention in wetlands, lakes and rivers. *Hydrobiologia* **2001**, *443*, 205–212. [CrossRef]

51. Pűtz, K.; Benndorf, J. The importance of pre-reservoirs for the control of eutrophication of reservoirs. *Water Sci. Technol.* **1998**, *37*, 317–324. [CrossRef]

52. Straškraba, M.; Tundisi, J.G.; Duncan, A. State-of-the-art of reservoir limnology and water quality management. *Comp. Reserv. Limnol. Water Qual. Manag.* **1993**, *77*, 213–288.

 *water*

Article

Functional Groups of Phytoplankton and Their Relationship with Environmental Factors in the Restored Uzarzewskie Lake

Anna Kozak *, Agnieszka Budzyńska, Renata Dondajewska-Pielka, Katarzyna Kowalczewska-Madura and Ryszard Gołdyn

Adam Mickiewicz University, Faculty of Biology, Department of Water Protection, Uniwersytetu Poznańskiego 6, 61-614 Poznań, Poland; ag.budzynska@gmail.com (A.B.); gawronek@amu.edu.pl (R.D.-P.); madura@amu.edu.pl (K.K.-M.); rgold@amu.edu.pl (R.G.)
* Correspondence: akozak@amu.edu.pl

Received: 2 December 2019; Accepted: 20 January 2020; Published: 21 January 2020

Abstract: Uzarzewskie Lake is a small, postglacial lake, located in western Poland. The lake is under restoration treatment since 2006. At first, iron treatment was done for 2 years. In the second stage, spring water was directed into the hypolimnion in order to improve water oxygenation near the bottom sediments. The purpose of our research was to determine changes in the contribution of functional groups to the total number of taxa and total biomass of phytoplankton due to changes in the physical and chemical characteristics of the restored lake. Phytoplankton composition was analyzed in three periods: (1) before restoration; (2) during the first method of restoration; and (3) when the second method was implemented in the lake. Epilimnetic phytoplankton was sampled every year monthly from March to November. The relationship between phytoplankton groups and environmental factors (water temperature, ammonium nitrogen, nitrate nitrogen, dissolved phosphorus, conductivity and pH) was examined, using the canonical analyses. The redundancy analysis indicated that the temperature, dissolved phosphates concentration, ammonium nitrogen and pH were the main determining factors of the phytoplankton community dynamics. During the study, 13 coda dominated the phytoplankton biomass. Cyanobacteria of the codon **H1** with such species as *Aphanizomenon gracile, Dolichospermum planctonicum, D. viguieri* dominated the phytoplankton community before restoration. **S1** group consisting of *Planktolyngbya limnetica, Limnothrix redekei* and *Planktothrix agardhii* mostly dominated during the period in which the first method was used. Improvement of water quality due to restoration efforts in the third period caused dominance of other groups, especially **J** (*Actinastrum hantzschii* and other Chlorococcales), **C** (*Asterionella formosa* and other diatoms), **Y** (*Cryptomonas marssonii* and other cryptophytes), **Lo** (*Peridiniopsis cunningtonii* and other dinophytes) and **X2** (*Rhodomonas lacustris*).

Keywords: functional groups of phytoplankton; innovative method of restoration; oxygenation near the bottom sediments; phytoplankton biomass; restoration treatment; small lake; sustainable restoration

---

1. Introduction

Due to the increasing eutrophication of waters, intensive algae blooms in lakes are noted worldwide, threatening freshwater biodiversity and supplying humans with water [1,2]. The restoration and preservation of aquatic ecosystems is a crucial issue in our society. To prevent the water blooms, attempts are being made to limit the anthropogenic deterioration of water quality. Functional groups were originally proposed to assess the ecological status of different types of lakes proposed by the European community [3]. If these protective measures do not bring the desired results within the required timespan, (in the European Union dictated by the requirements of the Directive 2000/60/EC of the European Parliament, WFD, 2000), restoration methods are being implemented to help improve water quality.

Despite lake restoration becoming more and more frequent, still little is known about the reaction of cyanobacteria blooms and the whole phytoplankton communities to restoration actions. Restoration activities do not only affect the long-term reduction in cyanobacteria abundance but contribute to phytoplankton adaptation to the new environmental conditions of a water body [4,5]. In this study, we focused on the changes in the phytoplankton community in a small lake restored first by phosphorus precipitation, and then by oxygenation of hypolimnetic waters by redirecting the inflows from two springs.

To analyze the impact of restoration on phytoplankton, the concept of functional groups [6] seems a suitable approach. Functional groups are used in several countries of the European Union to assess water quality and apply both to lakes and reservoirs, as well as rivers [7]. In each functional group, species with similar ecological characteristics are gathered [8], and thus their reactions to ecosystem changes caused by restoration activities should follow the same pattern. The use of this concept allows focusing on phytoplankton groups instead of individual species, which can be more useful from the application point of view. Thus, the purpose of our research was to determine changes in the phytoplankton functional groups due to changes in the physical and chemical characteristics of Uzarzewskie Lake during the restoration.

2. Materials and Methods

2.1. Uzarzewskie Lake and Its Restoration

Uzarzewskie Lake (52°26′53″ N, 17°08′00″ E) is a water body in the shape of a kettle, located on the River Cybina (an eastern tributary of the River Warta in western Poland, Wielkopolska Region). It is quite a small lake, covering 10.6 ha of area. The maximum depth of the lake is 7.3 m and the average depth is 3.4 m. It was assessed as a hypereutrophic lake [9]. This lake is dimictic with short mixing time. The bottom sediments of Uzarzewskie Lake provided a high internal load of nutrients [10].

The total catchment area of Uzarzewskie Lake is 160.8 km^2 and 79% of the direct catchment area is occupied by arable land [11]. The waters of Cybina River also supply the lake with nutrients, mainly from farmland and fish ponds. The excessive external load has resulted in very low water quality [12]. Very often cyanobacterial water blooms were noted, which excluded the lake from recreational use [13,14]. For this reason, in 2006, restoration treatments have been started. In the years 2006–2007, the method consisted of precise inactivation of phosphorus with iron sulfate (PIX-112) [10]. The coagulant was dosed 6 times in 2006 and 3 times in 2007 in quantities of 380 kg and 180 kg annually, respectively. These treatments did not bring the expected effect in the form of a clear reduction in the number of phytoplankton, especially cyanobacteria [15]. In 2008, the use of an innovative method of restoration was begun, which consisted of delivering water from two watercourses flowing from springs at the slope of the river valley directly to the hypolimnion, using two pipes (Figure 1). Water flowed to the hypolimnion by gravity and constantly, with a capacity of about 10 L per second. It was rich in nitrates and well oxygenated. The pipelines delivering this water were placed directly on the bottom of the lake and the water outflows were situated near the deepest place of the lake. Water spreading around this place caused improvement in the oxygenation of hypolimnetic water and increase in the redox potential over the bottom of Uzarzewskie Lake [16]. This procedure continues through the present.

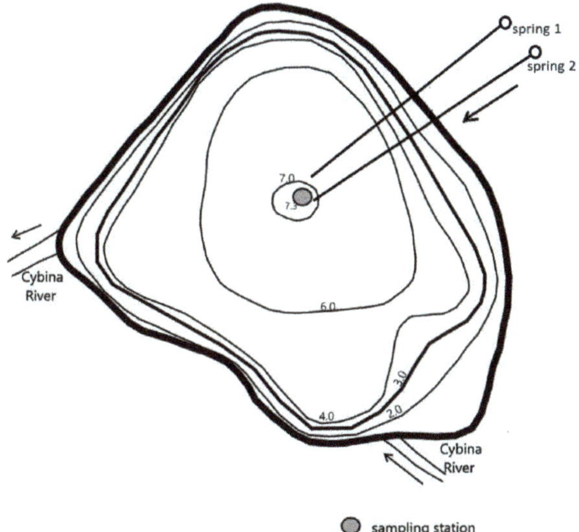

Figure 1. Uzarzewskie Lake and the location of the sampling station with the view of the system supplying hypolimnion with spring waters (according to Kowalczewska-Madura et al. [10], changed).

2.2. Sampling and Analysis

The study took place from 2005 to 2015, in three periods: (1) in the year 2005—before the restoration, (2) within 2006–2007—during the chemical treatment and (3) within 2008–2015—during the oxygenation of the hypolimnetic water layer. Phytoplankton was sampled usually once a month, except for some winter months, when the lack of a thick ice cover prevented sampling. Each time sample were collected at one station situated in the central part of the lake from the surface, 1 m and 2 m depths and fixed in situ with Lugol solution. The abundance of phytoplankton was calculated using the light microscope method [17]. Phytoplankton biomass was estimated using specific biovolumes obtained by geometrical approximations based on Hillebrandt et al. [18]. Phytoplankton functional group classification according to Reynolds et al. [6] and Padisák et al. [8] was used.

Together with phytoplankton sampling, the samples for chemical analyses were collected and Secchi depth, water temperature, electrolytic conductivity, pH, soluble oxygen and water saturation with oxygen were measured in situ using Secchi disc and YSI-meter. Results of the analysis of the physical and chemical variables of water quality were already published [11] and they were used here for statistical analyses as the restoration-dependent (PO_4-P, NO_2-N, NO_3-N and NH_4-N) and restoration-independent drivers of the phytoplankton community.

In order to determine changes in the composition of phytoplankton under the influence of restoration activities, functional groups with the highest share in the overall biomass of phytoplankton exceeding 25% were selected. The composition of functional groups among the three study periods mentioned above was compared.

The relation between the physico-chemical parameters of the water and functional groups of phytoplankton during three seasons (before restoration and when using the first and second methods) was analyzed by redundancy analysis (RDA). The relation between the most important functional groups of phytoplankton during three periods was analyzed by discriminant analysis (canonical variate analysis -CVA). The relative importance and statistical significance of each environmental factor in the ordination model were tested by a forward selection procedure and Monte Carlo permutation test. All ordination methods were applied with the use of the Canoco 4.5 package [19].

3. Results

3.1. The Number of Species in Functional Groups in the Three Periods of Study

The total number of phytoplankton species exceeded 300, then classified into 28 different functional groups. Not all coda were found in the three analyzed periods, and so in the pre-restoration period, representatives from 22 coda were found. During the restoration with the first and second method of restoration 25 and 27 coda were noted, respectively (Figure 2, Table 1). The most frequent coda were **J** (20%–22% in subsequent periods of research), **C** and **Y** (6%–8%, with the least during the period of using second method), **H1** (6%–7%), **X2** (6%–10%), **X1** (5%–7%), **F** (4%–6%), **S1** and **Lo** (4%–5%), and also **D** (4% each). Other functional groups were represented by fewer taxa. Representatives of several coda were found only in one research period, e.g., codon **K** was found only during the application of the first method. During the second method of restoration, representatives of coda **A**, **T** and **U** were found. They were not recorded in earlier periods of research.

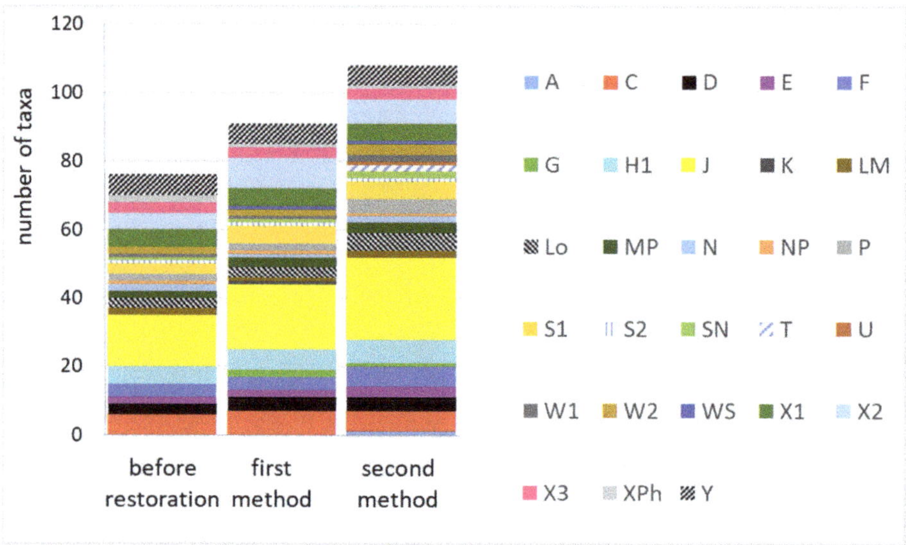

Figure 2. The contribution of the number of taxa from individual functional groups of phytoplankton to the total number of taxa in three study periods.

The number of functional groups, taking into account individual samples of the three research periods, was the highest during the period of restoration using the first method (14.7 on average). In the pre-restoration period, the average number of groups in subsequent months was 13.8 and in the period the second restoration method was 13.4. A similar trend was also observed for the number of phytoplankton taxa—the average number of taxa in individual samples was the highest during the period of using the first restoration method.

Table 1. Main representatives of the functional groups of phytoplankton noted in Uzarzewskie Lake.

Functional Group	The Most Abundant Representatives
A	*Acanthoceras zachariasii, Rhizosolenia longiseta*
C	*Asterionella formosa, Cyclotella meneghiniana, Stephanodiscus hantzschii*
D	*Nitzschia acicularis, Ulnaria acus*
E	*Dinobryon* sp.*, Mallomonas* sp.
F	*Elakatothrix gelatinosa, Keratococcus* sp.*, Oocystis* spp.
G	*Eudorina* sp.*, Pandorina morum*
H1	*Aphanizomenon gracile, Dolichospermum planctonicum, D. viguieri*
J	*Actinastrum hantzschii, Coelastrum astroideum, Crucigenia tetrapedia, Pediastrum* sp.*, Scenedesmus* sp.*, Tetraëdron* sp.*, Tetrastrum* sp.,
K	*Aphanocapsa* sp.
LM	*Microcystis aeruginosa, M. wessenbergii, Ceratium hirundinella*
Lo	*Peridiniopsis cunningtonii, Peridinium aciculiferum*
MP	*Achnanthes* sp.*, Amphora ovalis, Nitzchia palea, Gomphonema* sp.
N	*Cosmarium bioculatum, Staurastrum* sp.
NP	*Ulnaria ulna*
P	*Closterium aciculare, Closterium acutum, Fragilaria crotonensis*
S1	*Planktolyngbya limnetica, Limnothrix redekei, Planktothrix agardhii*
S2	*Raphidiopsis raciborskii*
SN	*Dolichospermum compactum, Raphidiopsis* sp.
T	*Binuclearia lauterbornii, Mougeotia* sp.
U	*Uroglena* sp.
W1	*Phacus longicauda, Euglena* sp. *,Lepocinclis acus*
W2	*Trachelomonas hispida*
WS	*Synura uvella*
X1	*Ankistrodesmus gracilis, Monoraphidium irregulare, M. contortum, M. minutum, Schroederia setigera*
X2	*Rhodomonas lacustris, Bicoeca planktonica, Ochromonas* sp.
X3	*Chrysochromulina parva, Chrysococcus* sp.
XPh	*Phacotus lenticularis*
Y	*Cryptomonas marssonii, C. ovata, C. reflexa*

3.2. Changes in the Biomass of Functional Groups in the Three Periods of Study

The phytoplankton biomass throught this study varied between 0.68 and 136.5 µg/mL and was mainly dominated by species belonging to coda **H1** (*Aphanizomenon gracile* (Lemm.) Lemm.*, A. flos-aquae* Ralfs ex Bornet and Flahault*, A. klebahnii* (Elenk.) Pechar et Kalina)*, Dolichospermum* (= *Anabaena*) *viguieri* (Denis and Frémy) Wacklin, L.Hoffmann and Komárek, *Cuspidothrix issatschenkoi* (Usachev) P.Rajaniemi, Komárek, R.Willame, P. Hrouzek, K.Kastovská, L.Hoffmann and K. Sivonen, *Dolichospermum planctonicum* (Brunnthaler) Wacklin, L.Hoffmann and Komárek*, D. spiroides* (Klebhan) Wacklin, L.Hoffmann and Komárek), **S1** (shade-adapted cyanobacteria: *Pseudanabaena limnetica* (Lemm.) Komárek, *Planktolyngbya limnetica* (Lemm.) Komárková-Legnerová and Cronberg, *Limnothrix redekei* (Goor) Meffert, *Planktothrix agardhii* (Gomont) Anagnostidis and Komárek), **Y** (*Cryptomonas* spp.), **C** (*Asterionella formosa* Hassall), **D** (*Ulnaria acus* (Kützing) Aboal and *Nitzschia* spp.), **X2** (*Chrysochromulina parva* Lackey) and codon **J** (*Actinastrum hantzschii* Lagerheim, *Crucigenia tetrapedia* (Kirchner) W. et G. S. West, *Stauridium tetras* (Ehrenberg) E. Hegewald in Buchheim et al. (Syn. *Pediastrum tetras*), *Coelastrum astroideum* De Notaris*, C. microporum* Nägeli in A. Braun) (Figure 3).

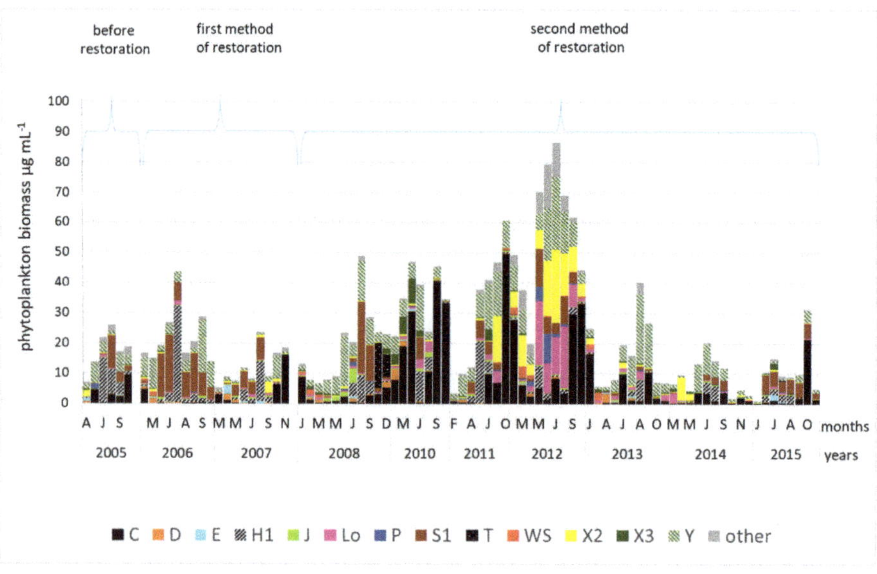

Figure 3. Distribution of phytoplankton functional groups in the three study periods (mean values from the surface layer up to 2 m).

In subsequent years of research, a different share of the biomass of particular functional groups in the total phytoplankton biomass was found. Cyanobacteria (coda **H1** and **S1**) dominated in the total biomass of phytoplankton in the first 3 years of our study (before restoration and during the first method of restoration). During water mixing period codon **Y** (*Cryptomonas erosa* Ehrenberg, *C. marssonii* Skuja, *C. ovata* Ehrenberg, *C. gracilis* Skuja, *C. reflexa* Skuja, *C. rostratiformis* Skuja) and codon **C** (*Asterionella formosa* and small centric diatoms, e.g., *Stephanodiscus hantzschii* Grunow) were the most abundant.

Coda **J** (green algae mentioned above) and **D** (*Fragilaria* spp., *Ulnaria ulna* (Nitzsch) Compère, *U. acus*) mostly dominated in spring or autumn. Such genera as *Cyclotella, Stephanodiscus, Ulnaria* and *Nitzschia* (from **D** group), especially dominated in 2010 while codon **X2** (*Rhodomonas lacustris* Pascher and Ruttner in Pascher, *Ochromonas* sp., *Rhodomonas lens* Pascher and Ruttner) especially dominated in the years 2011 and 2012 (Figure 3). Codon **P** (*Fragilaria crotonensis* Kitton and *Aulacoseira granulata* (Ehrenberg) Simonsen) dominated especially in spring 2005 and 2012.

3.3. The Relations of the Phytoplankton to Environmental Factors in Three Periods of Study

As shown by RDA analyses (Table 2, Figure 4), all the analyzed physical and chemical factors, except for NO_3-N, were significant for the distribution of phytoplankton coda. Water temperature was interrelated with the coda **S1, H1, Lo, J** and **P** (Figure 4), while pH had a significant impact on the distribution of taxa representing codon **T**.

Table 2. Results of the Monte Carlo permutation test (values of p and F are calculated using a test with 999 permutations). The overall percentage of explained variance was 22.98%. Bold = variables significantly adding to the model at $p \leq 0.05$ level. Explanations: EC—conductivity, NO_2—nitrite nitrogen, NO_3—nitrate nitrogen, NH_4—ammonium nitrogen, PO_4—orthophosphate phosphorus, Temp.—temperature.

Variable	Var.N	LambdaA	p	F
Temp.	6	0.11	0.002	29.98
EC	8	0.05	0.002	14.56
NO$_2$	12	0.02	0.002	5.39
PO$_4$	17	0.02	0.002	5.16
pH	7	0.01	0.032	2.75
NH$_4$	11	0.01	0.040	2.27
NO$_3$	13	0.01	0.056	2.09

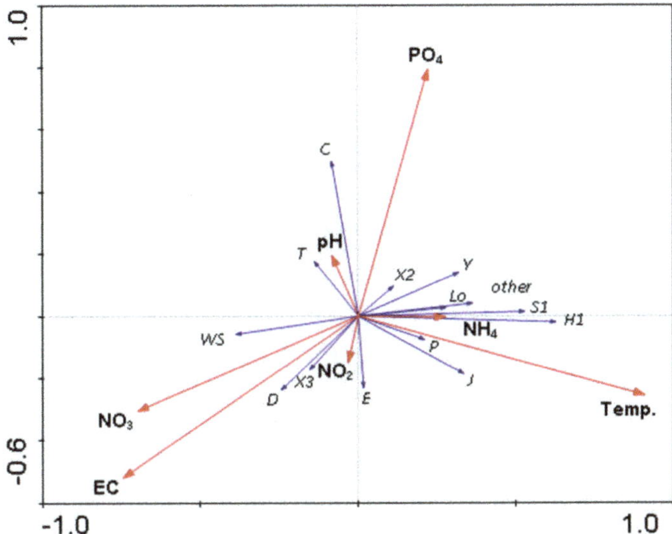

Figure 4. Results of redundancy analysis (RDA) illustrating the functional groups of phytoplankton distribution and environmental variables (physico-chemical parameters) in the Uzarzewskie Lake. Explanations: NO_2—nitrite nitrogen, NO_3—nitrate nitrogen, NH_4—ammonium nitrogen, PO_4—orthophosphate phosphorus, EC—conductivity, Temp.—temperature, phytoplankton dominating coda: **C, D, E, H1, J, Lo, S1, WS, X1, X3, Y,** other—the rest of phytoplankton coda.

The abundance of codon **C** was positively affected by PO_4-P concentration and pH, while NH_4-N positively affected coda **X2, Y** and **Lo**. Water conductivity along with NO_3-N positively affected the abundance of coda **WS, D** and **X3**, however, the correlation was not significant.

Analyzing the individual three periods of research (before, during the first and second restoration methods), significant decreases in the contribution of the **H1** group were found (Figure 5, Table 3). It was the highest before restoration and in subsequent years the share of this group decreased. Before restoration, a much lower share of **J** and **WS** groups was found. These groups increased their share in 2008 and 2013. In the period of using the first restoration method, especially codon **S1** was noted in a high percentage.

During the second method of restoration, the share of these groups was lower, while the share of taxa representing other groups increased, especially coda **J, Y** and **C**. Codon **C** was equally represented throughout the entire research period, a slight decrease was found during the use of chemical treatment (iron sulfate). Coda **S1** and **H1** decreased their share during the application of the second method (water from springs delivered to hypolimnion). Some groups were only noted in some years e.g., **T** (*Mougeotia* sp.) in 2008, or codon **Lo** (*Peridinium* sp.) in the year of 2012.

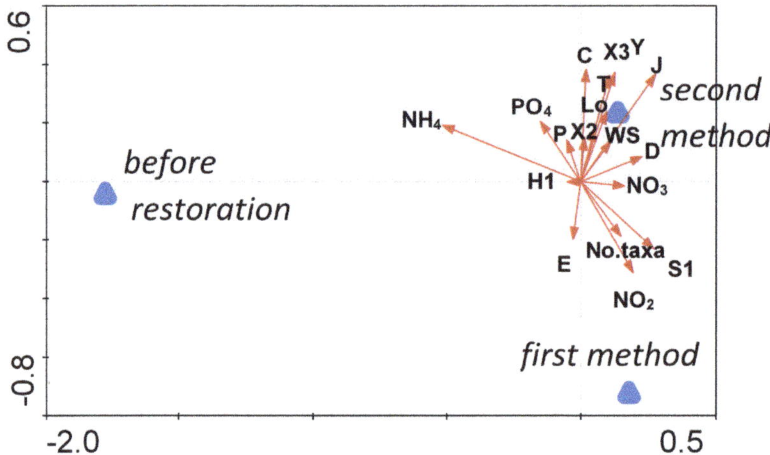

Figure 5. Canonical variate analysis (CVA), diagrams with axes 1 and 2 for the most important functional groups of phytoplankton and nutrients (ammonium nitrogen, nitrites and nitrates and dissolved phosphates) in three periods of investigations: before restoration and during the first and second method.

Table 3. Results of the Monte Carlo permutation test (values of p and F are calculated using a test with 999 permutations). The overall percentage of explained variance was 25.45%. Bold = variables significantly adding to the model at $p \leq 0.05$ level. Explanations: as in Figure 4, No. taxa—the number of taxa in the sample.

Variable	Var.N	LambdaA	p	F
NH$_4$	11	0.08	0.002	14.49
J	58	0.05	0.004	10.11
S1	66	0.05	0.004	8.69
C	52	0.02	0.020	4.75
NO$_3$	13	0.02	0.030	4.19
Y	78	0.02	0.030	4.50
No. taxa	83	0.02	0.030	4.10
NO$_2$	12	0.02	0.050	3.51
T	69	0.01	0.090	3.25
Lo	61	0.01	0.120	2.21
X3	76	0.01	0.206	1.57
E	54	0.01	0.108	2.10
WS	73	0.01	0.292	1.14
D	53	0.00	0.290	1.14
P	65	0.01	0.390	0.88
PO$_4$	17	0.00	0.560	0.50
H1	57	0.00	0.684	0.31
X2	75	0.00	0.902	0.08

4. Discussion

Restoration activities have a significant impact on the composition and abundance of phytoplankton species [20]. Cyanobacterial blooms have been found in Uzarzewskie Lake for years, which was largely due to unregulated water and sewage management in the catchment area [11]. The previous study showed that this lake was hypereutrophic [4].

In this study, the impact of restoration methods identified as sustainable on the development of phytoplankton was analyzed. In contrast to drastic methods (e.g., the use of high doses of chemical substances inactivating phosphorus), the methods analyzed in this paper only initiate and support naturally occurring processes of water quality improvement [16].

Different taxa numbers were found during the research periods, indicating environmental instability [21]. The highest taxonomic diversity was found during the period of application of the first method of restoration (Figure 5), which can be explained with the theory of intermediate disturbances [22,23]. The lowest biodiversity appears at the lowest level of disturbances because only the best competitors can persist under these conditions [24]. The lake is also diverse in terms of the number of functional phytoplankton groups. The species noted in the lake were classified into 28 functional groups. This is quite a significant number. In other water bodies, the number of coda was smaller, e.g., 12 functional groups in Lake Mogan [25], 16 in Liman Lake [26], 18 in a dam lake in Turkey [27] and 23 coda along the River Loire in France [7].

Phytoplankton biomass has undergone dynamic changes. The highest values of phytoplankton biomass were found during the restoration period with the second method, which aimed to oxygenate hypolimnetic waters and thus prevented the internal loading of nutrients. This stimulated the growth of large, long-living organisms, e.g., from coda **C** and **Lo**, not sensitive to periodic nutrient deficits [28]. Their growth was particularly intense in the years with high precipitation, e.g., in 2010 and 2012, which resulted in higher phytoplankton biomass. This was probably related to a greater supply of nutrients from the catchment. Periodic problems with the clogging of pipes supplying water from springs to hypolimnion may also have had some impact [11]. Also, the codon **X2** had high biomass in 2012, showing a clear correlation with PO_4-P concentration. Their numerous presence together with species from codon **Y** testifies to the high variability of the environmental conditions [28]. Coda **Y** (*Cryptomonas* spp.) and **X2** (mainly *Rhodomonas* spp.), correlated with PO_4-P concentration are rapidly developing pioneer algae groups. These highly flexible species are able to live in almost all lentic ecosystems when grazing pressure is low [8]. Their life strategy allows them to dominate the environment after the population of other algae collapse and they play a connecting role in the succession of summer phytoplankton. Due to their *r*-life strategy [29] and mixotrophic nutrition they are able to grow well or even outcompete cyanobacteria during colder seasons of the year, i.e., autumn, winter or spring [30].

The biomass of codon **C** (*Asterionella formosa* and centric diatoms) correlated not only with PO_4-P but also with pH. Centric diatoms belonging to this codon dominated mainly in the first and second periods owing to their low-light tolerance and fast growth [31,32].

Coda **H1** and **S1** representing cyanobacteria were closely correlated with the concentration of ammonium nitrogen in water (Figure 4). This is due to the fact that cyanobacteria representing these two coda prefer this form of nitrogen [33–40]. As the concentration of ammonium nitrogen was lower during the second method of restoration due to nitrification and denitrification processes in the hypolimnion [11], there was also a decrease in biomass of coda **H1** and **S1**.

The change of cyanobacteria belonging to codon **H1**, dominating before restoration, to **S1** during the period of the first restoration method and the numerous presence of green algae from codon **J** during the second method, is also associated with a decrease in dissolved phosphates concentrations as a result of restoration. According to Crossetti et al. [41] coda **S1** and **J** occurred in a shallow lake in Brazil under a lowered concentration of soluble phosphorus in the water. An additional reason for the intensive growth of phytoplankton, especially from the coda **H1** and **S1** (cyanobacteria), were very favorable temperature conditions in summer [42–44]. It is worth mentioning that the increasing trends of temperature in the last decades favor cyanobacterial blooms [45–48] and reconstruction of species composition towards an increase of the number of invasive taxa [49–51]. This was the cause of higher cyanobacteria biomass in 2015 than in other years of the second method application, as summer 2015 was extremely hot in Poland. Temperature is a factor that can very clearly affect the successful proliferation of most groups of phytoplankton [52], increasing their growth rates [53–55]. This is demonstrated by temperature correlations also with such coda as: **P** (*Fragilaria crotonensis* and *Aulacoseira granulata*) and **J** (*Actinastrum, Scenedesmus* and *Coelastrum*).

Coda **X3** (*Chrysococcus* and *Koliella*) and **D** (*Fragilaria* spp., *Ulnaria ulna, U. acus*), mostly found at a lower temperature, were correlated with EC and NO_2-N. Unlike cyanobacteria, these groups prefer other nitrogen forms such as nitrates and nitrites [20,56]. High biomass of codon **T** (*Mougeotia* sp.) in Uzarzewskie Lake was found especially in autumn of 2008 in the entire water column and was correlated with pH. This taxon may dominate in reservoirs from spring to early autumn [57,58]. There have also been cases when it caused intense blooms of water, associated with higher average annual temperatures, such as in Lake Geneva. *Mougeotia* sp. has lamellar chloroplasts that can move to present the largest surface area in relation to the light rays and this allows it to survive in low-light conditions [59].

Codon **Lo** (*Peridinium* sp.) was quite abundant in the year of 2012. This group prefers high concentration of dissolved oxygen in water, as the behavioral and physiological adaptation force dinoflagellates to remain in surface water, rich in oxygen [40]. The second restoration method helped to maintain aerobic conditions that can affect the development of this group of algae. It is worth noting that the average values of oxygen concentration in surface water from the entire research period were the highest in 2012 [11], in which organisms from codon **Lo** were the most abundant.

The biomass of phytoplankton and share of the functional groups in the period of the second method of restoration of Uzarzewskie Lake was very unstable. In the beginning, during the period 2008–2012, a higher biomass of phytoplankton could be observed. This trend was similar to the average values of both soluble reactive and total phosphorus in the water column [11] and was connected with quite high concentrations of phosphorus above the bottom sediments and its internal loading [60]. This increased internal loading was most likely associated with faster mineralization of organic matter as a result of oxygen supply to the bottom area. A similar increase in the first years of restoration was also found in Lake Durowskie [61]. The functional groups such as **C, T, Y** and **X2** were correlated with high phosphate concentration during those periods. That was also confirmed by the results of statistical analyses.

Summing up, the dominance of codon **H1** in the phytoplankton biomass in the Uzarzewskie Lake before restoration, and the change to codon **S1** during the first method of restoration and coda **C, J, Y** during the second method of restoration indicates the changes in environmental conditions, especially in trophic level. Phytoplankton groups have been replaced by others, which are indicators for the lower trophy. In turn, the share of cyanobacteria decreased. In such lakes, however, cyanobacterial blooms are decreasing gradually [62–64], and the stabilization of the ecosystem can last even many years.

5. Conclusions

The influence of particular physical and chemical properties of water on the qualitative and quantitative contribution of functional groups in the phytoplankton was demonstrated. The study performed from 2005 to 2015 in Lake Uzarzewskie allowed to analyze changes in the share of taxa representing individual functional groups in the total phytoplankton biomass. It was found that phytoplankton community was rebuilt. In the first study period i.e., before restoration, the **H1** group with such species as *Aphanizomenon gracile, Dolichospermum planctonicum, D. viguieri* predominated in phytoplankton biomass, although this trend was not statistically significant. During the use of the first restoration method (with iron-treatment), high biomass of the codon **S1** was found, mainly *Planktolyngbya limnetica, Limnothrix redekei* and *Planktothrix agardhii*. In the third period of research (oxygenation of hypolimnetic water layer), coda **J** (with *Actinastrum hantzschii* and other Chlorococcales), **C** (*Asterionella formosa* and other diatoms), **Y** (*Cryptomonas marssonii* and other cryptophytes) were dominant, indicating decreasing nutrient concentration and thus improving water quality.

Author Contributions: A.K.—analyzed the phytoplankton samples, performed statistical analysis, designed and wrote the manuscript, A.K. and A.B.—performed the phycological analyses, R.D.-P. and K.K.-M. prepared physico-chemical data, R.G.—reviewed the manuscript. All authors have read and agreed to the published version of the manuscript.

Funding: This work was supported by The Polish National Science Center as research project No. NN305 372838.

Conflicts of Interest: The authors declare no conflict of interest.

References

1. Paerl, H.W.; Fulton, R.S.; Moisander, P.H.; Dyble, J. Harmful freshwater algal blooms with an emphasis on cyanobacteria. *Sci. World J.* **2001**, *1*, 76–113. [CrossRef]
2. Huisman, J.; Codd, G.A.; Paerl, H.W.; Ibelings, B.W.; Verspagen, J.M.H.; Visser, P.M. Cyanobacterial blooms. *Nat. Rev. Microbiol.* **2018**, *16*, 471. [CrossRef] [PubMed]
3. WFD. Directive 2000/60/ec of the European Parliament and of the Council 22.12.2000. *Off. J. Eur. Communities* **2000**, *L327*, 1–72.
4. Kozak, A.; Gołdyn, R. Zooplankton versus phyto- and bacterioplankton in the Maltanski Reservoir (Poland) during an extensive biomanipulation experiment. *J. Plankton Res.* **2004**, *26*, 37–48. [CrossRef]
5. Kozak, A.; Gołdyn, R.; Dondajewska, R.; Kowalczewska-Madura, K.; Holona, T. Changes in phytoplankton and water quality during sustainable restoration of an urban lake used for recreation and water supply. *Water* **2017**, *9*, 713. [CrossRef]
6. Reynolds, C.S.; Huszar, V.; Kruk, C.; Naselli-Flores, L.; Melo, S. Towards a functional classification of the freshwater phytoplankton. *J. Plankton Res.* **2002**, *24*, 417–428. [CrossRef]
7. Abonyi, A.; Leitão, M.; Lançon, A.M.; Padisák, J. Phytoplankton functional groups as indicators of human impacts along the River Loire (France). *Hydrobiologia* **2012**, *698*, 233–249. [CrossRef]
8. Padisak, J.; Crossetti, L.O.; Naselli-Flores, L. Use and misuse in the application of the phytoplankton functional classification: A critical review with updates. *Hydrobiologia* **2009**, *621*, 1–19. [CrossRef]
9. Kozak, A.; Gołdyn, R. Variation in Phyto- and Zooplankton of Restored Lake Uzarzewskie. *Pol. J. Environ. Stud.* **2014**, *23*, 1201–1209.
10. Kowalczewska-Madura, K.; Gołdyn, R.; Dera, M. Spatial and seasonal changes of phosphorus internal loading in two lakes with different trophy. *Ecol. Eng.* **2015**, *74*, 187–195. [CrossRef]
11. Dondajewska, R.; Kozak, A.; Kowalczewska-Madura, K.; Budzyńska, A.; Gołdyn, R.; Podsiadłowski, S.; Tomkowiak, A. The response of a shallow hypertrophic lake to innovative restoration measures—Uzarzewskie Lake case study. *Ecol. Eng.* **2018**, *121*, 72–82. [CrossRef]
12. Gołdyn, R.; Kowalczewska-Madura, K.; Dondajewska, R.; Budzyńska, A.; Domek, P.; Romanowicz, W.; Sengupta, M.; Dalwani, R. (Eds.) Functioning of hypertrophic Uzarzewskie Lake ecosysytem. In *Proceedings of Taal 2007: The12th World Lake Conference, Andhra Pradesh, India, 8 July 2008*; ILEC: Jaipur, India, 2008; pp. 2233–2236.
13. Kozak, A. Community Structure and Dynamics of Phytoplankton in Lake Uzarzewskie. *Teka Komisji Ochrony Przyrody i Kształtowania Środowiska Przyrodniczego* **2009**, *6*, 146–152.
14. Kozak, A.; Gołdyn, R. Macrophyte response to the protection and restoration measures of four water bodies. *Int. Rev. Hydrobiol.* **2016**, *101*, 1–13. [CrossRef]
15. Budzyńska, A.; Gołdyn, R.; Zagajewski, P.; Dondajewska, R.; Kowalczewska-Madura, K. The dynamics of a Planktothrix agardhii population in a shallow dimictic lake. *Oceanol. Hydrobiol. Stud.* **2009**, *38*, 7–19.
16. Gołdyn, R.; Podsiadłowski, S.; Dondajewska, R.; Kozak, A. The sustainable restoration of lakes—Towards the challenges of the Water Framework Directive. *Ecohydrol. Hydrobiol.* **2014**, *14*, 68–74. [CrossRef]
17. Utermöhl, H. Zur Vervollkommnung der quantitativen Phytoplankton-Methodik. *Mitteilungen Internationale Vereinigung Theoretische und Angewandte Limnologie* **1958**, *9*, 1–38.
18. Hillebrand, H.; Dürselen, C.-D.; Kirschtel, D.; Pollingher, U.; Zohary, T. Biovolume calculation for pelagic and benthic microalgae. *J. Phycol.* **1999**, *35*, 403–424. [CrossRef]
19. Ter Braak, C.J.F.; Smilauer, P. *CANOCO Reference Manual and CanoDraw for Windows User's Guide: Software for Canonical Community Ordination*, version 4.5; Microcomputer Power: Ithaca, NY, USA, 2002.

20. Kozak, A.; Gołdyn, R.; Dondajewska, R. Phytoplankton composition and abundance in restored Maltanski Reservoir under the influence of physico-chemical variables and zooplankton grazing pressure. *PLoS ONE* **2015**, *10*, e0124738. [CrossRef]

21. Matsumura-Tundisi, T.; Tundisi, J. Plankton richness in a eutrophic reservoir (Barra Bonita Reservoir, SP, Brazil). *Hydrobiologia* **2005**, *542*, 367–378. [CrossRef]

22. Connell, J. Diversity in tropical rain forests and coral reefs. *Science* **1978**, *199*, 1304–1310. [CrossRef]

23. Padisak, J.; Reynolds, C.S.; Sommer, U. (Eds.) *Intermediate Disturbance Hypothesis in Phytoplankton Ecology*; Developments in Hydrobiology 81; Springer: Berlin/Heidelberg, Germany, 1991.

24. Li, Z.; Wang, S.; Guo, J.; Fang, F.; Gao, X.; Long, M. Responses of phytoplankton diversity to physical disturbance under manual operation in a large reservoir, China. *Hydrobiologia* **2012**, *684*, 45–56. [CrossRef]

25. Demir, A.M.; Fakioğlu, Ö.; Dural, B. Phytoplankton functional groups provide a quality assessment method by the Q assemblage index in Lake Mogan (Turkey). *Turk. J. Bot.* **2014**, *38*, 169–179. [CrossRef]

26. Soylu, E.N.; Gönülol, A. Functional Classification and Composition of Phytoplankton in Liman Lake. *Turk. J. Fish. Aquat. Sci.* **2010**, *10*, 53–60. [CrossRef]

27. Maraşlıoğlu, F.; Gönülol, A. Phytoplankton Community, Functional Classification and Trophic State Indices of Yedikır Dam Lake (Amasya). *J. Biol. Environ. Sci.* **2014**, *8*, 133–141.

28. Reynolds, C.S. *The Ecology of Freshwater Phytoplankton*; Cambridge University Press: Cambridge, UK, 1984.

29. Napiórkowska-Krzebietke, A. Diversity and dynamics of phytoplankton in lakes Licheńskie and Slesińskie in 2004–2005. *Arch. Pol. Fish.* **2009**, *17*, 253–265. [CrossRef]

30. Lenard, T.; Ejankowski, W.; Poniewozik, M. Responses of Phytoplankton Communities in Selected Eutrophic Lakes to Variable Weather Conditions. *Water* **2019**, *11*, 1207. [CrossRef]

31. Reynolds, C.S.; Descy, J.-P. The production, biomass and structure of phytoplankton in large rivers. *Large Rivers* **1996**, *10*, 161–187. [CrossRef]

32. Reynolds, C.S.; Descy, J.P.; Padisak, J. Are phytoplankton dynamics in rivers so different from those in shallow lakes? *Hydrobiologia* **1994**, *289*, 1–7. [CrossRef]

33. Dondajewska, R.; Kowalczewska-Madura, K.; Gołdyn, R.; Kozak, A.; Messyasz, B.; Cerbin, S. Long-term water quality changes as a result of a sustainable restoration-a case study of Dimictic Lake Durowskie. *Water* **2019**, *11*, 616. [CrossRef]

34. Dondajewska, R.; Kozak, A.; Rosińska, J.; Gołdyn, R. Water quality and phytoplankton structure changes under the influence of effective microorganisms (EM) and barley straw–Lake restoration case study. *Sci. Total Environ.* **2019**, *660*, 1355–1366. [CrossRef]

35. Grabowska, M.; Mazur-Marzec, H. The influence of hydrological conditions on phytoplankton community structure and cyanopeptide concentration in dammed lowland river. *Environ. Monit. Assess.* **2016**, *188*, 488. [CrossRef] [PubMed]

36. Herrero, A.; Muro-Pastor, A.M.; Flores, E. Nitrogen control in cyanobacteria. *J. Bacteriol.* **2001**, *183*, 411–425. [CrossRef] [PubMed]

37. Herrero, A.; Flores, E. Genetic responses to carbon and nitrogen availability in Anabaena. *Environ. Microbiol.* **2019**, *21*, 1–17. [CrossRef]

38. Kozak, A.; Celewicz-Gołdyn, S.; Kuczyńska-Kippen, N. Cyanobacteria in small water bodies: The effect of habitat and catchment area conditions. *Sci. Total Environ.* **2019**, *646*, 1578–1587. [CrossRef]

39. Pełechata, A.; Pełechaty, M.; Pukacz, A. Factors influencing cyanobacteria community structure in Chara-lakes. *Ecol. Indic.* **2016**, *71*, 477–490. [CrossRef]

40. Yatigammana, S.K.; Ileperuma, O.A.; Perera, M.B.U. Water pollution due to a harmful algal bloom: A preliminary study from two drinking water reservoirs in Kandy, Sri Lanka. *J. Natl. Sci. Found. Sri Lanka* **2011**, *39*. [CrossRef]

41. Crossetti, L.O.; Beckerb, V.; de Souza Cardosoc, L.; Rodriguesd, L.R.; da Costad, L.S.; da Motta-Marquesd, D. Is phytoplankton functional classification a suitable tool to investigate spatial heterogeneity in a subtropical shallow lake? *Limnologica* **2013**, *43*, 157–163. [CrossRef]

42. Kozak, A. Seasonal changes occurring over four years in a reservoir's phytoplankton composition. *Pol. J. Environ. Stud.* **2005**, *14*, 437–444.

43. Kozak, A.; Kowalczewska-Madura, K. Pelagic Phytoplankton of Shallow Lakes in the Promno Landscape Park. *Pol. J. Environ. Stud.* **2010**, *19*, 587–592.

44. Kozak, A.; Rosińska, R.; Gołdyn, R. Changes in the phytoplankton structure due to prematurely limited restoration treatments. *Pol. J. Environ. Stud.* **2018**, *27*, 1097–1103. [CrossRef]

45. Paerl, H.W.; Gardner, W.S.; Havens, K.E.; Joyner, A.R.; McCarthy, M.J.; Newell, S.E.; Qin, B.; Scott, J.T. Mitigating cyanobacterial harmful algal blooms in aquatic ecosystems impacted by climate change and anthropogenic nutrients. *Harmful Algae* **2016**, *54*, 213–222. [CrossRef] [PubMed]

46. Mantzouki, E.; Campbell, J.; Loon, E.V.; Visser, P.; Konstantinou, I.; Antoniou, M.; Giuliani, G.; Machado-Vieira, D.; Cromie, H.; Ibelings, B.W.; et al. A European Multi Lake Survey dataset of environmental variables, phytoplankton pigments and cyanotoxins. *Sci. Data* **2018**, *5*, 180226. [CrossRef]

47. Mantzouki, E.; Lürling, M.; Fastner, J.; De Senerpont Domis, L.; Wilk-Woźniak, E.; Koreivienė, J.; Seelen, L.; Teurlincx, S.; Carey, C.C.; Ibelings, B.W.; et al. Temperature effects explain continental scale distribution of cyanobacterial toxins. *Toxins* **2018**, *10*, 156. [CrossRef]

48. Pálffy, K.; Vörös, L. Phytoplankton functional composition shows higher seasonal variability in a large shallow lake after a eutrophic past. *Ecosphere* **2019**, *10*, e02684. [CrossRef]

49. Kozak, A. Changes in the structure of phytoplankton in the lowest part of the Cybina River and the Maltański Reservoir. *Oceanol. Hydrobiol. Stud.* **2010**, *39*, 85–94. [CrossRef]

50. Budzyńska, A.; Gołdyn, R. Domination of invasive Nostocales (Cyanoprokaryota) at 52° N latitude. *Phycol. Res.* **2017**, *65*, 322–332. [CrossRef]

51. Budzyńska, A.; Rosińska, J.; Pełechata, A.; Toporowska, M.; Napiórkowska-Krzebietke, A.; Kozak, A.; Messyasz, B.; Pęczuła, W.; Kokociński, M.; Szeląg-Wasielewska, E.; et al. Environmental factors driving the occurence of the invasive cyanobacterium *Sphaerospermopsis aphanizomenoides* (Nostocales) in temperate lakes. *Sci. Total Environ.* **2019**, *650*, 1338–1347. [CrossRef]

52. Hsieh, C.H.; Ishikawa, K.; Sakai, Y.; Ishikawa, T.; Ichise, S.; Yamamoto, Y.; Kuo, T.C.; Park, H.D.; Yamamura, N.; Kumagai, M. Phytoplankton community reorganization driven by eutrophication and warming in Lake Biwa. *Aquat. Sci.* **2010**, *72*, 467–483. [CrossRef]

53. De Senerpont Domis, L.N.; Elser, J.J.; Gsell, A.S.; Huszar, V.L.; Ibelings, B.W.; Jeppesen, E.; Kosten, S.; Mooij, W.M.; Roland, F.; Sommer, U.; et al. Plankton dynamics under different climatic conditions in space and time. *Freshw. Biol.* **2012**, *58*, 463–482. [CrossRef]

54. Lürling, M.; Eshetu, F.; Faassen, E.J.; Kosten, S.; Huszar, V.L.M. Comparison of cyanobacterial and green algal growth rates at different temperatures. *Freshw. Biol.* **2012**, *58*, 1–8. [CrossRef]

55. Singh, S.P.; Singh, P. Effect of temperature and light on the growth of algae species: A review. *Renew. Sustain. Energy Rev.* **2015**, *50*, 431–444. [CrossRef]

56. Donald, D.B.; Bogard, M.J.; Finlay, K.; Bunting, L.; Leavitt, P.R. Phytoplankton-Specific Response to Enrichment of Phosphorus-Rich Surface Waters with Ammonium, Nitrate, and Urea. *PLoS ONE* **2013**, *8*, e53277. [CrossRef] [PubMed]

57. Salmaso, N. Ecological patterns of phytoplankton assemblages in Lake Garda: Seasonal, spatial and historical features. *J. Limnol.* **2002**, *61*, 95–115. [CrossRef]

58. Sommer, U. The periodicity of phytoplankton in Lake Constance (Bodensee) in comparison to other deep lakes of central Europe. *Hydrobiologia* **1986**, *138*, 1–7. [CrossRef]

59. Tapolczai, K.; Anneville, O.; Padisák, J.; Salmaso, N.; Morabito, G.; Zohary, T.; Tadonléké, R.; Rimet, F. Occurrence and mass development of *Mougeotia* spp. (Zygnemataceae) in large, deep lakes. *Hydrobiologia* **2015**, *745*, 17–29. [CrossRef]

60. Kowalczewska-Madura, K.; Dondajewska, R.; Gołdyn, R.; Podsiadłowski, S. The influence of restoration measures on phosphorus internal loading from the sediments of a hypereutrophic lake. *Environ. Sci. Pollut. Res.* **2017**, *24*, 14417–14429. [CrossRef]

61. Kowalczewska-Madura, K.; Dondajewska, R.; Gołdyn, R.; .Kozak, A.; Messyasz, B. Internal Phosphorus Loading from the Bottom Sediments of a Dimictic Lake During Its Sustainable Restoration. *Water Air Soil Pollut.* **2018**, *229*, 280. [CrossRef]

62. Moustaka-Gouni, M.; Michaloudi, E.; Kormas, K.A.; Katsiapi, M.; Vardaka, E.; Genitsaris, S. Plankton changes as critical processes for restoration plans of lakes Kastoria and Koronia. *Eur. Water* **2012**, *40*, 43–51.

63. Rosińska, J.; Kozak, A.; Dondajewska, R.; Gołdyn, R. Cyanobacteria blooms before and during the restoration process of a shallow urban lake. *J. Environ. Manag.* **2017**, *198*, 340–347. [CrossRef]
64. Rosińska, J.; Kozak, A.; Dondajewska, R.; Kowalczewska-Madura, K.; Gołdyn, R. Water quality response to sustainable restoration measures-case study of urban Swarzędzkie Lake. *Ecol. Indic.* **2018**, *84*, 437–449. [CrossRef]

Article

The Effects of Sodium Percarbonate Generated Free Oxygen on *Daphnia*—Implications for the Management of Harmful Algal Blooms

Robin Thoo [1], Waldemar Siuda [2] and Iwona Jasser [1,*]

[1] Institute of Environmental Biology, Faculty of Biology, Biological and Chemical Research Centre, University of Warsaw, 02-089 Warsaw, Poland; r.thoo@student.uw.edu.pl
[2] Institute of Functional Biology and Ecology, Faculty of Biology, Biological and Chemical Research Centre, University of Warsaw, 02-089 Warsaw, Poland; w.siuda@biol.uw.edu.pl
* Correspondence: jasser.iwona@biol.uw.edu.pl

Received: 30 March 2020; Accepted: 1 May 2020; Published: 5 May 2020

Abstract: Increasing frequencies and durations of harmful algal blooms are a nuisance in many aquatic ecosystems. This has led to the use of a variety of control methods to prevent their appearance or to disperse them following their establishment. Most of these methods are not selective; consequently, research into alternative selective methods has been ongoing. Reactive oxygen species generated following the addition of hydrogen peroxide have been shown to selectively target the cyanobacterial component of harmful algal blooms in experimental and field settings. This study assesses the effects of increasing concentrations of reactive oxygen species from the addition of sodium percarbonate on zooplankton in a small experimental setting using a natural plankton sample. It was found that the genus *Daphnia* showed moderate sensitivity to sodium percarbonate. Preliminary evidence suggests that the size of an individual may affect the probability of survival, with larger individuals having a lower likelihood of survival. Lower survival rates of large *Daphnia* were hypothesized to have been caused by higher relative filtration rates of larger individuals. From the zooplankton data obtained, we suggest that a safe concentration of sodium percarbonate for *Daphnia* individuals would be below 10.0 mg·L^{-1} sodium percarbonate (2.8 mg·L^{-1} hydrogen peroxide).

Keywords: eutrophication; cyanobacteria; *Daphnia*; harmful algal bloom mitigation; phytoplankton; zooplankton; reactive oxygen specie

---

1. Introduction

Phytoplankton forms the base of the aquatic pelagic food web [1]. These photosynthetic organisms are a key component of freshwater aquatic ecosystems, and the majority of other aquatic environments [2]. Algae is a broad term that encompasses a wide range of organisms, some of which, due to increasing levels of pollution (among other factors) are the primary contributors to eutrophication. Eutrophication is undesirable for water bodies for a number of reasons, ranging from ecological and economic concerns to pure aesthetics [3]. It is not the only phytoplankton component that causes blooms to be a nuisance, but the proliferation of cyanobacteria are a particular cause of concern because of the large number of negative effects that can result when their abundance becomes too high. Such issues include but are not limited to: shading and outcompeting other phytoplankton species and macrophytes, interfering with large zooplankton feeding by clogging and providing poor quality food, forming scum on water surfaces, and eventually producing toxins which may be harmful for other organisms and humans [3–8]. Owing to the mixed or heterogeneous community of organisms within a fluid medium (notwithstanding stratification due to temperature, light and temperature gradients, and so on) a major complication in combatting harmful algal blooms (HABs), and an even greater problem in combatting the cyanobacterial component, is that of targeting a specific organism or group of organisms. This becomes more complex when we consider that cyanobacteria can be found not only as large colonies and scum, easily visible to the naked eye, but also as individual cells or clusters of cells (filaments and small colonies) dispersed within a varied algal community. Moreover, this varied algal community shares living space with numerous other biological entities from viruses to zooplankton to large, complex eukaryotes such as planktivorous and predatory fish, with each entity having the possibility of responding in undesirable or unexpected ways to any attempt made to mitigate a HAB.

Long-term preventative solutions are the primary goal of combatting eutrophication and of reducing the impact of HABs, with the restoration of water bodies to pristine or near pristine conditions via the reduction of nutrient inputs and regression to the former hydrological regime being the ideal endpoint [9]. Unfortunately, this is often not possible or takes a long time. Therefore, a large number of physical and chemical methodologies have been devised to provide short-term relief to affected bodies of water. The addition of chemicals has proven useful, with algaecides such as copper sulphate being particularly effective, but many such chemicals leave traces in the environment and cannot discriminate between organisms, with negative effects having been observed on all levels of the trophic chain [10,11].

Reactive oxygen species (ROS) are forms of oxygen with at least one unpaired electron and are therefore highly reactive, since atoms and molecules will 'seek' to return to an electrically neutral state. In essence they are a form of free radical with a high oxidative potential, and often cause chain reactions which damage biological molecules. Microscopic planktonic primary producers such as eukaryotic algae and cyanobacteria differ in their abilities to tolerate ROS as a result of different photosynthetic paths. Therefore, the addition of chemicals that generate ROS has been suggested as an effective tool to selectively remove cyanobacteria [12].

In eukaryotic algae and plants, ROS are continually produced as byproducts of various metabolic pathways localized in different cellular compartments. Under steady state conditions these byproducts are scavenged by anti-oxidative defense components to minimize possible oxidative damage, but this equilibrium can be disrupted [13]. Eukaryotic algae and plants also synthesize numerous oxidises and peroxidises in response to environmental changes. Within plants and eukaryotic algae, hydrogen peroxide (HP) is often created under natural conditions of high light. It is a waste product produced during the processes used to deal with surplus photons in the electron sinks associated with elevated photosynthesis [14].

In contrast, because of the flavoproteins flv1 and flv3 present in cyanobacteria, superoxide is not formed as part of the mechanisms used to deal with high light levels, thus cyanobacteria do not produce ROS internally in a Mehler-like reaction and so neither have nor need the same coping mechanisms to deal with their presence. It has been demonstrated, both experimentally in the laboratory and in the field, that at low doses ROS may have selectivity for the cyanobacteria within HABs because of the above [9]. Consequently, a large number of studies have been conducted using HP to generate ROS with the intention of targeting cyanobacteria while reducing the relative impact on other organisms. Hydrogen peroxide has the benefit of not only being readily available to facilitate the creation of ROS, but it also decays to water and oxygen within a maximum period of a few days [15,16]. Additionally, it is not considered to be carcinogenic [4] and it is a commonly encountered component of the aquatic environment [12], with typical concentrations in lakes of 1–30 $\mu g \cdot L^{-1}$ [15].

Far fewer studies are available on functionally similar compounds such as sodium percarbonate (SP), which decomposes in water to form sodium, carbonate and HP [16]. The main difference between these compounds is that the carbonate portion of SP can cause an increase in alkalinity and pH which depends on the buffering capacity of the water to which it is added [17].

$$2Na_2CO_3 \cdot 3H_2O_2 \rightarrow 3H_2O_2 + 4Na^+ + 2CO_3^{2-}$$
$$2H_2O_2 \rightarrow 2H_2O + O_2 \tag{1}$$

Interactions between the cyanobacterial component of phytoplankton, zooplankton and larger organisms such as fish are crucial to the understanding, maintenance and management of aquatic ecosystems and as a result have been widely studied [18]. Macroinvertebrates and zooplankton in particular are the main components of most pelagic food webs and are the major herbivores in freshwater lakes and ponds [19–21]. Aquatic macroinvertebrates have been shown to differ in sensitivity to changes in the physicochemical parameters of water and various toxic compounds [22]. In natural waterbodies, the pH fluctuates due to the photosynthetic activity of algae and the respiration of aquatic animals, varying daily between 7 and 9 [23]. Pedersen and Hansen [24], studying the marine plankton community, found that increases of pH above 9 induce unfavorable changes in the zooplankton community, while Yin and Niu [25] found that in experimental conditions, age-specific survival curves were not significantly different within five of the studied *Brachionus* rotifer species between pH values of 6 and 10. Among freshwater macroinvertebrates, *Daphnia* seems to be one of the most vulnerable species to decreases in pH. While fewer studies exist on pH increases, Beklioglu and Moss [26] found that increasing pH through the addition of NaOH did not negatively affect *D. hyalina* up until pH 11, and suggest an upper limit of approximately pH 10.5 to 11 for Cladocera survival. Taking this into account, sodium percarbonate increasing the alkalinity of water may begin to negatively influence zooplankton if the pH begins to rise above pH 9.

The aim of this investigation was therefore to gain a preliminary understanding of the possible effects of SP on the genus *Daphnia*, the keystone species in temperate freshwaters and thus one of the most commonly studied mesozooplankton [21,27], in controlled conditions using pond water with a natural plankton assemblage. If ROS generated from this substance are to be widely used to treat HABs, the possible effects on non-target organisms need to be established to facilitate making appropriate management decisions. It should be pointed out that *Daphnia*, especially large *Daphnia*, contribute significantly to the maintenance of clear water through filter feeding, in line with McQueen et al.'s [28] model of top-down control of phytoplankton. According to Christoffersen et al. [29], large *Daphnia* may prevent cyanobacteria from blooming, and therefore may strengthen the effect of SP on the aquatic community. A level of certainty is required so that the technologies used to maintain and protect ecosystem services, including the use of ROS, exert only a minimal amount of stress or damage on non-target organisms and do not negatively affect top-down control of algal growth.

2. Materials and Methods

A microcosm experiment was conducted where SP was added to ten plastic containers, each with 1 liter of lake water and approximately the same number of individual planktonic crustaceans. Water samples were collected on the 15th of May from the surface layer (0.5 m) of a shallow artificial lake (0.35 ha) with a maximum depth of 1.8 m and a mean depth of 1.4 m located in Dawidy Bankowe, a village in the Mazovian Voivodeship, Poland (52°07′0753″ N, 20°59′43.09″ E). In mid-April the lake was characterized by the following physiochemical characteristics: pH—8.1; NH_4^+—0.510 mg·L^{-1}; P-PO_4^{3-}—0.046 mg·L^{-1}; DOP—0.005 mg·L^{-1}; DOC—12.7 mg·L^{-1} and Chlorophyll a—83 µg·L^{-1}. At the time of sample collection for the experiment in May, the temperature of water was 17 °C and the phytoplankton biomass obtained by microscopic analysis was just over 3.0 mg fresh weight L^{-1}. Phytoplankton was dominated by Cryptophyceae and Chlorophyceae, with Cyanobacteria accounting for only about 3% of total phytoplankton biomass. In the laboratory the water sample was poured into a large mesocosm, from which the 1 liter experimental sub-samples were obtained. To ensure that approximately the same number of planktonic crustaceans were present in each treatment, the mesocosm was mixed slowly to ensure homogeneity and then 1 liter of lake water was extracted.

Three of the containers were set as controls and seven containers contained increasing concentrations of SP (5, 10, 20, 30, 40, 50 and 60 mg·L^{-1}). This kind of experimental design (three controls and a gradient experimental treatment) was chosen to ensure that the *Daphnia* which had been transferred to the mesocosms were not adversely affected under the base conditions of the experiment, allowing us to confidently attribute mortality to SP in the treatments along a gradient of increasing SP concentration. The SP was manufactured to a technical grade by envolab.pl (Długomiłowice, Poland) ($2Na_2CO_3 \cdot 3H_2O_2$ [30]). According to the manufacturer, the maximum solubility of the SP in water at 20 °C was 120 g·L^{-1}, giving a HP solution of approximately 3.2%. The addition of SP to lake water with pH 8.1 caused an increase of the pH to 8.4 at 60 mg·L^{-1} SP and 9.7 at 300 mg·L^{-1} SP (Supplementary Material Figure S1). Trace iron content was no greater than 15 ppm, with a minimum active oxygen content of 13%. To facilitate comparison of the results with studies using HP, information on the breakdown of SP to HP in water solutions is given in Table 1 and in brackets after the SP values in the text.

Table 1. Concentration (mg·L^{-1}) of sodium percarbonate added in the containers and concentration of hydrogen peroxide released in the water by its breakdown.

Treatment	Sodium Percarbonate (SP) ($2Na_2CO_3 \cdot 3H_2O_2$)	Hydrogen Peroxide (HP) (H_2O_2)
Control	0.0 mg·L^{-1}	0.0 mg·L^{-1}
T1	5.0 mg·L^{-1}	1.4 mg·L^{-1}
T2	10.0 mg·L^{-1}	2.8 mg·L^{-1}
T3	20.0 mg·L^{-1}	5.5 mg·L^{-1}
T4	30.0 mg·L^{-1}	8.3 mg·L^{-1}
T5	40.0 mg·L^{-1}	11.1 mg·L^{-1}
T6	50.0 mg·L^{-1}	13.8 mg·L^{-1}
T7	60.0 mg·L^{-1}	16.6 mg·L^{-1}

To test the immediate acute effect on *Daphnia* of SP addition to the aquatic environment, the experiment was left to run for 24 h at room temperature (approximately 20 °C) under natural light conditions, and without an additional supply of oxygen. This was to ensure a close approximation to the natural conditions of the sampling location in terms of: temperature (see above), light (a natural daily cycle) and oxygen availability (phytoplankton would have still provided an oxygen source in proportion to their response to SP concentrations). Moreover, the considerably large surface to volume ratio of the sample containers would allow diffusion to cover some of the additional oxygen demand in a manner similar to surface waters.

At the end of the experiment all zooplankton individuals were separated into living and dead fractions for each treatment, then preserved with formaldehyde (4%) in separate sample tubes. Zooplankton was identified using a binocular stereoscopic microscope with the aid of diagrams and keys modified from Chapman [31]. Copepoda were identified to the level of order with three orders being represented: (1) Calanoida, (2) Cyclopoida and (3) Harpacticoida. Cladocera were identified to genus or family level. All animals were measured along their major axes; for *Daphnia*, the length of the major axis from the upper eye to the tip of the tail spine was recorded. The size categories were 0.2 mm, 0.5 mm, 0.7 mm, 1.0 mm, 1.2 mm, 1.5 mm and 2.0 mm.

All statistical analyses were conducted using the statistics packages IBM SPSS statistics 23® (IBM Corporation, New York, NY, USA) and GraphPad Prism (GraphPad Software, San Diego, CA, USA) and Microsoft Excel® (Microsoft Corporation, Redmond, WA, USA). A 95% confidence level ($p \leq 0.05$) was chosen for all tests unless otherwise stated. To establish the strength of the relationship with SP concentration and zooplankton survival, a step-by-step approach was taken. Firstly, data on the percentage of individuals alive after 24 h were correlated with concentration using Spearman's rank to confirm variable association. In the second stage of analysis, linear, quadratic, cubic and four parameter logistic (4PL) regression were applied following McDonald [32]. The percentage survival data were obtained by transforming the raw binominal data (i.e., 'dead' or 'alive') into ratio variables (i.e., percentage values). Finally, binominal logistic regression was performed to obtain information on the probability of survival of *Daphnia* and to assess differences between the different sizes.

3. Results

A total of 1944 individuals belonging to seven different systematic groups were counted: four from the order of Cladocera, namely *Daphnia* (1610), Bosminidae (2), Chydoridae (1) and *Scapholeberis* (2), and three from subclass Copepoda, namely Calanoida (272), Cyclopoida (44) and Harpacticoida (14). *Daphnia* were the most numerous zooplankton group by far; therefore, it was decided that further detailed analyses were only to be performed on this genus (Table 2), though the actual number of animals from all identified groups are provided in the Table S1 in the Supplementary Materials.

Table 2. Total number of individuals per mesocosm. Concentration of sodium percarbonate (SP) ($mg \cdot L^{-1}$). Standard deviation for the control (SD).

Species	Control (Mean)	T1: 5 SP	T2: 10 SP	T3: 20 SP	T4: 30 SP	T5: 40 SP	T6: 50 SP	T7: 60 SP
Daphnia sp.	79.6 SD 47	280	292	150	185	104	95	265

Total Mortality and Concentration of SP

In the control, 97.9% of *Daphnia* were alive after 24 h. At a concentration of 5.0 $mg \cdot L^{-1}$ SP (1.4 $mg \cdot L^{-1}$ HP) the percentage of living individuals decreased slightly. The next treatment, 10.0 $mg \cdot L^{-1}$ SP (2.8 $mg \cdot L^{-1}$ HP), showed a further small decrease. This was followed by a more pronounced decrease at 20.0 $mg \cdot L^{-1}$ SP (5.5 $mg \cdot L^{-1}$ HP). Between 20.0 $mg \cdot L^{-1}$ SP (5.5 $mg \cdot L^{-1}$ HP) and 30.0 $mg \cdot L^{-1}$ SP (8.3 $mg \cdot L^{-1}$ HP) a dramatic decrease of almost 50 percentage points was observed. At 40.0 $mg \cdot L^{-1}$ SP (11.1 $mg \cdot L^{-1}$ HP) the percentage of living *Daphnia* dropped to 3%, while at 50.0 $mg \cdot L^{-1}$ SP (13.8 $mg \cdot L^{-1}$ HP), only 1.1% of *Daphnia* were still alive. The final treatment, 60.0 $mg \cdot L^{-1}$ SP (16.6 $mg \cdot L^{-1}$ HP), saw all individuals expire (Table 3). For zooplankton other than *Daphnia* a general downward trend was also observed. However, the small sample size and a number of outliers, particularly in the control and 50.0 $mg \cdot L^{-1}$ SP (13.8 $mg \cdot L^{-1}$ HP) treatments, prevented further analysis.

Table 3. Concentration of SP ($mg \cdot L^{-1}$), *Daphnia* survival percentage and actual numbers of living and dead individuals. Standard deviation for percentage of living *Daphnia* in the control (SD).

Species	Control (Mean)	T 1: 5 SP	T2: 10 SP	T3: 20 SP	T4: 30 SP	T5: 40 SP	T6: 50 SP	T7: 60 SP
Percentage living *Daphnia*	97.9 SD 1.23	93.9	91.1	66.0	16.8	2.9	1.1	0.0
Living	77	263	266	99	31	3	1	0
Dead	2.6	17	26	51	154	101	95	265

Using Spearman's rank correlation, a strong inverse relationship was found to exist between SP concentration and percentage of living *Daphnia* (r = −0.976, $p < 0.0005$, N = 8). Linear regression was significant ($R^2 = 0.901$, $p < 0.0003$, F = 54.826), and quadratic and cubic regression had a lower level of significance (both $p = 0.001$) and R^2 values of 0.942 and 0.978 (F = 40.909 and F = 58.970, respectively). Four parameter logistic regression showed the highest R^2 value (0.998) with $p = 0.0004$ and F = 97.15 in comparison to first order polynomial (straight line) linear regression (Figure 1).

For these data the dependent variable can be expressed as a percentage, or in two categories, i.e., whether or not an individual survived the experiment. Binominal logistic regression was used to estimate the probability of an event occurring (death of an organism following 24 h of exposure) based on the independent variables provided (concentration of SP ($mg \cdot L^{-1}$)).

Grouping all of the differently sized individuals together, the results indicate that the concentration of SP was a significant predictor of mortality (Chi-Square = 1429.918, df = 1 and $p = 0.001$). Sodium percarbonate concentration explains 91.3% of the variability of animal deaths, 35.9% greater than the preliminary Block 0 model which only bases the prediction on which category has a higher frequency (55.4% of variability was explained by this model). Concentration was a significant factor at the

0.001 level (Wald = 407.442, $p < 0.0005$, Odds Ratio (OR) = 0.839). The Hosmer–Lemeshow test produced the desired non-significant value ($p = 0.117$) indicating that there was not a significant difference between the observed value and the values predicted by this model. In Figure 2a the probability for each individual either surviving or dying during the 24-h experiment is displayed, with individuals scoring above 0.5 having a probability towards surviving and those below 0.5 towards dying.

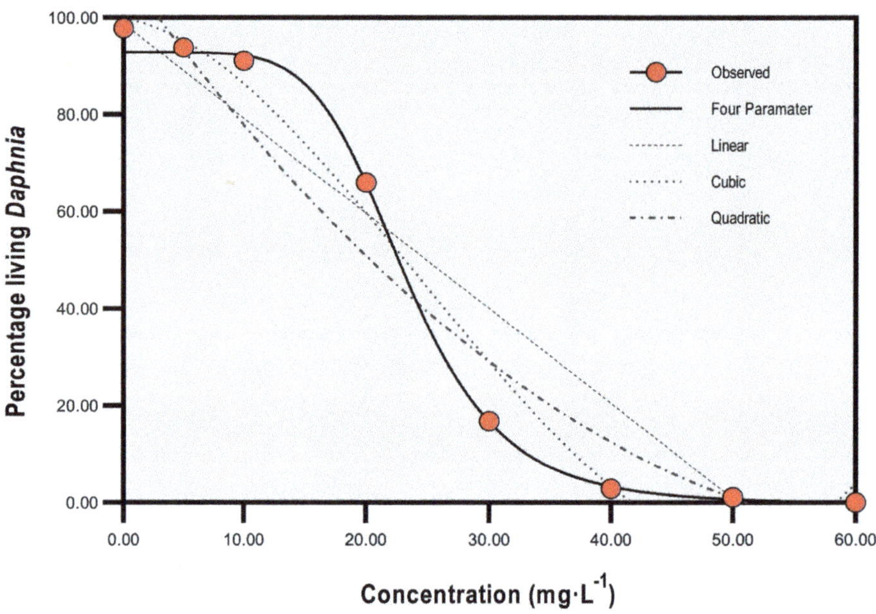

Figure 1. Regression lines for percentage survival of *Daphnia* and concentration of sodium percarbonate ($mg \cdot L^{-1}$).

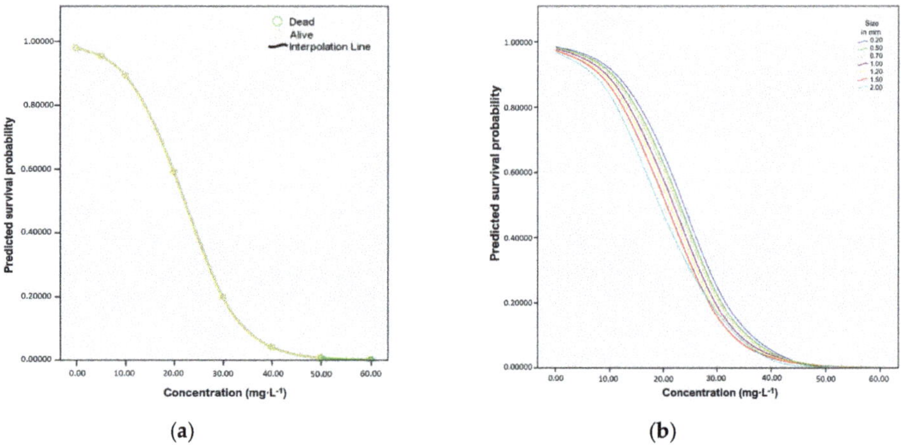

Figure 2. Predicted probability of survival dependent on sodium percarbonate concentration: (a) Grouped size; the line shows only values for dead individuals after 50 $mg \cdot L^{-1}$ (darker green) because there were no surviving individuals beyond that concentration; the prediction was based only on dead individuals. (b) The relationship between predicted probability of survival of different *Daphnia* size classes and sodium percarbonate concentration.

The analysis between size, concentration and mortality can be observed in Figure 2b, which presents the survival probability of each size category when included as a covariate scale variable alongside concentration. The justification for this is that size categories in themselves are artificial, since length is, in nature, a ratio variable (denoted as a 'scale variable' in the SPSS package (IBM Corporation, New York, NY, USA)).

In this case the accuracy of the prediction increased by 36.4% from 54.8% (Block 0) to 91.0%. Concentration was significant below the 0.0005 level (Wald $= 390.942$ $p \leq 0.0005$, Odds Ratio (OR) $= 0.839$), whereas size was not significant ($p = 0.111$, Wald $= 2.541$, Odds Ratio (OR) $= 0.651$). The pseudo R^2 value suggests that this model explains between 58.9% and 78.8% of the variance seen ($p = 0.0005$, Cox & Snell R Square $= 0.589$, Nagelkerke R Square $= 0.788$). The Hosmer–Lemeshow test showed a significant difference ($p = 0.001$) between the observed values and the values predicted by this model. All size classes showed a tipping point at around 10 mg·L^{-1}, when the probability of survival began to rapidly fall. Additionally, larger individual *Daphnia* appeared to show higher sensitivity to SP than smaller individuals, indicated by the largest *Daphnia* on the far left side of the curves and smaller individuals progressively shifting towards the right, with the smallest individual *Daphnia* in the furthest curve to the right.

4. Discussion

The experimental design applied in this study allowed changes in experimental treatments to be attributed to the increasing concentration of SP, because the three control replicates with 97.9% ± 1.23 of living *Daphnia* showed that enclosure effect was negligible. Furthermore, the gradient of SP concentrations allowed the non-linear response of *Daphnia* to be traced. According to Kreyling et al. [33], experiments with a gradient of factors outperform replication-based experiments in capturing non-linear responses to various environmental drivers. Accordingly, analysis of the results of our experiment showed that increasing concentrations of SP causes mortality to rise gradually but non-linearly. Additionally, both measured results and binominal regression showed a steep drop in survival in treatments containing between 10.0 mg·L^{-1} SP (2.8 mg·L^{-1} HP) and 20.0 mg·L^{-1} SP (5.5 mg·L^{-1} HP), with the LC$_{50}$ value falling within the range between 20.0 and 30.0 mg·L^{-1} SP (8.3 mg·L^{-1} HP).

The single, acute dosage had a very low threshold (the smallest amount of SP that caused a measurable effect). Although it is difficult to obtain the exact threshold dose based on our data, slight negative effects were observed at the lowest applied SP concentrations of 5.0 mg·L^{-1} SP (1.4 mg·L^{-1} HP; 93.9% living *Daphnia* as compared to 97.9% in the control samples). Meinertz et al. [34] found that chronic concentrations below 1.25 mg·L^{-1} HP did not significantly affect *D. magna*, while Reichwaldt et al. [35] showed that the LC$_{50}$ for *Daphnia* sp. was 5.6 mg·L^{-1} HP with no observed adverse effects at concentrations as high as 3.0 mg·L^{-1} HP. Similarly, in this study more than ninety percent of *Daphnia* remained alive after exposure to both 5.0 mg·L^{-1} SP (1.4 mg·L^{-1} HP) and 10.0 mg·L^{-1} SP (2.8 mg·L^{-1} HP) after 24 h, but at higher measured concentrations survival declined rapidly.

At much higher concentrations over a longer period, Wasielak [36] tested Envolab's SP formulation using the same artificial lake water sample at concentrations between 60.0 mg·L^{-1} SP (16.6 mg·L^{-1} HP) to 300 mg·L^{-1} SP (83.0 mg·L^{-1} HP). Following the addition of 60.0 mg·L^{-1} SP (16.6 mg·L^{-1} HP) crustacean zooplankton recovered to around 75% of their original abundance after 10 days, from a low point of around 50% at 5 days after addition. Zooplankton in the 180 mg·L^{-1} SP (49.8 mg·L^{-1} HP) and 300 mg·L^{-1} SP (83.0 mg·L^{-1} HP) treatments did not recover to more than 20% of their initial abundance within 10 days.

With regards to the sodium and carbonate components of sodium percarbonate and their effects on pH and alkalinity, pH values greater than 9.7 were not observed in mesocosm experiments following the addition of up to 300.0 mg·L^{-1} SP (83.0 mg·L^{-1} HP) [36], while with 60 mg·L^{-1} SP (the maximal concentration used in our experiment) the pH of 8.4 was well within natural daily fluctuations in eutrophic lake water. Furthermore, with a buffering capacity of 106 mg·L^{-1} HCO$_3^-$ (the mean value of 77 European rivers), the addition of 102 mg·L^{-1} of sodium carbonate would be needed to raise pH from

8.3 to 10.0 [37]. Regarding alkalinity, Bogart et al. [38] exposed *D. magna* to a combination of extreme shifts in hardness and alkalinity. Only extremely high (420 mg·L^{-1} CaCO$_3$ alkalinity and 600 mg·L^{-1} CaCO$_3$ water hardness) increases caused significant mortality after 72–96 h of exposure, albeit with calcium carbonate and not sodium carbonate. Warne & Schifko [39] analyzed the toxicity of numerous laundry detergent components, finding that sodium carbonate had a 48-h EC$_{50}$ of 199.82 mg·L^{-1} for the Cladocera *Ceriodaphnia dubia*. Given these results and the findings from Beklioglu and Moss [26], we suggest that the effects of changes in pH and alkalinity on *Daphnia* in our mesocosms would not be a significant contributor to mortality seen.

ROS can cause oxidative stress with the layers of cells becoming detached from tissues because of the formation of gas embolies [19]. Toxicity is likely to be because of ROS molecules reaching active sites within organisms [40]. Larger organisms must be exposed to higher concentrations to experience the same exposure per unit mass. Because of this it was initially thought that larger individual *Daphnia* would have benefited from a smaller surface area to volume ratio (SA:V) leading to less cell damage per unit of biomass. However, size differences were not shown to be statistically significant ($p = 0.111$) and based upon the pattern seen in Figure 2b there are some indications that large *Daphnia* may have experienced higher levels of mortality during the experiment. This non-significant finding may be explained by the fact that zooplankton are generally small enough to not require specialized respiratory surfaces and may acquire enough oxygen through integumental respiration [41,42]. The circulatory system of *Daphnia* also appears to lack specific sites for respiratory gas exchange and tissue transfer [43]. Therefore, the range of sizes of the *Daphnia* measured in this experiment (0.2–2.0 mm) may not have been wide enough for differences in mortality to be significant within 24 h at the measured levels of exposure to SP. *Daphnia* creates a dual-purpose feeding current that flows from the anterior to the posterior by beating its legs [41,44]. Pirow et al. [45] found that the partial pressure of oxygen in the water leaving the feeding chamber was significantly lower than that in water entering the chamber, indicating that oxygen was extracted between the inner and outer walls of the carapace. Burns [46] and Egloff and Palmer [47] found that the volume of filtered water and filtering rate increases with length of the body. Larger *Daphnia* may have filtered a greater proportion of water containing ROS than smaller individuals and may have increased their exposure to oxidative stress in this way. If larger *Daphnia* filter a proportionally greater volume of water, the transfer of ROS across the thin wall of the inner carapace was probably more effective in increasing exposure to ROS, which may have caused the higher levels of predicted mortality.

Additionally, *Daphnia*-sized aquatic animals have a layer of surrounding water which depends on the relative velocity of movement of their surroundings [41]. Pirow et al. [45] found a positive correlation between appendage beat rates and the flow rate of the medium around the animals, and Pirow and Buchen [43] suggest that oxygen supply may also be enhanced by ventilatory flow reducing fluid boundary layers, leading to greater levels of oxygen diffusion. Increased relative filtration rates would have served to reduce the fluid boundary layer between the epidermal layer and the ROS in the water, also resulting in a higher level of exposure for larger individuals. All of this may explain why it was not the case that smaller *Daphnia* were actually more affected by SP, with the opposite being suggested by these preliminary results. However, it must be kept in mind that because *Daphnia* were not identified to the species level, any differentiation in ROS tolerance between individuals that were not members of the same species would not have been accounted for in the analysis. To this point, Reichwaldt et al. [35] suggest that the differences observed in survival following HP exposure between *Moina* and *Daphnia* species were because of their different abilities to withstand oxidative stress, caused by higher levels of enzymes such as catalase, peroxidase and superoxide dismutase.

Geer et al. [40] assessed the effects of the HP-generating algaecide Phycomycins® SCP on various animals including *Ceriodaphnia dubia*, a member of the Daphniidae family, the cyanobacterium *Microcystis aeruginosa* and the chlorophyte *Pseudokirchneriella subcapitata*. They found that EC$_{50}$ values, based on cell density and chlorophyll *a* concentration for phytoplankton, were 0.9 to 1.0 mg·L^{-1} HP (over 96 h) for the cyanobacterium and 5.2 to 9.2 mg·L^{-1} HP for the chlorophyte (over 7 days). The LC$_{50}$ value over 96 h for the *C. dubia* was 1.0 mg·L^{-1} HP. Of the animals studied, *C. dubia* showed the highest sensitivity to HP equivalent concentrations with the authors suggesting that this was a result of the smaller size in comparison to other organisms under study (see above). Because the size structure of the *Daphnia* community was not uniform in all containers, the indication of higher vulnerability of larger *Daphnia* to ROS should be considered as preliminary results. Given the importance for lake management, further studies taking the effect of SP on *Daphnia* eggs and various sizes into consideration will be conducted to gain a more comprehensive understanding of the potential effects of SP on other stages of the lifecycle of *Daphnia*.

Crafton et al. [48] investigated the effects of the SP formulation PAK® 27, finding that 6.2 mg·L^{-1} HP equivalent led to statistically significant reductions in cyanobacterial populations and increases in the abundance of other classes of phytoplankton. Furthermore, results obtained by Kędzierska [49] showed that 20.0 mg·L^{-1} SP (5.5 mg·L^{-1}HP) of the SP formulation CYANOXIDE® reduced cyanobacteria from 63% of the phytoplankton assemblage to less than 5% 6 days after application in a 300 L experimental tank. Cyanobacteria did not recover over the 25-day duration of the experiment, while other phytoplankton groups increased in abundance. However, Sinha et al. [16] found that while 2.5 mg·L^{-1} HP did not negatively affect non-target algal and zooplankton communities, negative effects were observed in their 4.0 mg·L^{-1} HP treatment in comparison to the control and the 2.5 mg·L^{-1} HP treatment.

We find the above particularly significant because at concentrations greater than 10.0 mg·L^{-1} SP (2.8 mg·L^{-1} HP), we also began to see a rapid decline in *Daphnia* suggesting that concentrations below 10.0 mg·L^{-1} SP (2.8 mg·L^{-1} HP) may be able to reduce cyanobacterial abundance, while having a reduced impact on non-target zooplankton and phytoplankton. Moreover, numerous studies using direct HP application have shown that concentrations below 2.8 mg·L^{-1} HP have significant effects on cyanobacterial abundance in experimental laboratory conditions and in whole lake experiments [9,12,15]. As noted by Sinha et al. [16] and in line with our original concerns regarding the effects of top-down control, concentrations greater than 10.0 mg·L^{-1} SP (2.8 mg·L^{-1} HP) may also impact non-cyanobacterial phytoplankton, thereby reducing potential food sources for *Daphnia* and other zooplankton.

5. Conclusions

In conjunction with these cited works, our findings show that ROS released following the degradation of SP have a non-linear effect on the mortality of zooplankton over a very short period, with an apparent tipping point followed by a steep drop in survival in treatments containing 10.0 mg·L^{-1} (2.8 mg·L^{-1} HP) or more of SP. For individual *Daphnia* ranging in size from 0.2 to 2.0 mm no significant difference in survival was observed, with the suggestion that large individuals may be more vulnerable to the SP than the smaller individuals, which was in contradiction to our initial speculations, as explained above.

Given the importance of *Daphnia* as a keystone species in exercising control over the biomass and structure of phytoplankton in lakes and many other water bodies through top-down control, we suggest that the amount of SP that should be added to a lake to combat cyanobacterial occurrence while limiting collateral damage to zooplankton should be below 10.0 mg·L^{-1} SP (2.8 mg·L^{-1} HP) and possibly less than 5.0 mg·L^{-1} SP (1.4 mg·L^{-1} HP).

Supplementary Materials: The following are available online at http://www.mdpi.com/2073-4441/12/5/1304/s1, Table S1: Number of individuals representing various taxonomic groups per microcosm in three controls and seven experimental treatments, Figure S1: The changes of pH at various concentrations of Envolab's sodium percarbonate.

Author Contributions: Conceptualization, I.J., W.S.; formal analysis, R.T., methodology, I.J.; R.T., investigation, R.T.; visualization, R.T.; writing—original draft preparation, R.T.; writing—review and editing, R.T.; I.J.; W.S., supervision, I.J. All authors have read and agreed to the published version of the manuscript.

Funding: This work was supported by the statutory program of the Department of Microbial Ecology and Environmental Biotechnology, Faculty of Biology, University of Warsaw.

Acknowledgments: The authors thank colleagues from the Department of Microbial Ecology and Environmental Biotechnology and from the Environmental Management Master's Program at the Biology Faculty UW for support during the experimental works.

Conflicts of Interest: The authors declare no conflict of interest.

References

1. Shurin, J.B.; Gruner, D.S.; Hillebrand, H. All wet or dried up? Real differences between aquatic and terrestrial food webs. *Proc. R. Soc. B Biol. Sci.* **2005**, *273*, 1–9. [CrossRef] [PubMed]
2. Field, C.B.; Behrenfeld, M.J.; Randerson, J.T.; Falkowski, P. Primary production of the biosphere: Integrating terrestrial and oceanic components. *Science* **1998**, *281*, 237–240. [CrossRef] [PubMed]
3. Anderson, D.M. Prevention, control, and mitigation of harmful algal blooms: Multiple approaches to HAB management. In *Harmful Algae Management mitigate, Heng Mui Keng, Terrace, Singapore, 21 December 2004*; Etheridge, S., Anderson, D., Kleindinst, J., Zhu, M., Zou, Y., Eds.; Asia Pacific Economic Cooperation: Heng Mui Keng Terrace, Singapore, 2004; pp. 123–130.
4. Barrington, D.; Xiao, X.; Coggins, L.; Ghadouani, A. Control and management of harmful algal blooms. In *Climate Change and Marine and Freshwater Toxins Botana*; Louzao, L.M., Vilariño, M.C., Eds.; N. De Gruyter: Berlin, Germany, 2015; pp. 313–358. ISBN 978-3-11-033303-9.
5. Meriluoto, J.; Blaha, L.; Bojadzija, G.; Bormans, M.; Brient, L.; Codd, G.A.; Drobac, D.; Faassen, E.J.; Fastner, J.; Hiskia, A.; et al. Toxic cyanobacteria and cyanotoxins in european waters—Recent Progress Achieved through the CYANOCOST Action and Challenges for Further Research. *Adv. Oceanogr. Limnol* **2017**, *8*, 161–178. [CrossRef]
6. Medrano, E.A.; Uittenbogaard, R.; Wiel, B.V.D.; Pires, L.D.; Clercx, H. An alternative explanation for cyanobacterial scum formation and persistence by oxygenic photosynthesis. *Harmful Algae* **2016**, *60*, 27–35. [CrossRef]
7. Tiling, K.; Proffitt, C.E. Effects of *Lyngbya Majuscula* blooms on the seagrass *Halodule Wrightii* and resident invertebrates. *Harmful Algae* **2017**, *62*, 104–112. [CrossRef]
8. Ghadouani, A.; Pinel-Alloul, B.; Prepas, E.E. Effects of experimentally induced cyanobacterial blooms on crustacean zooplankton communities. *Freshw. Biol.* **2003**, *48*, 363–381. [CrossRef]
9. Weenink, E.F.J.; Luimstra, V.M.; Schuurmans, J.M.; Herk, M.J.V.; Visser, P.M.; Matthijs, H.C.P. Combatting Cyanobacteria with Hydrogen Peroxide: A laboratory study on the consequences for phytoplankton community and diversity. *Front. Microbiol.* **2015**, *6*. [CrossRef]
10. Dokulil, M.T.; Teubner, K. Cyanobacterial dominance in lakes. *Hydrobiologia* **2000**, *438*, 1–12. [CrossRef]
11. Matthijs, H.C.P.; Jančula, D.; Visser, P.M.; Maršálek, B. Existing and emerging cyanocidal compounds: New perspectives for cyanobacterial bloom mitigation. *Aquat. Ecol.* **2016**, *50*, 443–460. [CrossRef]
12. Drábková, M.; Admiraal, W.; Maršálek, B. Combined exposure to hydrogen peroxide and light selective effects on cyanobacteria, green algae, and diatoms. *Environ. Sci. Technol.* **2007**, *41*, 309–314. [CrossRef]
13. Apel, K.; Hirt, H. Reactive oxygen species: Metabolism, oxidative stress, and signal transduction. *Annu. Rev. Plant Biol.* **2004**, *55*, 373–399. [CrossRef] [PubMed]
14. Veljovic-Jovanovic, S. Active oxygen species and photosynthesis: Mehler and ascorbate peroxidase reactions. *Iugoslav Physiol. Pharmacol. Acta* **1998**, *34*, 503–522.
15. Matthijs, H.C.; Visser, P.M.; Reeze, B.; Meeuse, J.; Slot, P.C.; Wijn, G.; Talens, R.; Huisman, J. Selective suppression of harmful cyanobacteria in an entire lake with hydrogen peroxide. *Water Res.* **2012**, *46*, 1460–1472. [CrossRef] [PubMed]
16. Sinha, A.K.; Eggleton, M.A.; Lochmann, R.T. An environmentally friendly approach for mitigating cyanobacterial bloom and their toxins in hypereutrophic ponds: Potentiality of a newly developed granular hydrogen peroxide-based compound. *Sci. Total Environ.* **2018**, *637–638*, 524–537. [CrossRef]
17. Massachusetts Department of Environmental Protection—Hydrogen Peroxide, Peracetic Acid and Sodium Percarbonate. Available online: http://www.mass.gov/eea/docs/agr/pesticides/aquatic/sodium-carbonate-peroxyhydrate-and-hydrogen-peroxide.pdf (accessed on 10 October 2016).
18. Wilson, A.E.; Sarnelle, O.; Tillmanns, A.R. Effects of cyanobacterial toxicity and morphology on the population growth of freshwater zooplankton: Meta-analyses of laboratory experiments. *Limnol. Oceanogr.* **2006**, *51*, 1915–1924. [CrossRef]

19. Lampert, W. Laboratory studies on zooplankton-cyanobacteria interactions. *N. Z. J. Mar. Freshw. Res.* **1987**, *21*, 483–490. [CrossRef]

20. Lampert, W. Zooplankton research: The contribution of limnology to general ecological paradigms. *Aquat. Ecol.* **1997**, *31*, 19–27. [CrossRef]

21. Sarnelle, O. *Daphnia* as keystone predators: Effects on phytoplankton diversity and grazing sesistance. *J. Plankton Res.* **2005**, *27*, 1229–1238. [CrossRef]

22. Wogram, J.; Liess, M. Rank ordering of macroinvertebrate species sensitivity to toxic compounds by comparison with that of *Daphnia Magna. Bull. Environ. Contam. Toxicol.* **2001**, *67*, 360–367.

23. Wetzel, R. *Limnology*, 3rd ed.; Academic Press: Cambridge, MA, USA, 2001; p. 1006.

24. Pedersen, F.; Hansen, P. Effects of high pH on the growth and survival of six marine heterotrophic protists. *Mar. Ecol. Prog. Ser.* **2003**, *260*, 33–41. [CrossRef]

25. Yin, X.W.; Niu, C.J. Effect of pH on Survival, Reproduction, Egg Viability and Growth Rate of Five Closely Related Rotifer Species. *Aquat. Ecol.* **2007**, *42*, 607–616. [CrossRef]

26. Beklioglu, M.; Moss, B. The Impact of pH on interactions among phytoplankton algae, zooplankton and perch (*Perca Fluviatilis*) in a Shallow, Fertile Lake. *Freshw. Biol.* **1995**, *33*, 497–509. [CrossRef]

27. Reynolds, C. *The Ecology of Phytoplankton*; Cambridge University Press: Cambridge, UK; Melbourne, Australia; Madrid, Spain; Cape Town, South Africa; Singapore; São Paulo, Brazil, 2006; p. 551.

28. McQueen, D.J.; Post, J.R.; Mills, E.L. Trophic relationships in freshwater pelagic ecosystems. *Can. J. Fish. Aquat. Sci.* **1986**, *43*, 1571–1581. [CrossRef]

29. Christoffersen, K.; Riemann, B.; Klysner, A.; Søndergaard, M. Potential role of fish predation and natural populations of zooplankton in structuring a plankton community in eutrophic lake water. *Limnol. Oceanogr.* **1993**, *38*, 561–573. [CrossRef]

30. Envolab Fine Chemicals Nadwęglan sodu. Available online: https://envolab.pl/sklep/nadweglan-sodu-12kg/ (accessed on 1 January 2016).

31. Chapman, M.A.; Lewis, M.H.; Stout, V.M. *Introduction to the freshwater crustacea of New Zealand*; Collins: Auckland, New Zealand, 1976.

32. McDonald, J.H. Regressions. In *Handbook of Biological Statistics*; Sparky House Publishing: Baltimore, MD, USA, 2014; Volume 2, pp. 173–181.

33. Kreyling, J.; Schweiger, A.H.; Bahn, M.; Ineson, P.; Migliavacca, M.; Morel-Journel, T.; Christiansen, J.R.; Schtickzelle, N.; Larsen, K.S. To replicate, or not to replicate—that is the question: How to tackle nonlinear responses in ecological experiments. *Ecol. Lett.* **2018**, *21*, 1629–1638. [CrossRef]

34. Meinertz, J.R.; Greseth, S.L.; Gaikowski, M.P.; Schmidt, L.J. Chronic toxicity of hydrogen peroxide to *Daphnia magna* in a continuous exposure, flow-through test system. *Sci. Total Environ.* **2008**, *392*, 225–232. [CrossRef]

35. Reichwaldt, E.S.; Zheng, L.; Barrington, D.J.; Ghadouani, A. Acute toxicological response of *Daphnia* and *Moina* to hydrogen peroxide. *J. Environ. Eng.* **2011**, *138*, 607–611. [CrossRef]

36. Wasielak, K. *The Influence of Atomic Oxygen on Chemical Properties and Microbial Activity in Aquatic Ecosystem*; The University of Warsaw: Warsaw, Poland, 2016.

37. OECD SIDS Initial Assessment Report for SIAM 15—Sodium Carbonate. Available online: https://hpvchemicals.oecd.org/ui/handler.axd?id=5A6538BE-AA30-4A72-AD1C-906D9B5413BD (accessed on 19 March 2020).

38. Bogart, S.J.; Woodman, S.; Steinkey, D.; Meays, C.; Pyle, G.G. Rapid changes in water hardness and alkalinity: Calcite formation is lethal to *Daphnia magna. Sci. Total Environ.* **2016**, *559*, 182–191. [CrossRef]

39. Warne, M.S.D.; Schifko, A.D. Toxicity of laundry detergent components to a freshwater cladoceran and their contribution to detergent toxicity. *Ecotoxicol. Environ. Saf.* **1999**, *44*, 196–206. [CrossRef]

40. Geer, T.D.; Kinley, C.M.; Iwinski, K.J.; Calomeni, A.J.; Rodgers, J.H. Comparative toxicity of sodium carbonate peroxyhydrate to freshwater organisms. *Ecotoxicol. Environ. Saf.* **2016**, *132*, 202–211. [CrossRef]

41. Pirow, R.; Wollinger, F.; Paul, R.J. The sites of respiratory gas exchange in the planktonic crustacean *Daphnia magna*: An in vivo study employing blood haemoglobin as an internal oxygen probe. *J. Exp. Biol.* **1999**, *202*, 3089–3099. [PubMed]

42. Broönmark, C.; Hansson, L.-A. *The Biology of Lakes and Ponds*; Oxford University Press: Oxford, UK, 2018.

43. Pirow, R.; Buchen, I. The dichotomous oxyregulatory behaviour of the planktonic crustacean *Daphnia Magna. J. Exp. Biol.* **2004**, *207*, 683–696. [CrossRef] [PubMed]

44. Ebert, D. *Ecology, Epidemiology, and Evolution of Parasitism in Daphnia*; National Library of Medicine (US): Bethesda, MD, USA; National Center for Biotechnology Information: Basel, Switzerland, 2004.
45. Pirow, R.; Wollinger, F.; Paul, R.J. The importance of the feeding current for oxygen uptake in the water flea *Daphnia* magna. *J. Exp. Biol.* **1999**, *202*, 553–562. [PubMed]
46. Burns, C.W. Relation between filtering rate, temperature, and body size in four Species of *Daphnia*. *Limnol. Oceanogr.* **1969**, *14*, 693–700. [CrossRef]
47. Egloff, D.A.; Palmer, D.S. Size relations of the filtering area of two *Daphnia* species. *Limnol. Oceanogr.* **1971**, *16*, 900–905. [CrossRef]
48. Crafton, E.A.; Cutright, T.J.; Bishop, W.M.; Ott, D.W. Modulating the effect of iron and total organic carbon on the efficiency of a hydrogen peroxide-based algaecide for suppressing cyanobacteria. *Water Air Soil Pollut.* **2019**, *230*, 56. [CrossRef]
49. Kędzierska, A. Methods of Mitigation of Cyanobacterial Blooms Basing on Various-Scale Experiments. Master's Thesis, University of Warsaw, Warsaw, Poland, 2016.

MDPI

St. Alban-Anlage 66

4052 Basel

Switzerland

Tel. +41 61 683 77 34

Fax +41 61 302 89 18

www.mdpi.com

Water Editorial Office

E-mail: water@mdpi.com

www.mdpi.com/journal/water